Springer Series on Fluorescence
Methods and Applications

O. Wolfbeis
Editor-in-Chief

Springer-Verlag Berlin Heidelberg GmbH

Ruud Kraayenhof · Antonie J. W. G. Visser ·
Hans C. Gerritsen (Eds.)

Fluorescence Spectroscopy, Imaging and Probes

New Tools in Chemical, Physical and Life Sciences

With 175 Figures

Springer

Fluorescence spectroscopy, fluorescence imaging and fluorescent probes are indispensible tools in numerous fields of modern medicine and science, including molecular biology, biophysics, biochemistry, clinical diagnosis and analytical and environmental chemistry. Applications stretch from spectroscopy and sensor technology to microscopy and imaging, to single molecule detection, to the development of novel fluorescent probes, and to proteomics and genomics. The *Springer Series on Fluorescence* aims at publishing state-of-the-art articles that can serve as invaluable tools for both practitioners and researchers being active in this highly interdisciplinary field. The carefully edited collection of papers in each volume will give continuous inspiration for new research and will point to exciting new trends.

– Springer WWW home page: http://www.springer.de

ISSN 1617-1306
ISBN 978-3-642-62732-3

Library of Congress Cataloging-in-Publication Data applied for

Die Deutsche Bibliothek – Cip-Einheitsaufnahme

Fluorescence spectroscopy, imaging and probes : new tools in chemical, physical and life sciences / Ruud Kraayenhof ... (ed.). – Berlin ; Heidelberg ; New York ; Barcelona ; Hong Kong ; London ; Milan ; Paris ; Tokyo : Springer, 2002
(Springer series on fluorescence ; Vol. 2)
ISBN 978-3-642-62732-3   ISBN 978-3-642-56067-5 (eBook)
DOI 10.1007/978-3-642-56067-5

http://www.springer.de
© Springer-Verlag Berlin Heidelberg 2002
Originally published by Springer-Verlag Berlin Heidelberg New York in 2002
Softcover reprint of the hardcover 1st edition

Typesetting: Data delivered by editors
Cover: design & production, Heidelberg
Printed on acid-free paper   SPIN: 10993422   52/3111 – 5 4 3 2 1

Editor-in-Chief

Professor Dr. Otto Wolfbeis
University of Regensburg
Institute of Analytical Chemistry, Chemo- and Biosensors
93040 Regensburg
Germany

Volume Editor

Professor Ruud Kraayenhof
Institute of Molecular Biological Sciences, Amsterdam
Vrije Universiteit
De Boelelaan 1087
1081 HV Amsterdam
The Netherlands
e-mail: kr@bio.vu.nl

Professor Antonie J. W. G. Visser
MicroSpectroscopy Center
Wageningen University
Dreijenlaan 3
6703 HA Wageningen
The Netherlands

Professor Hans C. Gerritsen
Dept. of Molecular Biophysics
University of Utrecht
Princetonplein 1
3584 CC Utrecht
The Netherlands

Preface

Fluorescence techniques enjoy ever-increasing interest from scientists of a multitude of disciplines: physics, chemistry, biology, geology, pharmacology, toxicology and medicine. Besides widespread fundamental and applied research of fluorescence in university laboratories, one observes a substantially enhanced effort by smaller and larger companies towards the development of new fluorescence-based diagnostic tools. In particular, the application of fluorescence in high-throughput screenings in genomics and proteomics is evidently successful.

This increased use of fluorescence techniques is greatly enhanced by the improved instrumentation pioneered by inventive scientists and now made available commercially by several high-tech companies. In recent years we observed a vast improvement of microscopic imaging techniques, such as the introduction of confocal scanning microscopy and multi-photon excitation microscopy. Moreover, the design and development of many new molecular probes with higher selectivity for specific micro-environmental properties has stimulated many new researchers to employ fluorescence techniques for solving their problems.

Probably the most significant breakthrough in fluorescence is its use in detection of single molecules and even of their real-time dynamics. Also, probing inside living cells has become a hot topic in the life sciences.

This topic book reflects the updates of scientific progress as presented by invited lecturers and other participants at the *7th Conference on Methods and Applications of Fluorescence: Spectroscopy, Imaging and Probes (MAF)*, held in Amsterdam, The Netherlands, 16–19 September 2001. The previous conference series on Fluorescence Microscopy and Fluorescent Probes, so successfully organized by Jan Slavik in Prague, was merged with the MAF series after Jan Slavik's untimely death in January 1999. This is emphasized by increased attention for new microscopic techniques and new fluorescent probes at the Amsterdam meeting. Also, the 360 participants from 35 countries at the 7th MAF conference demonstrate the growing interest in fluorescence techniques. The devastating events of 11 September 2001 in the USA prevented several speakers to come. We are grateful to those who replaced them at short notice, as speakers and contributors to this book.

We are very grateful to our sponsors (see next page) for their generous support and to Jeannet Wijker and the team of *Lidy Groot Congress Events* for their efficiency and enthousiasm. We also acknowledge the editorial assistance of Nina Visser and Wilfried van Sark in formatting this topic book.

Amsterdam, May 2002

Ruud Kraayenhof
Antonie J.W.G. Visser
Hans C. Gerritsen

Traditionally, the MAF conferences seek to integrate the active participation of fluorescence-related companies (lectures, exhibition, sponsoring), stimulating fruitful interactions between fundamental and applied research in a pleasant setting.

The Scientific and Organizing Committee of the *7th Conference on Methods and Applications of Fluorescence: Spectroscopy, Imaging and Probes* gratefully acknowledges the sponsors and exhibitors, that financially supported the conference:

Amersham Pharmacia Biotech, The Netherlands
ATTO-TEC GmbH, Germany
BERTHOLD TECHNOLOGIES GmbH & Co KG, Germany
Beun-de Ronde BV, The Netherlands
BFi OPTiLAS BV, The Netherlands
BIO-RAD Microscience Ltd, United Kingdom
BMG Labtechnologies GmbH, Germany
Byk Nederland BV, The Netherlands
Carl Zeiss NV, Belgium, and Carl Zeiss Jena GmbH, Germany
City of Amsterdam, The Netherlands
Coherent BV, The Netherlands
Chroma Technology Corp, USA
DSM Food Specialties, The Netherlands
Evotec OAI, Germany
IBH, United Kingdom
ISS, USA
Jobin Yvon SA, France
Kluwer Academic Publishers, The Netherlands
Lambert Instruments, The Netherlands
LaVision BioTec GmbH, Germany
Leica Microsystems, The Netherlands
L'ORÉAL Recherche Avancée, France
Molecular Devices, United Kingdom
Molecular Probes Europe BV, The Netherlands
Netherlands Society for Biochemistry and Molecular Biology
Organon NV, The Netherlands
Netherlands Organization for Scientific Research
Photon Technology International Inc. (PTI), USA
PicoQuant GmbH, Germany
Scientific Volume Imaging, The Netherlands
Unilever Research, The Netherlands
Unilever Research, United Kingdom
Varian Analytical Instruments, The Netherlands
Vrije Universiteit Amsterdam, Institute of Molecular Biological Sciences

Contents

Part 2 Fluorescence Spectroscopy of Single Molecules and Molecular Assemblies

Part 3 Application of Fluorescence in Biological Membrane and Enzyme Studies

17 Emission Spectroscopy of Complex Formation between *Escherichia coli* Purine Nucleoside Phosphorylase (PNP) and Identified Tautomeric Species of Formycin Inhibitors Resolves Ambiguities Found in Crystallographic Studies ... 277

B. KIERDASZUK

Part 4 Microscopic Imaging Techniques and their Application for the Study of Living Cells

18 Fluorescence Lifetime Imaging Implemented with Resonant Galvanometer Scanners ... 297

J. J. BIRMINGHAM

Contributors[1]

M.A.H. ASSELBERGS
Department of Molecular Biophysics,
Debye Institute, Utrecht University,
P.O. Box 80000, 3508 TA Utrecht,
The Netherlands
e-mail: m.a.h. asselbergs@phys.uu.nl

S. D'AURIA
University Maryland School of Medicine,
Center for Fluorescence Spectroscopy,
Dept. of Biochemistry and Molecular Bio-
logy, 725 W. Lombard Street, Baltimore,
Maryland 21201, USA
e-mail: sabato@cfs.umbi.umd.edu

J. BEUTHAN
Freie Universität Berlin, Institut für Medizinische Physik/Lasermedizin, and Laser-
und Medizin- Technologie GmbH Berlin,
Fabeckstr. 60-62, 14195 Berlin, Germany

D.J.S. BIRCH *
The Photophysics Research Group, Depart-
ment of Physics and Applied Physics,
Strathclyde University, Glasgow G4 ONG,
UK
e-mail: djs.birch@strath.ac.uk

J.J. BIRMINGHAM *
Unilever Research, Port Sunlight Laboratory, Quarry Road East, Bebington, Wirral,
Merseyside L63 3JW, UK
e-mail: john.birmingham@unilever.com

M. BÖHMER
Forschungszentrum Jülich, Institute for Biological Information Processing I,
D-52425 Jülich, Germany
e-mail: m.boehmer@fz-juelich.de

M. VAN BORREN
Department of Physiology, Academic
Medical Center, University of Amsterdam,
Academic Medical Center, Meibergdreef
9,1105 AZ Amsterdam, The Netherlands
e-mail: m.m.vanborren@amc.uva.nl

M. BÖRSCH *
Institut für Physikalische Chemie, Albert-
Ludwigs-Universität Freiburg, Albertstr.
23a, 79104 Freiburg, Germany
e-mail: boersch@uni-freiburg.de

J.W. BORST
MicroSpectroscopy Center, Wageningen
University, Dreijenlaan 3, 6703 HA Wageningen, The Netherlands
e-mail: janwillem.borst@laser.bc.wau.nl

N.R. BRADY
Department of Molecular Cell Physiology,
BioCentrum Amsterdam, Vrije Universiteit,
De Boelelaan 1087, 1081 HV Amsterdam,
The Netherlands
e-mail: brady@bio.vu.nl

C.K.D. BREEK
Clusius Laboratory, Leiden University,
Wassenaarseweg 64 2333 AL, Leiden,
The Netherlands

A. CHATTOPADHYAY *
Centre for Cellular and Molecular Biology,
Uppal Road, Hyderabad 500 007, India
e-mail: amit@gene.ccmbindia.org

M. COLE
Evotec OAI AG, Schnackenburgallee 114,
22525 Hamburg, Germany
e-mail: mary.cole@evotecoai.com

R. CORNELL
Department of Molecular Biology and Biochemistry, Simon Fraser University, Burnaby, British Columbia, V5A 1S6, Canada
e-mail: cornell@sfu.ca

M. COTLET
Department of Chemistry, Katholieke Universiteit Leuven, Celestijnenlaan 200 F,
3001 Heverlee, Belgium

---

[1] * = corresponding author

J.C. CRONEY
Department of Cell and Molecular Biology,
John A. Burns School of Medicine, University of Hawaii, 1960 East-West Rd., Honolulu, HI 96822, USA
e-mail: croney@gold.chem.hawaii.edu

S.M.A. DAVIES
Department of Molecular Biology and Biochemistry, Simon Fraser University, Burnaby, British Columbia, V5A 1S6, Canada;
present address: Department of Preclinical Veterinary Sciences, R.(D).S.V.S., Summerhall, University of Edinburgh, Edinburgh EH9 1QH, UK

A.P. DEMCHENKO *
TUBITAK Research Institute for Genetic
Engineering and Biotechnology, Gebze-
Kocaeli 41470, Turkey; present address:
A.V. Palladin Institute of Biochemistry,
National Academy of Sciences of Ukraine,
Kiev 01030, Ukraine
e-mail: adem@biochem.kiev.ua

M. DIEZ
Institut für Physikalische Chemie, Albert-
Ludwigs-Universität Freiburg, Albertstr.
23a, 79104 Freiburg, Germany

L. DOWAL
Department of Physiology and Biophysics,
State University of New York at Stony
Brook, Stony Brook, New York, 11794-
8661, USA

A. DÜRKOP
University of Regensburg, Institute of Analytical Chemistry, Chemo- and Biosensors,
D-93040 Regensburg, Germany
e-mail: axel.duerkop@chemie.uni-
regensburg.de

C. EGGELING
Evotec OAI AG, Schnackenburgallee 114,
22525 Hamburg, Germany
e-mail: christian.eggeling@evotecoai.com

J. ENDERLEIN
Forschungszentrum Jülich, Institute for Biological Information Processing I,
D-52425 Jülich, Germany
e-mail: j.enderlein@fz-juelich.de

R.M. EPAND *
Department of Biochemistry, McMaster
University Health Sciences Centre,
1200 Main Street West, Hamilton, ON,
L8N 3Z5, Canada
e-mail: epand@mcmaster.ca

S. ERCELEN
TUBITAK Research Institute for Genetic
Engineering and Biotechnology, Gebze-
Kocaeli 41470, Turkey
e-mail: sebnem@rigeb.gov.tr

P.L.T.M. FREDERIX
Department of Molecular Biophysics, Debye Institute and Department of Medical
Physiology, Faculty of Medicine, Utrecht
University, P.O. Box 80000, 3508 TA
Utrecht, The Netherlands
e-mail: patrick.frederix@unibas.ch

K. GALL *
Evotec OAI AG, Schnackenburgallee 114,
22525 Hamburg, Germany
e-mail: karsten.gall@evotecoai.com

C.D. GEDDES
The Photophysics Research Group, Department of Physics and Applied Physics,
Strathclyde University, Glasgow G4 ONG,
UK; present address: University Maryland
School of Medicine, Center for Fluorescence Spectroscopy, Department of Biochemistry and Molecular Biology, 725 W.
Lombard Street, Baltimore, Maryland
21201, USA
e-mail: chris@cfs.umbi.umd.edu

H.C. GERRITSEN *
Department of Molecular Biophysics, Debye Institute, Utrecht University, P.O. Box
80000, 3508 TA Utrecht, The Netherlands
e-mail: h.c.gerritsen@phys.uu.nl

I. GOSSE
Laboratoire d'Analyse Chimique par Reconnaissance Moléculaire, Ecole Nationale Supérieure de Chimie et de Physique de Bordeaux, 16 Avenue Pey-Berland, 33607 Pessac cedex, France
e-mail: gosse@enscpb.u-bordeaux.fr

P. GRÄBER
Institut für Physikalische Chemie, Albert-Ludwigs-Universität Freiburg, Albertstr. 23a, 79104 Freiburg, Germany
e-mail: graeberp@uni-freiburg.de

E. GRATTON
Laboratory for Fluorescence Dynamics, Department of Physics, University of Illinois at Urbana-Champaign, Urbana, Illinois 61801, USA
e-mail: lfd@uiuc.edu

K.O. GREULICH *
Institute of Molecular Biotechnology, Postfach 100 813, D 07708 Jena Germany
e-mail: kog@imb-jena.de

M. GRUBER
University of Regensburg, Institute of Analytical Chemistry, Chemo- and Biosensors, D-93040 Regensburg, Germany
e-mail: michaela.gruber@chemie.uni-regensburg.de

I. GRYCZYNSKI
University Maryland School of Medicine, Center for Fluorescence Spectroscopy, Dept. of Biochemistry and Molecular Biology, 725 W. Lombard Street, Baltimore, Maryland 21201, USA
e-mail: ignazy@cfs.umbi.umd.edu

Z. GRYCZYNSKI
University Maryland School of Medicine, Center for Fluorescence Spectroscopy, Dept. of Biochemistry and Molecular Biology, 725 W. Lombard Street, Baltimore, Maryland 21201, USA
e-mail: karol@cfs.umbi.umd.edu

C.V. HENKEL
Institute of Molecular Plant Sciences and Leiden Center for Natural Computing, Leiden University, Wassenaarseweg 64, 2333 AL Leiden, The Netherlands
e-mail: henkel@rulbim.leidenuniv.nl

D.J. VAN DEN HEUVEL
Department of Molecular Biophysics, Debye Institute, Utrecht University, P.O. Box 80000, 3508 TA Utrecht, The Netherlands
e-mail: d..j.vandenheuvel@phys.uu.nl

M. HOF *
J. Heyrovsky Institute of Physical Chemistry, ASCR, and Center for Complex Molecular Systems and Biomolecules, 18223 Prague 8, Czech Republic
e-mail: hof@jh-inst.cas.cz

J. HOFKENS
Department of Chemistry, Katholieke Universiteit Leuven, Celestijnenlaan 200 F, 3001 Heverlee, Belgium
e-mail: johan.hofkens@chem.kuleuven.ac.be

O. HOLUB
Laboratory for Fluorescence Dynamics, UIUC, Department of Physics, 1110 W. Green St., Urbana, IL 61801, USA
e-mail: ollihol@web.de

R. HUTTERER
Institute of Analytical Chemistry, Chemo- and Biosensors, University of Regensburg, D-93040 Regensburg, Germany
e-mail: rudolf.hutterer@chemie.uni-regensburg.de

F. VAN IREN
Clusius Laboratory, Leiden University, Wassenaarseweg 64 2333 AL, Leiden, The Netherlands

T. ISHII
Department of Cell and Molecular Biology, John A. Burns School of Medicine, University of Hawaii, 1960 East-West Rd., Honolulu, HI 96822, USA

D.M. JAMESON *
Department of Cell and Molecular Biology,
John A. Burns School of Medicine, University of Hawaii, 1960 East-West Rd.,
Honolulu, HI 96822, USA
e-mail: djameson@hawaii.edu

J. KAROLIN
The Photophysics Research Group, Department of Physics and Applied Physics,
Strathclyde University, Glasgow G4 ONG,
UK

P. KASK
Evotec OAI AG, Schnackenburgallee 114,
22525 Hamburg, Germany and Institute of
Experimental Biology, Instituudi tee 11,
Harku 76902, Estonia
e-mail: peet.kask@evotecoai.com

B. KIERDASZUK *
University of Warsaw, Institute of Experimental Physics, Department of Biophysics,
93 Zwirki i Wigury Street, 02-089 Warsaw,
Poland
e-mail: borys@asp.biogeo.uw.edu.pl

I. KLIMANT
Technical University of Graz, Institute of
Analytical Chemistry, 8010 Graz, Austria
e-mail: klimant@analytchem.tu-graz.ac.at

A.S. KLYMCHENKO
TUBITAK Research Institute for Genetic
Engineering and Biotechnology, Gebze-
Kocaeli 41470, Turkey
e-mail: andrey@rigeb.gov.tr

R. KRAAYENHOF
Department of Structural Biology,
Institute of Molecular Biological Sciences,
Vrije Universiteit Amsterdam, De Boele-
laan 1087, 1081 HV Amsterdam,
The Netherlands
e-mail: kr@bio.vu.nl

C. KRAUSE
University of Regensburg, Institute of Analytical Chemistry, Chemo- and Biosensors,
D-93040 Regensburg, Germany
e-mail: christian.krause@chemie.uni-regensburg.de

J. KÜRNER
University of Regensburg, Institute of Analytical Chemistry, Chemo- and Biosensors,
D-93040 Regensburg, Germany
e-mail: jens.kuerner@chemie.uni-regensburg.de

J.R. LAKOWICZ *
University Maryland School of Medicine,
Center for Fluorescence Spectroscopy,
Dept. of Biochemistry and Molecular Biology, 725 W. Lombard Street, Baltimore,
Maryland 21201, USA
e-mail: lakowicz@cfs.umbi.umd.edu

R. LAPOUYADE *
Laboratoire d'Analyse Chimique par Reconnaissance Moléculaire, Ecole Nationale
Supérieure de Chimie et de Physique de
Bordeaux, 16 Avenue Pey-Berland, 33607
Pessac cedex, France
e-mail: lapouyad@enscpb.u-bordeaux.fr

R.P. LEARMONTH *
Centre for Rural and Environmental Biotechnology and Department of Biological
and Physical Sciences, University of
Southern Queensland, Toowoomba 4350
Australia
e-mail: learmont@usq.edu.au

R. LEISHMAN
The Photophysics Research Group, Department of Physics and Applied Physics,
Strathclyde University, Glasgow G4 ONG,
UK

K. LICHA
Schering AG Berlin, Müllerstrasse 178,
13342 Berlin, Germany
e-mail: kai.licha@schering.de

G. LIEBSCH
University of Regensburg, Institute of Analytical Chemistry, Chemo- and Biosensors,
D-93040 Regensburg, Germany
e-mail: gregor.liebsch@chemie.uni-regensburg.de

Z. LIN
University of Regensburg, Institute of Analytical Chemistry, Chemo- and Biosensors, D-93040 Regensburg, Germany
e-mail: zhihong.lin@chemie.uni-regensburg.de

C. MAHNKE
Laser- und Medizin- Technologie GmbH Berlin, Fabeckstr. 60-62, 14195 Berlin, Germany

J. MALICKA
University Maryland School of Medicine, Center for Fluorescence Spectroscopy, Dept. of Biochemistry and Molecular Biology, 725 W. Lombard Street, Baltimore, Maryland 21201, USA
e-mail: joanna@cfs.umbi.umd.edu

J.-P. MALVAL
Laboratoire d'Analyse Chimique par Reconnaissance Moléculaire, Ecole Nationale Supérieure de Chimie et de Physique de Bordeaux, 16 Avenue Pey-Berland, 33607 Pessac cedex, France
e-mail: malval@enscpb.u-bordeaux.fr

M. MAUS
Department of Chemistry, Katholieke Universiteit Leuven, Celestijnenlaan 200 F, 3001 Heverlee, Belgium

A. MEIJERINK
Department of Physics and Chemistry of Condensed Matter, Debye Institute, Utrecht University, P.O. Box 80000, 3508 TA Utrecht, The Netherlands
e-mail: a.meijerink@phys.uu.nl

Ü. METS
Evotec OAI AG, Schnackenburgallee 114, 22525 Hamburg, Germany
e-mail: ylo.mets@evotecoai.com

O. MINET *
Freie Universität Berlin, Institut für Medizinische Physik/Lasermedizin, and Laser- und Medizin- Technologie GmbH Berlin, Fabeckstr. 60-62, 14195 Berlin, Germany
e-mail: minet@zedat.fu-berlin.de

J.-P. MORAND
Laboratoire d'Analyse Chimique par Reconnaissance Moléculaire, Ecole Nationale Supérieure de Chimie et de Physique de Bordeaux, 16 Avenue Pey-Berland, 33607 Pessac cedex, France
e-mail: morand@enscpb.u-bordeaux.fr

B. NASANSHARGAL
Institute of Molecular Biotechnology, Postfach 100 813, D 07708 Jena Germany
e-mail: tulga@imb-jena.de

B. OSWALD
University of Regensburg, Institute of Analytical Chemistry, Chemo- and Biosensors, D-93040 Regensburg, Germany
e-mail: bernhard.oswald@chemie.uni-regensburg.de

K. PALO
Evotec OAI AG, Schnackenburgallee 114, 22525 Hamburg, Germany
e-mail: kaupo.palo@evotecoai.com

V.G. PIVOVARENKO
Kiev Taras Shevchenko University, Department of Chemistry, Kiev 01017, Ukraine
e-mail: pvg@blizzard.sabbo.kiev.ua

J. RAVESLOOT
Department of Physiology, Academic Medical Center, University of Amsterdam, Academic Medical Center, Meibergdreef 9,1105 AZ Amsterdam, The Netherlands
e-mail: jh.ravesloot@amc.uva.nl

R. REUTER
Institut für Physikalische Chemie, Albert-Ludwigs-Universität Freiburg, Albertstr. 23a, 79104 Freiburg, Germany

O.J. ROLINSKI
The Photophysics Research Group, Department of Physics and Applied Physics, Strathclyde University, Glasgow G4 ONG, UK

G. ROZENBERG
Leiden Institute of Advanced Computer
Science and Leiden Center for Natural
Computing, Leiden University, Niels-Bohr-
weg 1, 2333 CA Leiden, The Netherlands
e-mail: rozenber@liacs.leidenuniv.nl

W.G.J.H.M. VAN SARK
Department of Molecular Biophysics, Debye Institute, Utrecht University, P.O. Box
80000, 3508 TA Utrecht, The Netherlands
e-mail: w.g.j.h.m.vansark@phys.uu.nl

S. SCARLATA *
Depart. of Physiology and Biophysics,
State Univ. of New York at Stony Brook,
Stony Brook, New York, 11794-8661, USA
e-mail: suzanne@dualphy.pnb.sunysb.edu

B. SCHÄFER
Institute of Molecular Biotechnology, Postfach 100 813, D 07708 Jena, Germany
e-mail: schaefer@imb-jena.de

K.A. SCHMIDT *
Institute of Molecular Plant Sciences and
Leiden Center for Natural Computing, Lei-
den University, Wassenaarseweg 64, 2333
AL Leiden, The Netherlands
e-mail: schmidt@rulbim.leidenuniv.nl

F.C. DE SCHRYVER *
Department of Chemistry, Katholieke Universiteit Leuven, Celestijnenlaan 200 F,
3001 Heverlee, Belgium
*e-mail: frans.deschryver@chem.kuleuven.
ac.be*

S.E. SEIFRIED
Department of Cell and Molecular Biology,
John A. Burns School of Medicine, University of Hawaii, 1960 East-West Rd., Honolulu, HI 96822, USA
e-mail: seifried@hawaii.edu

Y. SHEN
University Maryland School of Medicine,
Center for Fluorescence Spectroscopy,
Dept. of Biochemistry and Molecular Bio-
logy, 725 W. Lombard Street, Baltimore,
Maryland 21201, USA
e-mail: yibing.shen@probes.com

H.P. SPAINK *
Clusius Laboratory , Institute of Molecular
Plant Sciences and Leiden Center for
Natural Computing, Leiden University,
Wassenaarseweg 64, 2333 AL Leiden,
The Netherlands
e-mail: spaink@rulbim.leidenuniv.nl

N. STUURMAN
University of California at San Francisco,
Department of Cellular and Molecular
Pharmacology, 513 Parnassus Avenue,
S-1210, San Francisco, CA-94143-0450,
USA

M. TRAMIER
Institut Jacques Monod, 4 Place Jussieu,
Tour 43, 75251 Paris Cedex 05, France
e-mail: tramier@ijm.jussieu.fr

M.A. USKOVA
Enzymology Department, Chemistry Fa-
culty, Moscow State University, 119899
Moscow, Russian Federation

N.V. VISSER
MicroSpectroscopy Center, Wageningen
University, Dreijenlaan 3, 6703 HA
Wageningen, The Netherlands
e-mail: nina.visser@laser.bc.wau.nl

A.J.W.G. VISSER *
MicroSpectroscopy Center, Wageningen
University, Dreijenlaan 3, 6703 HA Wage-
ningen, The Netherlands; also affiliated at:
Department of Structural Biology, Institute
of Molecular Biological Sciences, Vrije
Universiteit, 1081 HV Amsterdam, The
Netherlands
e-mail: ton.visser@laser.bc.wau.nl

H.V. WESTERHOFF *
Department of Molecular Cell Physiology,
BioCentrum Amsterdam, Vrije Universiteit,
De Boelelaan 1087, 1081 HV Amsterdam,
The Netherlands
e-mail: hw@bio.vu.nl

S. WIJTING
Clusius Laboratory, Leiden University,
Wassenaarseweg 64 2333 AL, Leiden,
The Netherlands

O.S. WOLFBEIS *
University of Regensburg, Institute of Analytical Chemistry, Chemo- and Biosensors, D-93040 Regensburg, Germany
e-mail: otto.wolfbeis@chemie.uni-regensburg.de

M. WU
University of Regensburg, Institute of Analytical Chemistry, Chemo- and Biosensors, D-93040 Regensburg, Germany
e-mail: wu.meng@chemie.uni-regensburg.de

B. ZIMMERMANN
Institut für Physikalische Chemie, Albert-Ludwigs-Universität Freiburg, Albertstr. 23a, 79104 Freiburg, Germany

Part 1
Fluorescence Spectroscopy: New Approaches and Probes

Advanced Luminescent Labels, Probes and Beads and their Application to Luminescence Bioassay and Imaging

O. S. WOLFBEIS, M. BÖHMER, A. DÜRKOP, J. ENDERLEIN, M. GRUBER, I. KLIMANT, C. KRAUSE, J. KÜRNER, G. LIEBSCH, Z. LIN, B. OSWALD, AND M. WU

The design of fluorescent probes (and labels) is as challenging as it ever was. Such probes enable studies on the molecular dimensions and dynamics of even complex (bio)matter, but also bioanalytical and screening assays whose sensitivity can reach the single molecule level. The design of advanced labels for bioassays is paralleled by developments in (laser) fluorescence spectroscopy, opto-electronics and data processing. Light-emitting diodes (LEDs) and diode lasers (DLs) are particularly attractive light sources and we therefore have focused our research (a) on labels that are LED- or DL-compatible, and (b) on applications of such labels to various analytical formats.

In this article, we give an overview of our recent activities in the following areas: (1) a general logic for designing fluorescent probes and labels; (2) new diode laser-excitable probes for non-cocalent protein detection; (3) diode laser-compatible amino-reactive covalent labels; (4) diode laser-assisted fluorescent single molecule detection of dyes and labeled proteins; (5) new labels for flow cytometric determination of HSA; (6) new DNA labels; (7) fluorescence resonance energy transfer gene assays; (8) reactive ruthenium ligand complexes as markers for bioassays; (9) diode laser-excitable fluorescent polymer beads; (10) polyaniline-coated nanobeads as fluorescent pH probes; (11) phosphorescent poly(acrylonitrile) nanospheres as markers for optical assays; (12) competititve binding of streptavidin to biotinylated nanobeads as studied by resonance energy transfer; (13) nanobeads as reference dyes in luminescent lifetime imaging using DLR; (14) phosphorescent nanospheres for use in advanced time-resolved multiplexed bioassays; (15) beads dyed with a europium-based label and excitable with the 405-nm diode laser; and (16) a europium(III)-based probe for use in oxidase-associated reactions.

1.1
Introduction

The current popularity and success of fluorescent methods in bioanalysis and bio-physics [1–4] is based on several grounds including the versatility of fluorescence spectroscopy and the availability of molecular probes, of opto-electronic compo-nents (including LEDs and diode lasers) and of microprocessors. The former two are closely related in that certain labels and probes require specific light sources (and vice versa). The design of the label is, however, also determined by the spec-tral properties of the material to be probed. Fig. 1.1 shows the absorption spectra of common biological materials and it is obvious that the intrinsic absorption of blood, for example, prevents any spectroscopy to be performed on whole blood samples at wavelengths below 500 nm.

Diode lasers and light-emitting diodes cover most of the near UV, visible and near infrared. Table 1.1 gives an overview of currently available semiconductor light sources. Hence, both absorption and emission spectroscopy can be performed over a wide spectral range using such light sources which often are inexpensive, have small size and low power consumption.

The background luminescence of biological material is particularly strong in the ultraviolet and the near visible. Fig. 1.2 shows the intrinsic fluorescence of human blood serum in 2-dimensional format, demonstrating that in the UV there is an extremely strong fluorescence peaking at the emission maximum of HSA-tryptophane (at 287/340 nm) even though albumin contains one Trp unit only per 10 tyrosine units. This, in fact, is due to almost complete resonance energy trans-fer from Tyr to Trp, so that no peak resulting from Tyr emission (expected at 276/303 nm) is visible.

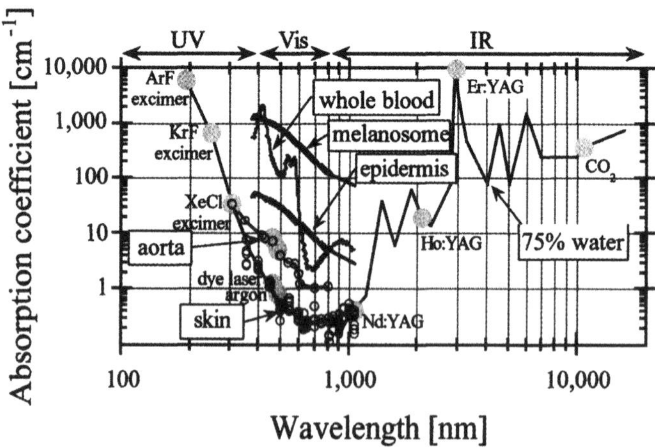

Fig. 1.1. Absorption spectra of typical biological matter; note that both scales are logarith-mic. Courtesy S. L. Jacques, University of Dallas

Table 1.1. Properties of presently available semiconductor (SC) light sources

peak wavelength (nm)	emitter type	SC material
340	LED[a]	ZnS
370	LED	InGaN/GaN
405	LED and DL	InGaN
430	LED	GaN on SiC
430–445	LED	announced (Nichia)
385 and 450	LED	InGaN/AlGaN, Zn doted
470	LED	SiC
450–500	LED, DL [a]	InGaN/GaN
525+	LED	announced (Marl)
515	DL [a]	ZnCdSe/ZnSSe
555+	LED	announced (Sharp; Toshiba)
555–605	LED	GaP
590–630	LED	AlGaInP
630–690	DL and LED	AlGaInP (many)
660–880	DL and LED	AlGaAs
940	DL	GaAs
980	DL	InGaAs
1300–1550	DL	InGaAsP

[a] prototype only

Also shown is the strong fluorescence in the near visible which is dominated by peaks at 344/460 nm (NADH), 370/500 nm (flavine nucleotide), 410/507 nm (pyridoxal phosphate Schiff base), and 455/515 and 465/515 nm (the bilirubin double peak). The intrinsic fluorescence of blood [5] and urine [6] is the major factor with respect to the limits of detection in fluorescence intensity-based bioassays.

The limitations can be overcome in two ways. The first is to make use of time-resolved gated measurements as, for example, in delayed fluorescence immunoassay [7–10]. The other is to shift the analytical wavelength of labels and probes into the red (or near-infrared) part of the spectrum where background fluorescence is much weaker [11]. We have pursued both approaches, and the respective results are presented here.

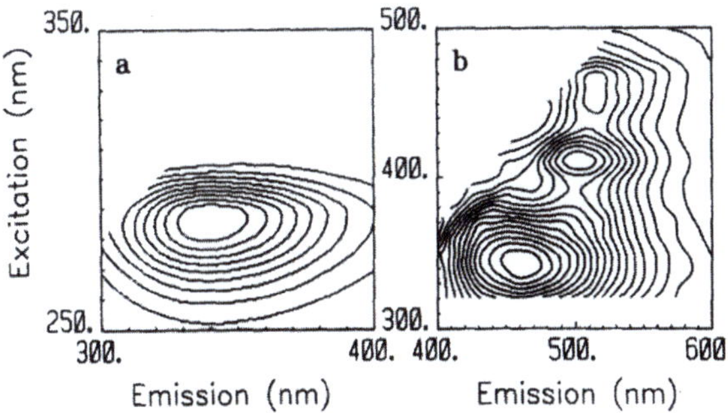

Fig. 1.2. The intrinsic fluorescence of human blood serum (from [5])

1.2
A General Logic for Designing Fluorescent Cyanine Type Probes and Labels of Defined Color

Cyanines form a group of dyes that combine relatively long-wave absorption with comparatively small molecular size, a feature that is desirable for labels in order not to disturb the system to be probed. In addition, the color of cyanines is fairly predictable from its molecular structure. The chemical structure of cyanine dyes can be represented in general form by structure $X(-CH=CH)_n-CH=Y$, where X and Y typically are nitrogen substituents like those shown in Table 1.2. One of the two substitutents (here X) has to be present in quaternized (cationic) form (A to G in Table 1.2).

The parameter n in $X(-CH=CH)_n-CH=Y$ has the largest effect on the absorption maximum (λ_{max}). The number of n typically varies from 0 to 3. It is known [12] that the λ_{max} values of cyanines increase almost linearly by 100 nm with n. However, if $n = 0$, the absorption of the cyanines does not exceed 600 nm. In this work, dyes are presented where $n = 1$ or 2. Such dyes are referred to as tricyanines and pentacyanines, respectively [12, 13].

While the number for n exerts a massive effect on the absorption maxima, spectral fine-tuning can only be accomplished by variation of substitutents X and Y. Table 1.1 gives the kind of substitutents standing for X (A–G) and Y (A'–G'), respectively, in this study. Symmetrical merocyanines (i.e., those where heterocycle X is of the same type as is Y) have been described rather often, and their absorption maxima have been measured [12–14], while those of unsymmetrical dyes are widely unknown.

It was found, to our surprise, that the arithmetic average of the λ_{max} values of the respective symmetrical dyes gave λ_{max} values that were in excellent (+/– 3–5 nm) agreement with the experimental findings, despite the fact that λ_{max} values are not linear with energy. Table 1.3 compiles the data for a whole set of trimethine dyes and in our eyes represents a general approach to fine-tune diode laser-compatible dyes.

The approach has been extended to the pentamethines ($n = 2$). The results show that the absorbances of the pentamethines extend far into the near infrared (from 645 to 808 nm) for which numerous diode laser lines are known. The approach presented here enables the absorption maxima of labels to be adjusted (in most cases) to the desired wavelength by better than +/– 6 nm.

Table 1.2. Typical substituents in cyanine dyes of the general structure X(-CH=CH)$_n$-CH=Y

X	Y
A	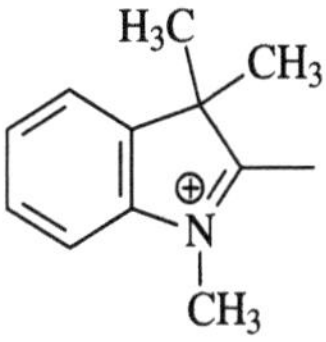 A' 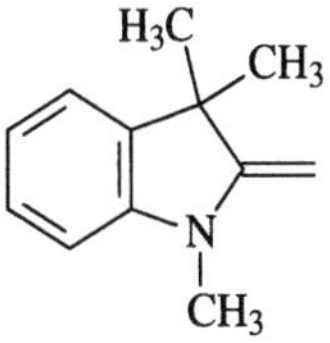
B	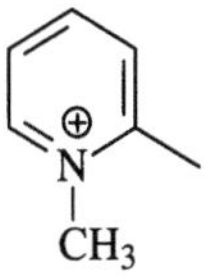 B' 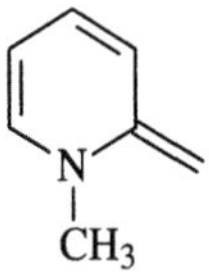
C	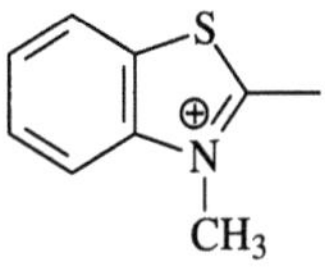 C' 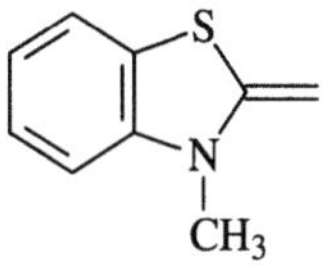
D	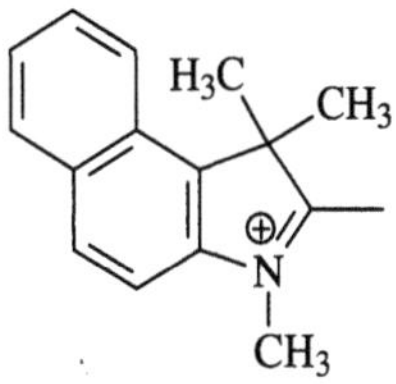 D' 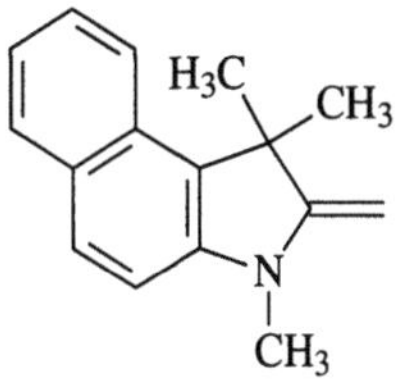
E	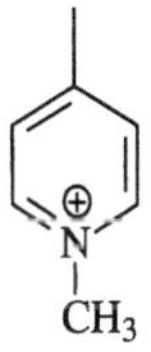 E' 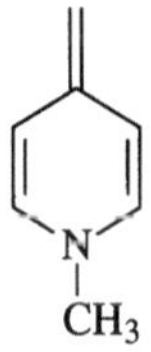
F	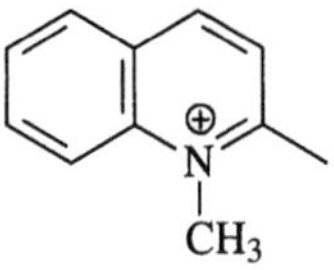 F' 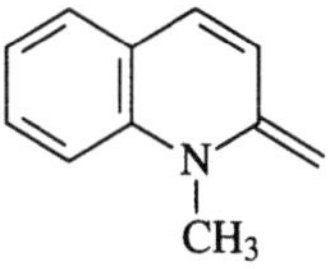
G	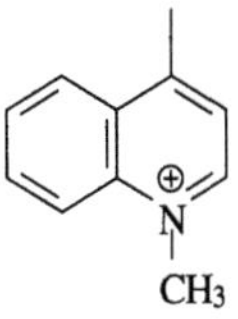 G' 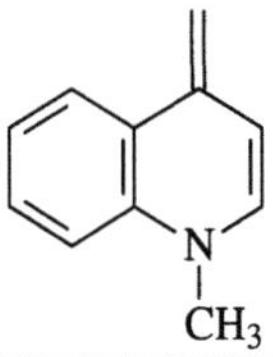

Table 1.3. Calculated absorption maxima (in nm) of trimethine dyes of type X-CH=CH-CH=Y, showing that by proper variation of substituents (A–G and A'–G', respectively; see Table 1.2) the whole long-wave part of the visible spectrum can be covered. Other substituents are known as well.

X↑ Y→	A'	B'	C'	D'	E'	F'	G'
A	543						
B	547	552					
C	549	551	555				
D	565	567	572	590			
E	525	590	579	595	603		
F	551	578	577	598	597	610	
G	615	630	630	642	651	657	710

1.3
Diode Laser-excitable Probes for General Protein Detection

Based on the above logic, we have prepared [15] various functional dyes. They contain (a) a chromophore of predetermined color, (b) a functional group imparting water solubility, and (c) a spacer with a terminal functional group (such as COOH) to enable conjugation to biomolecules. Fig. 1.3 gives the chemical structures of 3 typical dyes.

It is found that many of the fluorophores obtained in this way display weak fluorescence in aqueous solution but undergo a large increase in quantum yield (QY) on addition of protein. For example, the fluorescence intensity of the dye RB-627 in a 1 g L^{-1} HSA solution in phosphate-buffered saline rises about 27-fold (Fig. 1.4), that of dye RG-702 about 16-fold. This may be explained by the electrostatic and hydrophobic interactions between dye and protein, leading to a rigidization of the fluorophore and to better shielding of water molecules (which quench fluorescence) by the protein.

From plots of fluorescence intensities versus the concentration of HSA (on a log scale), the binding constants can be calculated to be 245 mg L^{-1} for RB-627 and 290 mg L^{-1} for RG-702. These findings may be exploited for purposes of protein detection (e.g., in proteomics) using a 635-nm or a 710-nm diode laser, respectively. The dyes reported here bind rather unspecifically, while others bind rather selective to certain proteins, e.g., the albumins [16].

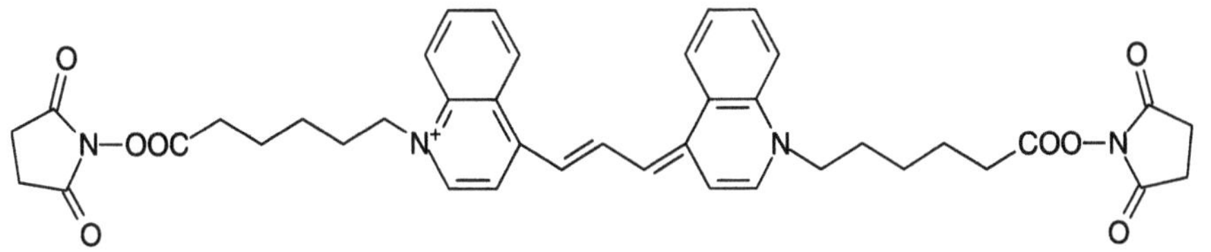

RG-702

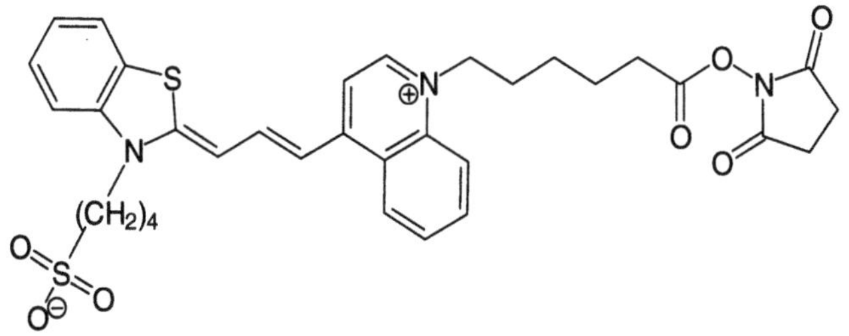

RB-627

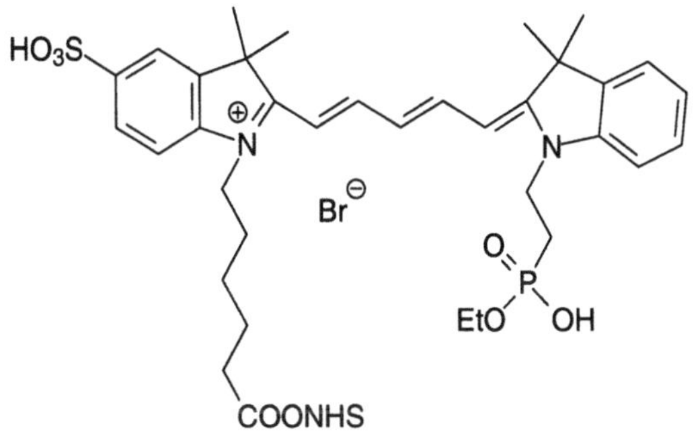

FO-646

Fig. 1.3. Chemical structures of typical diode laser-compatible fluorophores. In the form of their *N*-hydroxysuccinimide (NHS) esters they can be linked to amino groups of proteins, modified oligomers, or polymers and beads

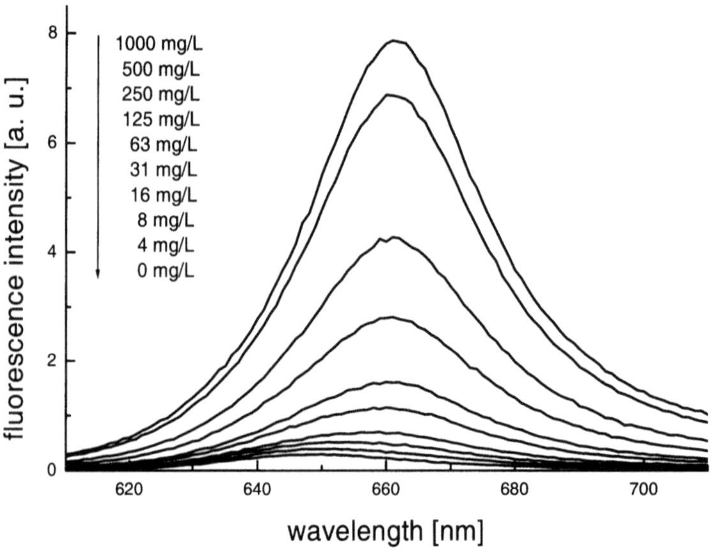

Fig. 1.4. Effect of HSA concentration on the fluorescence of RB-627 free acid ($c = 1\ \mu\text{mol}\cdot\text{L}^{-1}$; excitation by a 635-nm diode laser) on exposure to varying concentrations of HSA

1.4
Diode Laser-compatible Amino-Reactive Covalent Labels

The dyes described before have been converted into amino-reactive cyanine labels for *covalent* linkage to proteins and amino-modified nucleic acid oligomers. This was accomplished by converting them into the NHS esters. Their high absorbances ($\varepsilon > 100.000$) and adequate fluorescence quantum yields (up to 0.68 if bound to proteins) make them viable labels for proteins and in fluorescence energy transfer immunoassay which is demonstrated here for the system HSA/anti-HSA. In this assay, the donor dye was covalently coupled to HSA, and the acceptor dye (RB-627) to the antibody. Fig. 1.5 shows the change in the emission spectrum of donor-labelled HSA as increasing quantities of the acceptor-labeled anti-HSA are added.

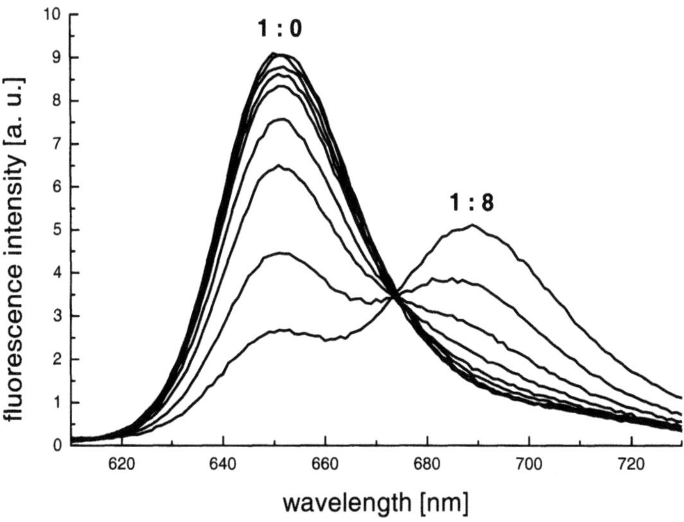

Fig. 1.5. Energy transfer study where HSA, labelled with RB-634 at a dye-to-protein ratio of 0.8 was titrated with anti-HSA labelled with RG-667. Excitation wavelength 635 nm). HSA concentration 2.1 μMol·L^{-1}, molar ratios 1:8, 1:4, 1:2, 1:1, 1:0.5, 1:0.25, 1:0.125, 1:0.063, and 1:0 respectively (from [15])

Recently, we have developed [17] near-infrared fluorescent squaraine dyes, and have exxamined their spectra, their covalent linkage to proteins, and their use as donors and acceptors, respectively, in fluorescence resonance energy transfer immunoassay based on the use of red lasers. The dyes show quantum yields of around 10% in the free form and up to 68% when covalently bound to proteins.

1.5
Diode Laser-assisted Fluorescent Single Molecule Detection of Dyes and Labeled Proteins

The probes described before are ideally suited for diode laser-based single molecule detection. A confocal laser-scanning microscope for ultrasensitive fluorescence lifetime-imaging was used. It is based on pulsed diode laser excitation and piezo scanning of the sample [18]. In confocal laser-induced fluorescence detection of single molecules, a dye or labeled protein is illuminated by a laser beam and thereby repeatedly cycled between the ground electronic state and the excited electronic state. When scanned across a focused laser beam in a confocal microscope set-up, the single fluorophores produce *bursts* of photons that can be detected, e.g., by a single photon avalanche diode. These bursts are generated as the fluorophores absorb and emit photons during their transit through the laser beam. The duration of the bursts is given by the velocity of the scanning equipment.

The photo-electrons generated by the detector are amplified and registered by a compact electronic system for time-correlated single-photon counting (TCSPC) allowing for measuring fluorescence lifetime with 40 ps time resolution, and for continuously recording photon arrival times with 100 ns time resolution [19]. Additionally, a driver electronics can be applied to synchronize the steps of scanning and data acquisition, which is essential for achieving high spatial image resolution.

To demonstrate the lifetime imaging capabilities on a single molecule level, a sample with a solution of two different dyes was prepared: the commercially available dye Light Cycler Red (LCR; from Roche Molecular Biochemical) and the new dye RB-646, both diluted to a concentration of 10^{-12} mol·L^{-1}. A drop of the solution was dried on a glass cover slip as used in microscopy, thus immobilizing the molecules at a glass/air interface.

Fig. 1.6 shows image intensity data of this mixture. The scanner driver was set to perform 500 scan steps in each direction, with a step size of 50 nm and a step time of 500 µs, corresponding to a scanned area of 25×25 µm^2. The fluorescence intensity image was calculated by first sorting the photon arrival times into time bins of 500 µs width, and then ordering the time bins into a 500×500 array, paying attention to the alternating scan directions of subsequent scan lines. Single molecules are clearly visible as circular fluorescing spots. Blinking of single molecules (due to temporary transition of molecules into a non-fluorescent state) and photobleaching also can be seen as white vertical streaks within the dark shapes of single molecule spots. Small residual line-to-line misalignment effects are also discernable in some images, due to the limited precision of the table positioning at high scan speeds.

Once knowing the exact correlation between detected photons and image position, the time-correlated single photon counting times of the photons can be used for calculating lifetime images. In Fig. 1.7, the "lifetime" of an image pixel was calculated as the average lifetime of all photons hitting the pixel, taking into account only photons arriving within a time window *after* the exciting laser pulse. More sophisticated evaluation procedures of the lifetime data are also possible, e.g., by fitting lifetime decay values to every pixel [20]. However, such a method is only reasonable for sufficiently large count rates per pixel, which is rarely the case in single molecule experiments. Two populations of molecules can be discerned: dark spots which correspond to LCR molecules with ~3.6 ns lifetime; and bright spots that are assigned to RB-646 molecules and having an ~1.4 ns lifetime. Thus, the lifetime image clearly shows the difference in lifetimes between the LCR and RB-646. More sophisticated data evaluation techniques allow even for distinguishing between these two species with nearly 100% certainty [21].

Additionally, the time-correlated single photon counting instrumentation of the time-resolved confocal microscope was used to determine the fluorescence lifetime of the dyes. Specifically, the traces of the decay profile of the fluorescence of dyes RB-627, RB-634, RB-646 and RB-661 have been recorded for aqueous solutions. In all cases, two decay times were found to describe the profiles adequately. The short-decaying components decay within typically 0.2 to 0.7 ns, with weighing factors ranging from 0.5 to 0.97. The longer components decay within 1.4 to 1.7 ns, but their relative contributions usually are much smaller.

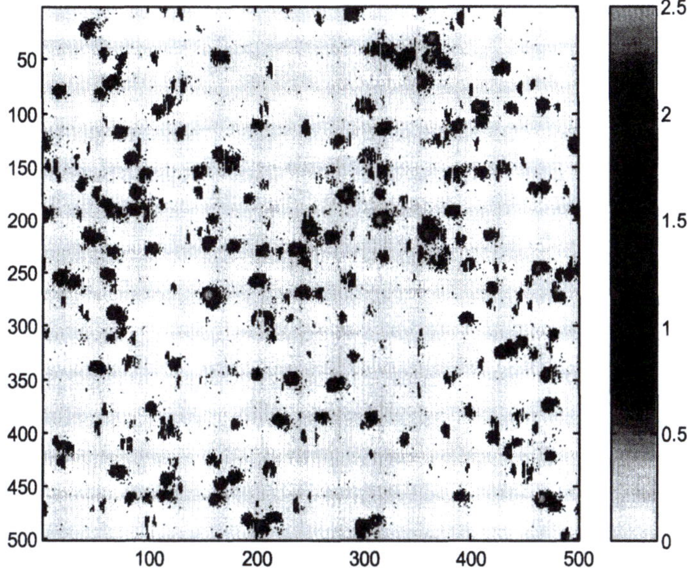

Fig. 1.6. Fluorescence *intensity* image of dyes RB-646 and Light Cycler Red on a glass surface. The gray scale visualizes the decadic logarithm of photon number per pixel. Axes labels on the image give pixel number, one pixel represents an area of 50×50 nm^2

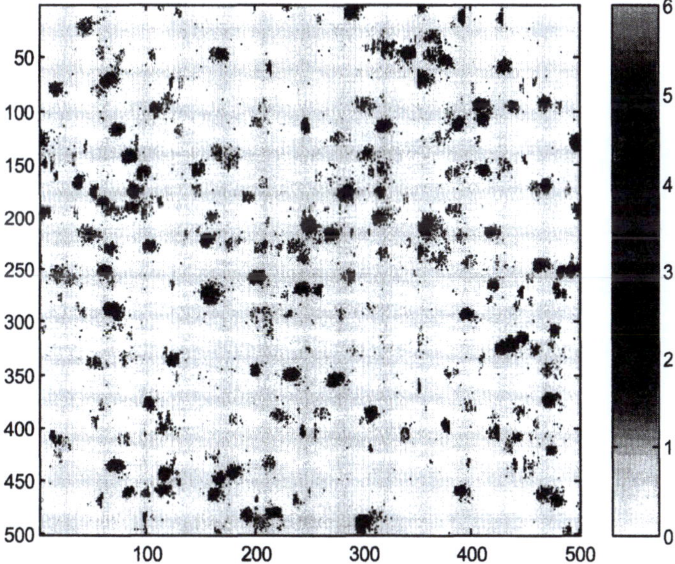

Fig. 1.7. Fluorescence *lifetime* image, corresponding to the intensity image in Fig. 1.6. The gray scale visualizes the average lifetime per pixel (in nanoseconds)

1.6
New Labels for Flow Cytometric Determination of HSA

Flow cytometry allows the simultaneous quantitative measurement of up to 100 different diagnostic parameters in a single drop of a sample. At the same time it may even save time when compared to a single parameter measurement using conventional methods. Due to its high flexibility, the technology can be used to measure numerous diagnostic analytes if it undergoes a specific molecular interaction with a partner molecule, e.g., an antibody an enzyme or a substrate, any ligand or receptor, or a complementary nucleic acid.

The Ab-Ag interaction takes place on the surface of the microspheres. The extent to which the Ag has interacted with the immobilized Ab is measured using a fluorescent label. Usually, polystyrene particles serve as a solid phase. Thousands of microspheres are analyzed per second individually. The kind of assay may be encoded as well by rendering the beads fluorescent using a second (encoding) color that typically has a red fluorescence. Flow cytometric assays involve less washing steps, enable working with whole blood, and can be fully programmed. Hence, they represent a substantial advantage over previous assays protocols.

The assay was demonstrated for the system HSA/anti-HSA. Polystyrene beads of 5.4 μm diameter were loaded with HSA (via NHS coupling). Anti-HSA was labeled with Fluorescent Orange (FO-548; Fluka; λ_{exc} 548 nm) following a standard labeling protocol (in bicarbonate buffer). The beads were incubated with the antigen solution, washed, and submitted to the flow cytometer. HSA is detectable in the 0–120 mg/L concentration range as can be seen from Fig. 1.8.

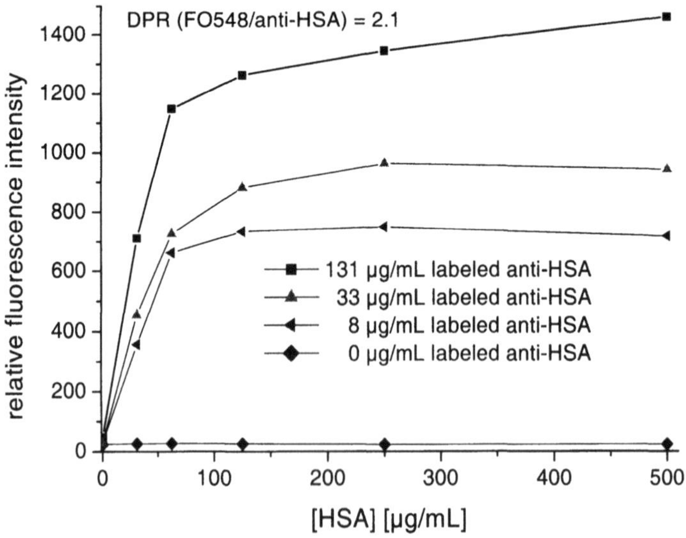

Fig. 1.8. Flow cytometric assay of HSA using the label Fluorescent Orange 548 (FO-548)

1.7
Diode Laser-excitable DNA Labels

There is a large interest in labeling oligonucleotides with fluorophores because the resulting fluorescent oligonucleotides are needed for DNA sequencing and DNA hybridization studies. The label can be introduced into the oligonucleotide via a reactive group such as an NHS ester which binds to amino groups [22, 23]. Hence, the NHS labels described before may be used to render amino-modified oligonucleotides fluorescent. An amino group may be incorporated onto the 5-end of a synthetic oligonucleotide in the last step during synthesis. Most amino-reactive labels contain spacer groups (C_4–C_6) in order to reduce the interaction of the label with the oligonucleotide.

However, the most important tagging method for DNA is based on the use of a phosphoramidite derivative of a fluorophore. In order to obtain phosphoramidite labels, a fluorophore containing a hydroxy group is reacted with a phosphine to give the corresponding phosphoramidite (Fig. 1.9). The dye activated in this way is capable of coupling to the hydroxy group of the (desoxy)ribose of an oligonucleotide. First, the hydroxylated fluorophore (F-OH) is reacted with a phosphine I to give the phosphoramidite label II. The latter is reacted with a deoxynucleoside to give III which, on oxidation with iodine, yields the labeled nucleotide IV.

Fig. 1.9 gives the chemical structures of two new phosphoramidite labels for oligomers. Both can be excited by conventional diode lasers. More importantly, they form a matched pair of labels for the kind of FRET studies described in Sect. 1.7. Table 1.4 also documents the relatively high molar absorbance and quantum yields of these squaraine dyes. Finally, it makes obvious that the introduction of a dicyanomethylene group causes a 40-nm bathochromic shift in absorption.

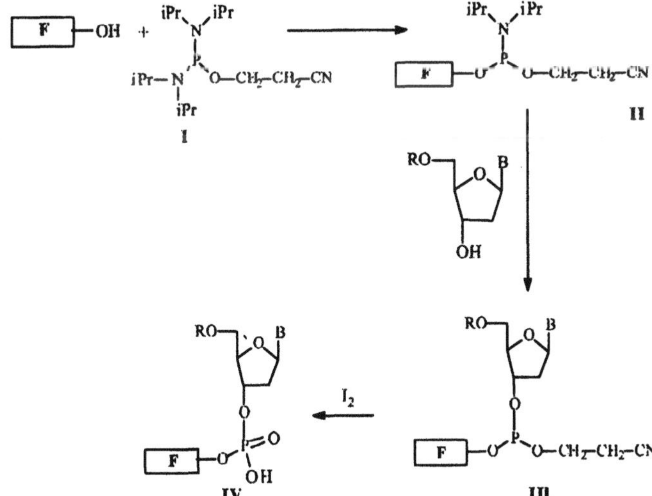

Fig. 1.9. Schematic of the synthesis of fluorescently labeled nucleotides. F denotes a fluorophore, B a nucleobase

Table 1.4. Chemical structures, absorption and emission maxima and quantum yields of diode-laser compatible phosphoramidite labels of the squaric acid type. PA stands for the phosphoramidite rest (see II in Fig. 1.9)

Structure	Absorption / emission maxima
OligoBlue 630-P	630/648 nm in EtOH $\varepsilon \approx 180,000$ QY ~ 0.6
OligoGreen 670-P	670/691 nm in EtOH $\varepsilon \approx 120,000$ QY ~ 0.5

1.8
New Resonance Energy Transfer Gene Assays

We have used both the NHS esters and the phosphoramidites to study the interaction of complementary oligomers via fluorescence resonance energy transfer (FRET). This section demonstrates the applicability of the new dyes in fluorescence resonance energy transfer assays. Amino-modified complementary oligonucleotides (15-mer) are labeled with pairs of dyes (via NHS ester coupling) which act as donors and acceptors, respectively. One label was attached to the 3'-end, the other to the 5'-end so to warrant close spatial proximity of the two labels once the duplex is formed (see Fig. 1.10).

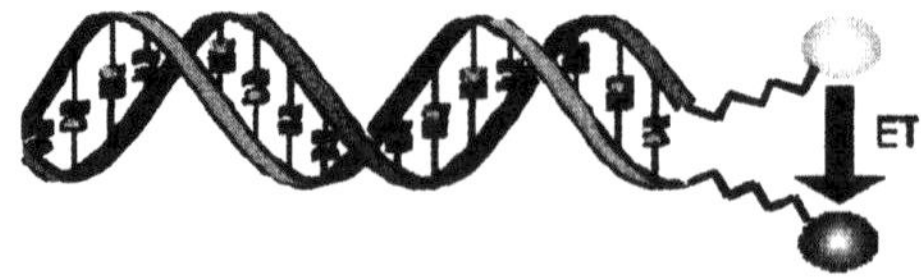

Fig. 1.10. Labeling of two complementary strands on the 3'-end and the 5'-end, respectively, results in a close proximity of the two labels in the duplex, thus resulting in efficient energy transfer

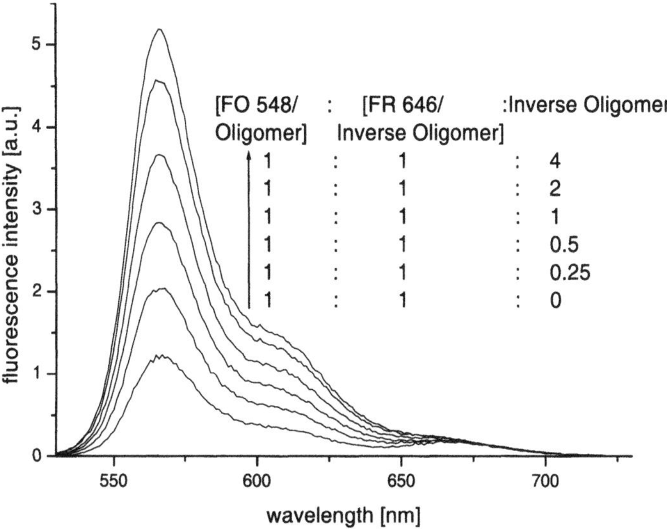

Fig. 1.11. Competitive binding of labeled (FR-646) and unlabeled inverse oligomer to a 15-mer (labeled with FO-548) as a function of the ratio between labeled and unlabeled inverse oligomer

Fig. 1.11 shows the results of a typical FRET hybridization assay. The 15-oligomer was labeled with donor dye FO-548. The inverse oligomer was labeled with acceptor dye FR-646. The inverse oligomers (labeled and unlabeled) bind *competitively* to the first strand. The efficiency of ET decreases with increasing quantities of non-labeled inverse oligomer (for quantities, see Fig. 1.11). This leads to an increase in the fluorescence intensity of the donor-labeled strand. It is noted that the second dye acts as a quencher in this case and does not display a strong fluorescence of its own.

1.9
Reactive Ruthenium Ligand Complexes as Markers for Bioassays

So far, labels have been presented with decay times in the order of ns. These can hardly be discriminated from the luminescence of biomatter if occurring at the same wavelength. Slow decaying probes, in contrast, enable time-resolved (or gated) measurements and thus allow for a superb method for improving selectivity and sensitivity.

We have synthesized the label Ru(bpy)-COOH (see Fig. 1.12) which, after activation to an NHS ester, was conjugated to HSA. In parallel, the label FR-642 was covalently linked to anti-HSA, again via an NHS ester. Fig. 1.12 also shows the result of a titration of the labeled HSA with the labeled antibody. In the absence of antibody, the typical luminescence of the ruthenium complex (with a

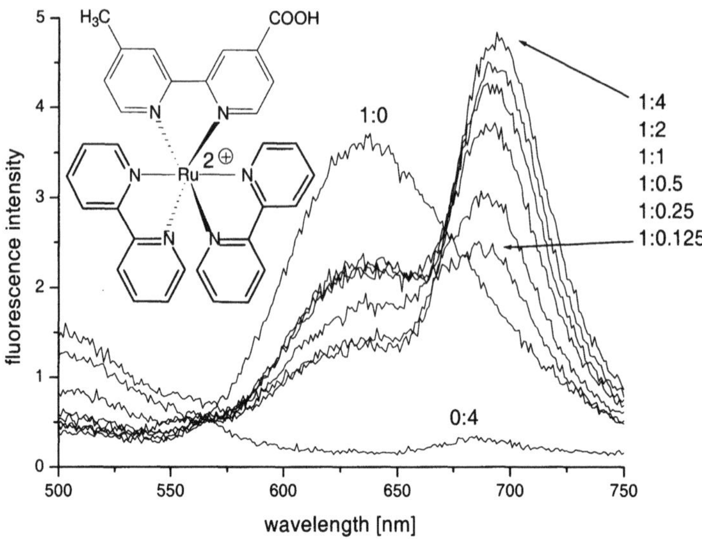

Fig. 1.12. Luminescence resonance energy transfer study on the system HSA/anti-HSA. A ruthenium label with a decay time of ≈ 450 ns acts as the donor (on HSA), a cyanine dye (on anti-HSA) as the acceptor

peak at 632 nm after excitation at 460 nm) is visible. On addition of labeled antibody, the emission of the ruthenium label is increasingly quenched, while that of the label FR-642 increases steadily. In the complete absence of labeled HSA, virtually no fluorescence is observed (see the bottom graph in Fig. 1.12) because FR-642 hardly absorbs light at the excitation wavelength.

1.10
Diode Laser-excitable Fluorescent Polymer Beads

Polystyrene beads (of 0.1–5 µm diameter) are widely used in immunoassay and in studies on receptor-ligand interactions, often in combination with flow cytometry as was shown in section 1.6. In recent years, beads have been fluorescently dyed for purposes of encoding. The color of the fluorescence of the bead, the ratio of two fluorescences of a bead, or the decay time of the fluorophore can serve for identification purposes. It is rather surprising to see that no beads have been described so far whose fluorescence can be excited by semiconductor light sources with emissions at above 600 mm. The respective beads are highly desirable, though, in view of the advantages of diode laser-based assays as outlined in earlier sections, and for encoding.

The lipophilic cyanine and squaraine dyes of Table 1.5 were synthesized and studied [24]. They display blue or green color, and we refer to them as the Lipo-Blue and LipoGreen dyes, respectively. Their chemical structures and spectral maxima are given in Table 1.5 and it is evident that they can be excited by the

Table 1.5. Chemical structures, spectral data, and quantum yields of the lipophilic dyes used for dyeing polystyrene beads

Chemical Structure	Absorption / emission maxima (solvent); quantum yields
LipoBlue 631 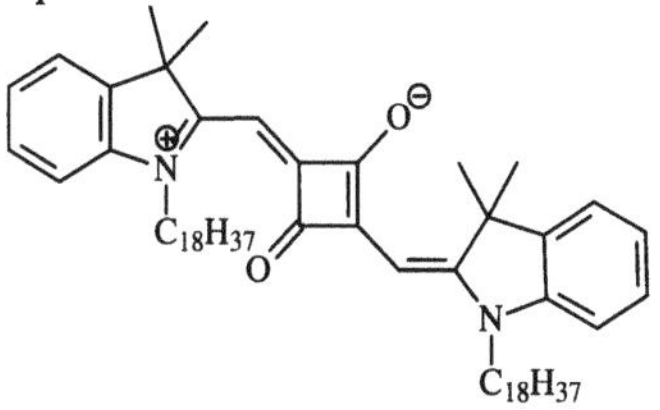	631/647 nm (ethanol) ε ~190,000 QY ~0.6 (in bead)
LipoBlue 644 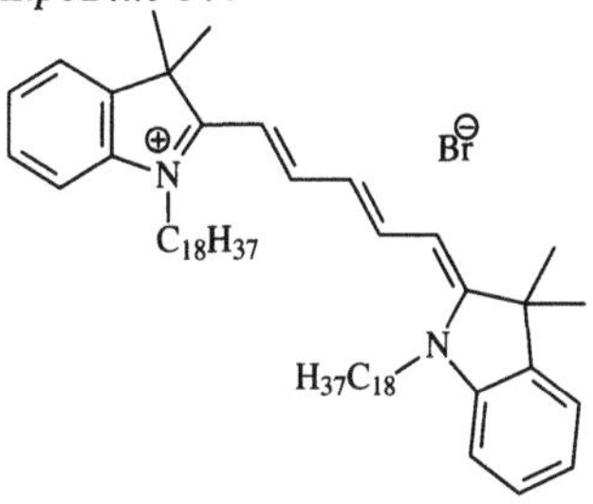	644/665 (ethanol) ε ~175,000 QY ~0.4
LipoGreen 671 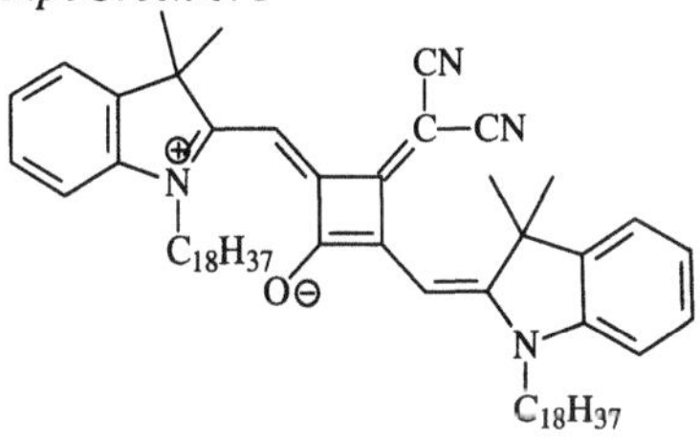	671/692 nm (ethanol) ε ~120,000 QY ~0.45
LipoGreen 783 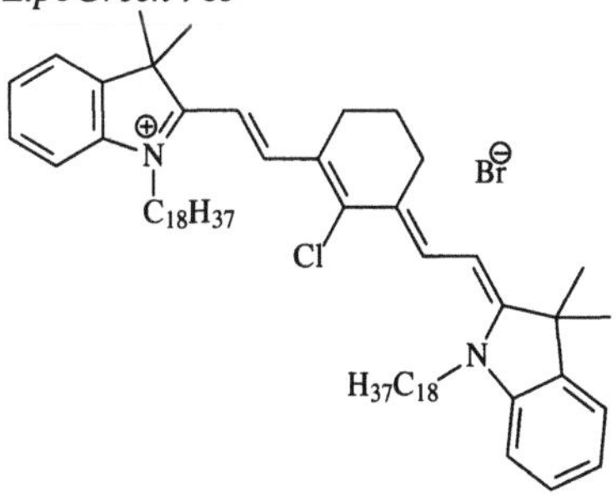	783/804 nm (chloroform) ε ~ 380,000 QY not determined

635-nm, 670-nm or the 780-nm diode laser, respectively. A C_{18} side chain renders them highly lipophilic. They have been used to dye 5-µm polystyrene particles which were first suspended in water/methanol (1:1; *v/v*), swollen by addition of 2% dichloromethane, and dyed by slow addition of the lipodye in dichloromethane. The resulting beads are weakly blue or green and display a strong fluorescence that is not quenched by oxygen, proteins, or ions such, as halides. Very recently, we also have prepared nm-sized beads (by co-precipitation of polymer and dye from solutions in dimethylformamide) using carboxy-modified poly-(acrylonitriles), with the aim to use such nanobeads as fluorescent markers for biomolecules.

1.11
Polyaniline-Coated Nano-beads (~200 nm in Diameter) with pH-dependent Fluorescence

We have previously described optical chemical sensors [25, 26] and biosensors [27] that are based on thin films of polypyrrole films obtained by chemical oxidation. Such films represent an interesting alternative to indicator-based sensor films because they exploit the intrinsic optical properties of the polymer (i.e., an additional indicator is not required), are compatible with LED and diode laser sources, and can easily be prepared. However, the pK_a values are outside the physiological pH range (which is the most important one in practice).

The polyanilines (PANIs) were found to represent an alternative class of polymers. They display pK_a values that enable optical sensing in the neutral and weakly acidic pH range [28–30]. Like polypyrrole, PANIs have absorption spectra that extend to above 1000 nm, but are nonfluorescent. Since, however, fluorescence is the preferred method in bioanalytical sciences, we were looking for a fluorescent sensing scheme.

We have designed fluorescent nanobeads that can be used for optical sensing of pH. The sensing scheme is based on the finding that aniline, if oxidized in presence of fluorescent polystyrene beads, is being deposited on the beads as a thin film of polyaniline (PANI). The resulting coated beads, schematically shown in Fig. 1.13 and typically 360 nm in diameter, have been characterized by fluorescence spectroscopy, atomic force microscopy and flow cytometry.

The fluorescence intensity of the PANI-coated beads undergoes pH-dependent changes even though the fluorophore is inert to pH (Fig. 1.14). This is due to an inner filter effect caused by the pH-sensitive PANI coating which modulates fluorescence intensity. The beads thus can act as fluorescent "bead probes" for physiological pHs.

Fig. 1.15 gives typical pH titration plots of 0.2 μm beads (from commercial sources) that were coated with a 80-nm layer of polyaniline [31]. This demonstrates that the approach is applicable to beads of various fluorescence colors. The "pK_a" of the beads may be fine-tuned by making use of substituted anilines rather than plain aniline.

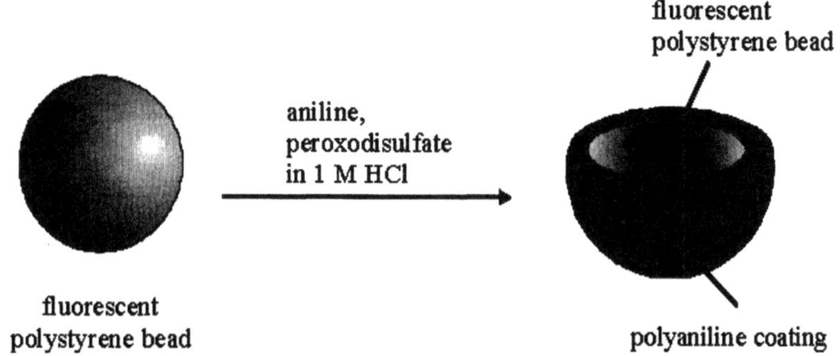

Fig. 1.13. Schematic of a fluorescent bead coated with polyaniline (PANI)

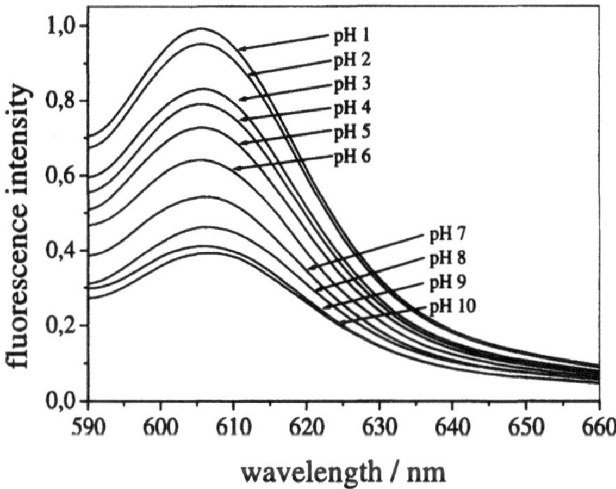

Fig. 1.14. pH-dependence of the fluorescence emission of inert fluorescent polystyrene beads coated with a thin layer of polyaniline

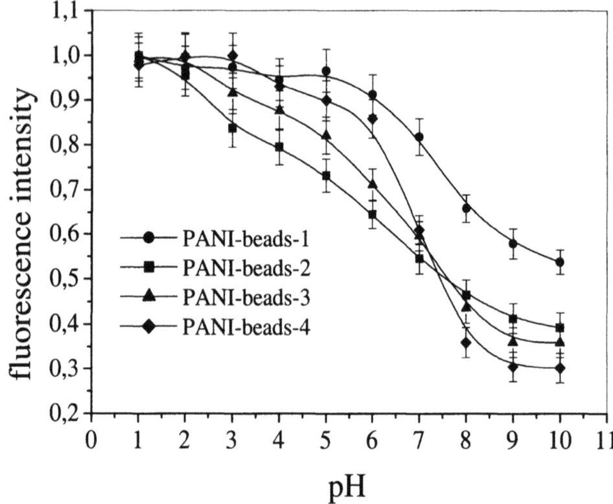

Fig. 1.15. Fluorescence titration plots of PANI-coated beads of varying fluorescence emission maxima over the pH 1 to 10 range. PB-1: λ_{em} 515 nm, PB-2: λ_{em} 560 nm, PB-3: λ_{em} 580 nm, PB-4: λ_{em} 606 nm). From [31]

1.12
Phosphorescent Poly(acrylonitrile) Nanospheres (10–100 nm in ∅) as Markers for Optical Assays

Micro- and nanospheres are useful markers in optical assays because they can largely increase sensitivity [32, 33]. In fact, thousands of fluorophores can be attached to a biomolecule via a nanobead, while conventional markers label to an extent of 1–10 markers per biomolecule only (depending, of course, on the size of the biomolecule).

In previous sections, the problems associated with the background luminescence of biological samples were overcome by shifting the analytical wavelengths of the labels (or beads) into the red or infrared. In this section, we describe an alternative approach, viz. the use of beads with a decay time in the order of several µs. The decay time of practically all natural fluorophores (causing the background) is in the order of < 20 ns. Hence, by making use of gated or time-resolved measurements, background luminescence can be widely suppressed. We use certain ruthenium(II) bipyridyl complexes (λ_{exc} 460 nm, λ_{em} 610–630 nm) as fluorescent markers.

Poly(acrylonitrile) (PAN) and its derivatives are attractive polymeric matrices for the encapsulation of phosphorescent dyes [34]. They display an extraordinarily poor permeability for gases and ionic as well as uncharged species [35]. As a result, they can protect luminescent dyes against potential quenchers such as oxygen. PAN is soluble in dimethylformamide (DMF) and acts itself as a solvent for

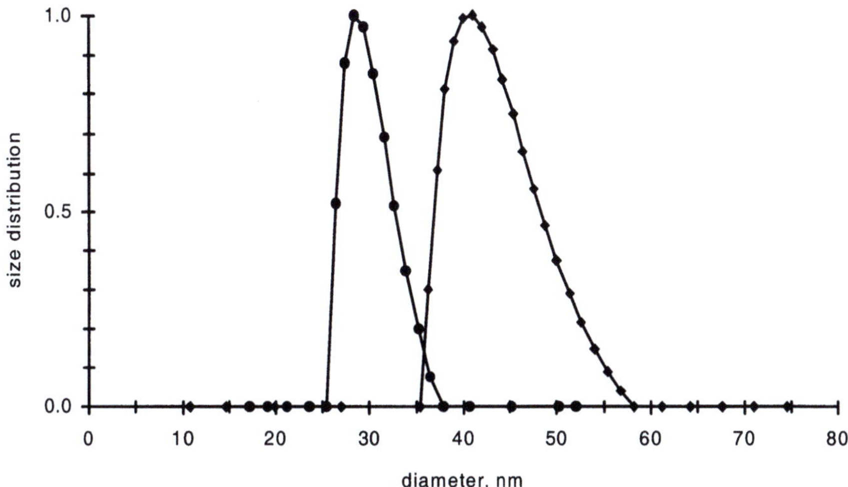

Fig. 1.16. Radii of two types of PAN nanospheres as determined by dynamic light scattering at a detection angle of 90°

lipophilic dyes. In addition to plain PAN, we have employed several functional copolymers. Among those, copolymers of PAN and acrylic acid (5%, *w/w*) proved to be most suitable as far as quenching by oxygen, quantum yields, decay times, and size of the nanosphere are concerned. PAN is amphiphilic in a sense that it is both hydrophilic and lipophilic. Nanospheres were prepared by a precipitation process: on dropwise addition of water to a dilute solution of PAN in DMF, a stable dispersion of nanoscale aggregates is formed.

If the spheres are precipitated from DMF solutions containing ruthenium-tris(diphenyl-phenanthroline) – in the following referred to as Ru(dpp) – the dye is co-precipitated with the spheres. This is an elegant way to stain nanospheres in a defined manner [36]. Solvents other than water may also be used provided that they are miscible with DMF, and that the polymer is not soluble in the binary mixture. Since PAN and its copolymers are soluble in DMF only, the nanospheres may be suspended in almost any other solvent. As can be seen from Fig. 1.16, the spheres are rather homogeneous in terms of size distribution.

The requirements for a dye to be ideally suited for incorporation into beads include (a) good solubility in both PAN (the polymer matrix) and DMF (the solvent), (b) insolubility in water for precipitation of the nanospheres, and (c) a positive charge so to bind to the negatively charged carboxy groups of the matrix. The phosphorescent ruthenium(II)-tris-polypyridyl complexes were selected as dyes since they exhibit these features. First, they yield brightly luminescent nanospheres with a Stokes' shift as large as almost 150 nm (λ_{exc} 465 nm, λ_{em} 610 nm). Due to their positive charge they electrostatically bind to copolymers containing negatively charged carboxy groups. Furthermore, the dyes can be rendered lipophilic by using proper ligands and counterions. The lipophilic dyes are extracted quantitatively into the nanospheres during the above preparation process because they are very well soluble in the polymer. Even in a lipophilic environment, e.g., if

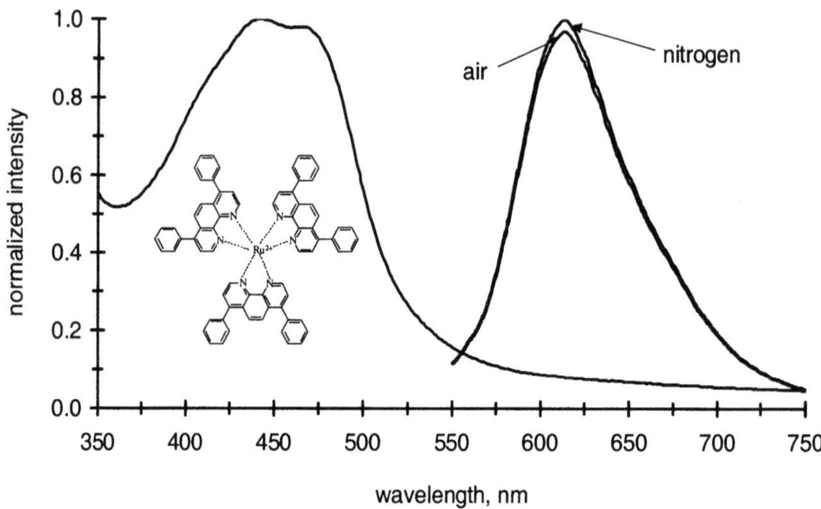

Fig. 1.17. Absorption and emission spectra (λ_{exc} 488 nm) of Ru(dpp) encapsulated into nanobeads of poly(acrylonitrile)

certain proteins are added, no leaching is observed. The high quantum yields (> 40%) and the fairly large molar absorbances ($\varepsilon \approx 30,000$ L mol^{-1} cm^{-1}) are further advantages. They are excitable by the argon ion laser at 488 nm or by blue light-emitting diodes (LEDs with emision peaks at either 450 nm or 470 nm). Last but not least, the complexes are stable against the loss of ligands, and its emission spectrum is broad enough to overlap with the absorbance spectra of various luminescence acceptor dyes which is of interest in the context of resonance energy transfer (also see Sect. 1.9).

Fig. 1.17 shows the excitation and emission spectra of beads dyed with Ru(dpp) and suspended in buffer of pH 7.0 for both aerated and de-aerated solutions. Quenching by oxygen is virtually absent (3–5%), a fact that indicates that the dye is fully incorporated into the beads, since quenching of the luminescence of Ru(dpp) is much stronger in water solution.

The luminescence frequency spectra (Fig. 1.18) reveal the dependence of the phase angle and the modulation on the modulation frequency applied. An evaluation of the data points leads to a bi-exponential fit as the best match which is typical for incorporated dyes. The apparent decay time varies between 6–7 μs for the main component (~95%) and 1–2 μs for a second component (~5%). The minor component is assumed to result from surface-bound dye which is susceptable to quenching by oxygen.

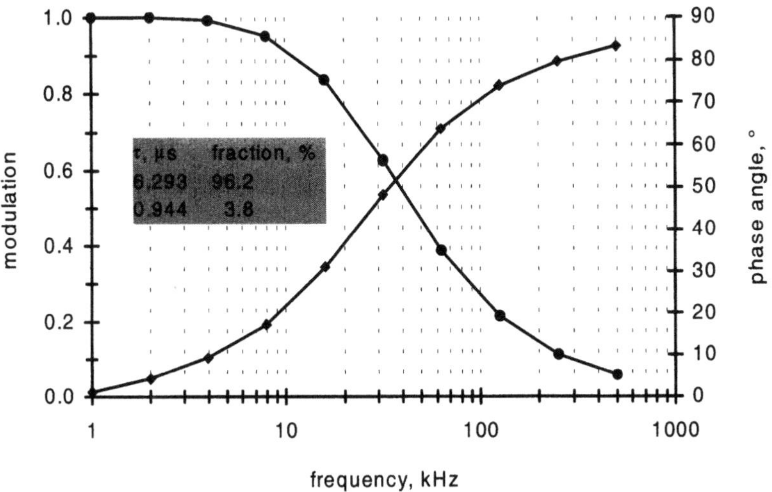

Fig. 1.18. Luminescence frequency spectra of Ru(dpp) nanobeads displaying modulation (●) and phase angle (♦)

1.13
Competitive Binding of Streptavidin to Biotinylated Nanobeads as Studied by Resonance Energy Transfer

In the previous section a technique has been presented that leads to highly phosphorescent, inert nanospheres which can act as luminescent markers due to the presence of carboxy groups. These nanospheres offer the possibility to create a novel scheme for a homogeneous bioassay based on the long decaying luminescence of ruthenium polypyridyl complexes. In order to prove this, biotin was covalently linked to the Ru(dpp)/PAN nanobeads (with COOH groups on the surface) via amide coupling. Streptavidin (SA), fluorescently labeled with AlexaFluor 633 (absorption/emission maxima at 633/645 nm), was used as acceptor dye. The beads were then contacted with solutions containing avidin in varying concentrations and labeled SA in constant concentration (see Fig. 1.19). At low fractions of avidin, most of the surface will be covered with labeled SA, while the opposite is the case if high fractions of avidin are employed.

Upon binding of labeled SA to the biotinylated surface of the nanobeads, resonance energy transfer (RET) occurs between Ru(dpp) (contained in the beads), and the acceptor (AlexaFluor 633 attached to SA). As a result of RET, the luminescence of Ru(dpp) excited at 470 nm is affected in two ways, namely in terms of intensity and in terms of decay time. The ratio of the intensity of the emission of labeled SA (with its peak at 645 nm) and the Ru(dpp) emission (with its peak at 610 nm) is reduced by 40% on complete coverage of the surface with avidin. The intensity of the emission of AlexaFluor 633 (peaking at 645 nm) increases by a factor of 1.8 [37].

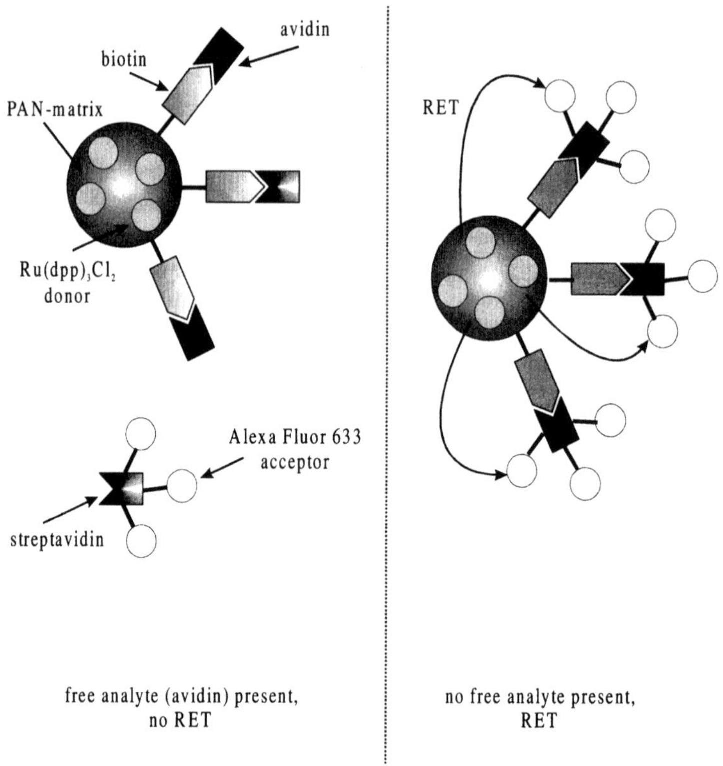

Fig. 1.19. Competitive binding assay using nanobeads labeled with Ru(dpp). Labeled streptavidin and free avidin compete for the biotin binding site on the surface of the bead. Left: situation in case of a large excess of avidin; right: no unlabeled avidin present, so that all binding sites are blocked with labeled streptavidin; as a result, efficient RET occurs [37]

Similar effects can be seen in the time domain. The decay time of nanobeads dyed with Ru(dpp) is 4.4 µs in presence of a large concentration of avidin, but is reduced to 2.5 µs if loaded with labeled SA and no avidin present, as can be seen from Fig. 1.20. It should be noted that in these experiments the concentrations of both the Ru(dpp) and labeled SA were kept constant (33.1 mg of beads per liter; corresponding to 700 nMol dye per liter), while the concentration of unlabeled avidin was varied.

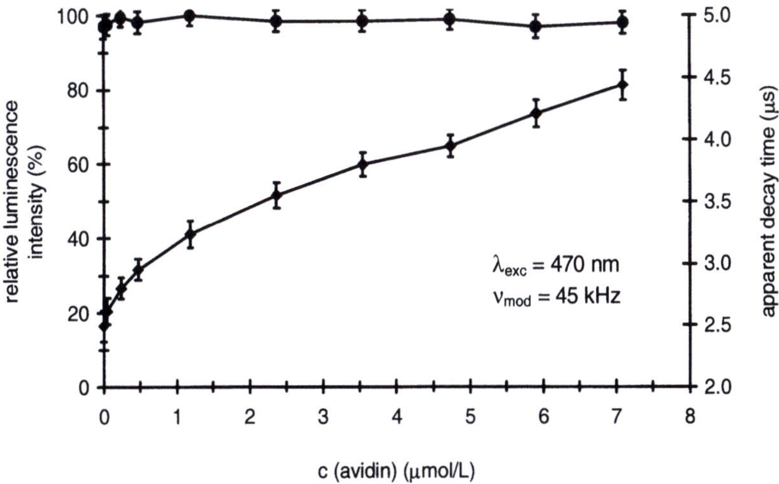

Fig. 1.20. Effects of competitive binding of avidin and fluorescently labeled streptavidin (SA) to biotinylated nanobeads dyed with Ru(dpp). The relative luminescence intensity of the overall luminescence (•) remains constant. In contrast, the decay time (♦) of the emission of Ru(dpp) increases with the concentration of avidin, because with increasing fractions of avidin in the solution less SA binds to the beads, so that less energy transfer does occur

1.14
Nanobeads as Reference Dyes in Luminescent Lifetime Imaging Using DLR

Recently, we have introduced a new method for time-resolved luminescence lifetime imaging of µs-decaying optical sensors [38, 39]. Sensor materials with µs decay times have been presented for imaging of oxygen, carbon dioxide, pH [40], and temperature [41]. The opto-electronic system is composed of a CCD camera (with a fast electronic shutter) and a pulsed LED light source. Both are fast enough to measure decay times in the µs time regime without employing expensive components such as blue lasers or fast gateable image intensifiers. The system also makes use of reference dyes (with µs decay times) which – in contrast to the indicator dyes – remain totally inert and serve as a generator for a reference luminescence signal only. By making use of this so-called DLR technique [39], the rather large number of fluorescent probes that decay within nanoseconds are susceptible to µs decay time imaging.

DLR is an intrinsically referenced scheme for imaging, i.e., it can compensate for fluctuations in light source intensity and for positional variations. It is based on the use of a fluorophore (indicator) and an inert phosphor (the reference) and can convert fluorescence intensity into a time-dependent parameter. We use the dye

Ru(dpp) almost exclusively. It is added to the sensing layer in the form of nano-beads prepared from poly(acylonitrile) as described before. Both the reference and the indicator probe (e.g., for pH) are excited simultaneously by a blue LED, and an overall luminescence is imaged. In the time-resolved imaging method, two images are taken at different time gates (= delay times) by the CCD-camera. The first image is recorded during excitation (A_{exc}) and reflects the luminescence signal of both the fluorophore (the pH probe) and the phosphor (reference). The second image (A_{em}) which is measured after a certain delay (after switching off the light source), is solely caused by the long-lived phosphorescent dye. Since the intensity of the fluorophore contains the information on pH, whereas phosphorescence is pH-independent, the ratio of the images displays a referenced intensity distribution that reflects the pH at each picture element (pixel). The scheme is particularly compatible with LED light sources and cameras that can be gated with square pulses in the microsecond range.

For a successful implementation of the time-resolved DLR scheme (t-DLR), it is mandatory that (a) reference luminophore and indicator fluorophore have largely different decay times, (b) both the decay time and quantum yield of the reference luminophore are not affected by the sample; (c) the indicator fluorophore changes its fluorescence intensity as a function of the analyte concentration; (d) the excitation spectra of reference and indicator overlap in order to allow the simultaneous excitation at a single wavelength; (e) the luminescence of both the reference and the indicator are detectable at a common wavelength (or range of wavelengths); (f) both luminophores are in tight spatial proximity; and (g) the ratio of the two dyes remains constant.

When using an ideal t-DLR sensor (where both dyes display identical absorption and emission), all disturbing influences having the same effect on both dyes are referenced out, no matter whether the effect is generated within the sample (e.g., filtering effects), in the sensor foil (e.g., inhomogeneous dye distribution) or by the system components (e.g., inhomogeneous excitation lightfield). This makes t-DLR imaging clearly superior over standard 2-color encoding methods, especially when working with poorly defined heterogeneous systems (e.g., tissue), and therefore provides a unique method in terms of quality of assays and images. In addition, applying nanobeads as a reference extends the parameters to be quantitatively sensed with our "slow" microsecond resolving imaging system from temperature to pO_2, pCO_2, and pH. Fig. 1.21 shows pH images obtained by the t-DLR technique using Ru(dpp)-based nanobeads as reference dyes in order to correct for the heterogeneous intensity pattern of a planar pH sensor foil.

The utility of t-DLR imaging is demonstrated for the visualization of a pH gradient on a pH sensor foil [39]. A planar optical sensor foil (1×5 cm^2) sensor strip was placed on a transparent plastic support contained in a plastic tube as shown in Fig. 1.22B. The sensor was exposed to a pH gradient inside the tube which was created by filling the tube with buffer solution of pH 6.0 on one side and with buffer of pH 8.0 on the other. The higher liquid level at the side of the more acidic pH forced the fluid to migrate through a barrier (consisting of filter paper) to the side of the basic pH. The movement of the migration front was imaged in time-intervals of 1 minute.

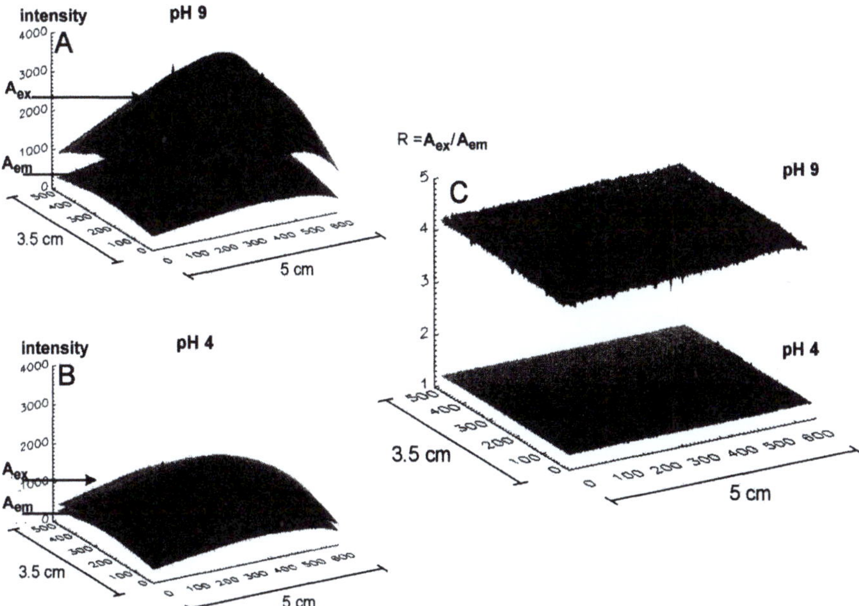

Fig. 1.21. Nanobeads can serve as reference dyes in time-resolved imaging for intrinsically correction of the heterogeneous intensity pattern of a planar pH sensor foil. **A** Surface plots A_{exc} (reflecting the sensor response towards pH) and A_{em} (pH independent response of the nanobeads) at pH 9, shown in one graph. **B** Plots A_{exc} and A_{em} at pH 4, both pictures again combined in one plot. **C** Intrinsically referenced parameter R at pH 4 and pH 9 (after correction for light field inhomogeneities and other perturbers) combined into plots for pH 4 and pH 9, respectively

Contour plots of the pH distribution were combined into on image, given in Fig. 1.22A. The first data were acquired 1 min after filling the tube with buffers. The image shows a homogeneous pH distribution of about pH 8.0 all over the sensor area and the pH 6.0 buffer does not affect the image. From minute 2 to 11, the front of the pH 6.0 buffer migrates from the right to the left side of the sensor. Due to a non-homogeneous front of the diffusion barrier, the pH front is not rectangular to the sensor expansion. However, this does not affect migration. The pH drop at the front is surprisingly sharp. After 12 min, the whole sensor area reports pH 6.0.

In order to quantify the speed of migration, the sensor area was divided into 20 segments, as shown in Fig. 1.22B. Each segment is 2.5 mm long. The average pH in the respective segment was calculated from the pixel data and plotted against the segment number. Then, the migration distance of the diffusion front at 50% of ΔpH is calculated and plotted versus time, as shown in Fig. 1.22C. A linear relation is obtained that can be fitted to a linear regression. Its slope reflects the migration speed which in our case is approximately 5.5 mm per min.

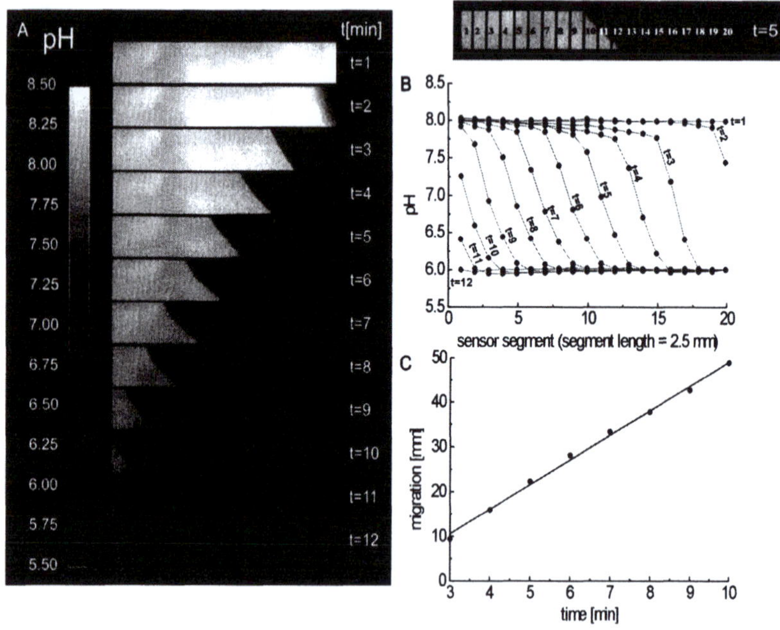

Fig. 1.22. Time course of the migration of a pH gradient along a planar t-DLR-sensor. **A** data plotted as contour plots and combined into one image. **B** plot of the average pH of defined sensor segments for different times. **C** plot of migration distance of the pH front versus time

1.15
Phosphorescent Nanospheres for Use in Advanced Time-resolved Multiplexed Bioassays (λ, τ)

There is a tremendous need for so-called multiplexing beads with clearly distinguishable optical properties [32, 42–47]. Such beads enable the differentiation of a large number of species such as oligonucleotides or proteins. The use of beads increases the sensitivity of luminescence assays [5–9]. In practice, the number of clearly and quickly distinguishable labels can be 100 or more. This is particularly true for applications such as cell separation, flow cytometry, DNA-chips and immunochips, fluorescence microscopy, and in so-called nano-devices. Here, we introduce a scheme leading to a vast number of markers that can be differentiated in terms of both emission wavelength and decay time.

The scheme is based on the finding that efficient resonance energy transfer (RET) can occur in modified poly(acrylonitrile) beads (PAN) between the donor dye Ru(dpp) (λ_{exc} 465 nm, λ_{em} 610 nm) if a second dye is added with a significant absorption within the emission band of Ru(dpp). Since this emission band is rather broad (extending from 550 to 750 nm), a number of acceptor dyes can be em-

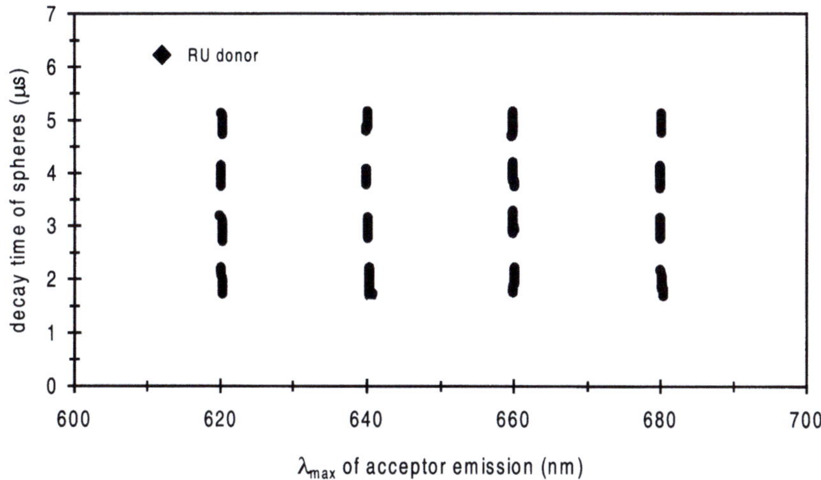

Fig. 1.23. Two-dimensional field showing classification of 16 nanosphere markers based on simultaneous analysis of both the emission wavelength of the cyanine acceptor and the apparent decay time of spheres [48]

ployed. In our experiments [48], cyanine dyes were used as acceptors because they have narrow absorption and emission bands, high absorbances, and can be fine-tuned (in terms of absorption maxima) rather easily as has been shown in Sect. 1.1.

Fig. 1.23 shows a two-dimensional dot plot of luminescence markers that can be clearly assigned by both the emission wavelength of the cyanine acceptor (x-axis) and the apparent decay time of the nanospheres (y-axis). By varying the concentration of the acceptor dye, the luminescence decay time of the donor can be modulated. Consequently, the decay time of the fluorescence of the acceptor dye is governed by the decay time of the donor. It therefore is possible to use the decay time of the microspheres as a parameter for identification along with the spectral maxima.

The ruthenium complex Ru(dpp) was chosen as a donor label because it has a decay time in the order of several µs ($\tau = 6.2$ µs in PAN) and is well soluble in DMF and in the polymer matrix, but not in water. Due to its positive charge, it strongly interacts with polymers containing negatively charged groups. It is extracted quantitatively into the nanospheres during the preparation process that starts from DMF solutions of the dye and the (acrylonitrile)/(acrylic acid) copolymer. Rather high concentrations of Ru(dpp) are tolerable without self-quenching, thus leading to brightly luminescent beads. The quantum yield is 0.38, and the molar absorbance ($\varepsilon = 30,000$ L mol^{-1} cm^{-1}) fairly high.

The choice of cyanines as acceptor is based on the following reasons: Cyanines absorbing between 550 and 750 nm show almost no intrinsic absorbance at wavelengths shorter than 500 nm. Consequently, the donor dye Ru(dpp) can be selectively excited with the blue LED or the argon ion laser. Many cyanines have molar absorbances exceeding 200,000 L mol^{-1} cm^{-1} and large quantum yields. Their

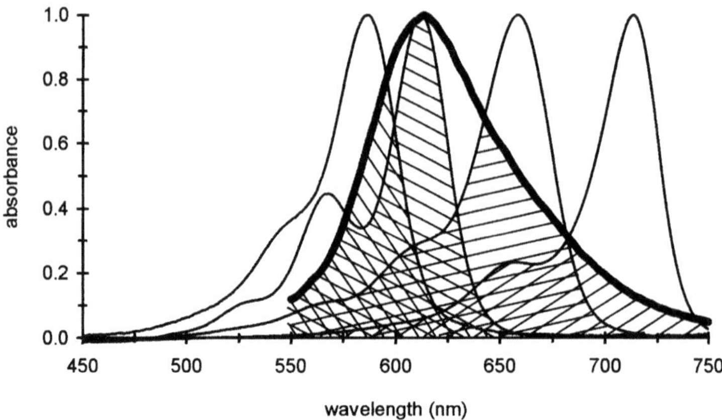

Fig. 1.24. Normalized absorbance spectra of the cyanines in DMF. From left to right: dyes CY582, CY604, CY655, CY703 (from Aldrich). The hatched areas indicate the overlap of the emission of the Ru(dpp) donor with the absorption of the four cyanines

lipophilic character along with the usually positive charge (which results in a strong interaction with the carboxy groups of the respective copolymers) simplifies the incorporation into nanospheres. The large overlap with the absorption spectra of the luminescence of Ru(dpp) along with the small fluorescence emission peaks of the cyanines are further attractive features for RET applications.

Fig. 1.24 shows the absorbance spectra of the cyanines in DMF solution. The hatched areas indicate the overlap of the phosphorescence of the ruthenium complex with the cyanine absorption spectra. In fact, up to eight different cyanines can serve as energy acceptors, provided that their excitation wavelengths match the donor emission wavelength range (which is from approximately 550 nm to 750 nm).

Fig. 1.25 shows the emission spectra (at λ_{exc} 488 nm) of the nanospheres containing Ru(dpp) (in constant concentration) and the carbocyanine dye CY655 at various concentrations. The relative luminescence intensity of the nanospheres rises with the increase of the acceptor concentration. Since the excitation of the cyanines cannot result directly from the argon ion laser light source due to negligible absorbance at 488 nm, it must result from a RET from the excited ruthenium complex to the cyanine dye.

Fig. 1.25 gives the phase angles and its modulation over the frequency range from 1 kHz to 1 MHz. The apparent decay time is composed of both that of the ruthenium donor and the delayed decay time of the cyanine acceptor. In the frequency spectra, the modulation drops from 1 to almost 0 with rising modulation frequencies. Simultaneously, the phase angle increases from 0° to approximately 90°. The slight decrease of the phase angles at frequencies close to 1 MHz is caused by a small fraction of direct excitation of the cyanines at those high frequencies. If a mono-exponential fit is applied, the intersection of the modulation and phase angle curves lead to the proper modulation frequency for the ruthenium complex of about 45 kHz.

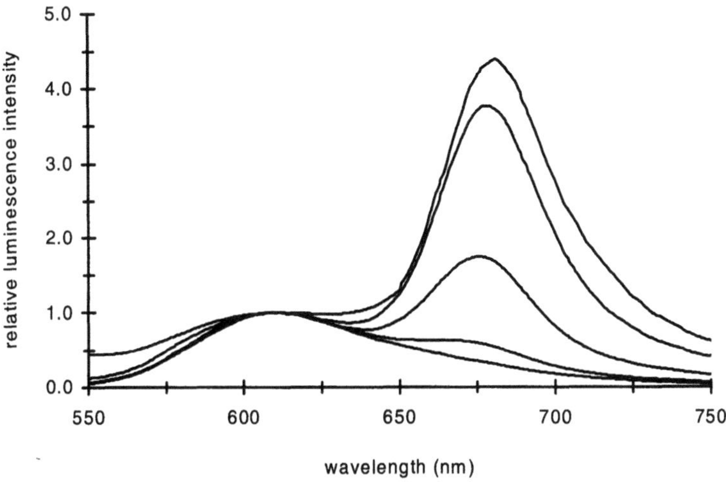

Fig. 1.25. Emission spectra of CY655 in phosphate buffer at increasing acceptor dye concentrations. From top to bottom: 2.43, 0.89, 0.32, 0.08, 0 mmol dye per kg of beads [48]. The emission of the Ru(dpp) is normalized to 1.0 at 610 nm

In each of the four arrays, the cyanine absorbance at the appropriate wavelength increases with dye concentration at a constant level of donor concentration. The underlying principle of all arrays is that the luminescence intensity (and decay time) of the ruthenium complex decreases due to resonance energy transfer to the cyanine acceptor within the nanosphere.

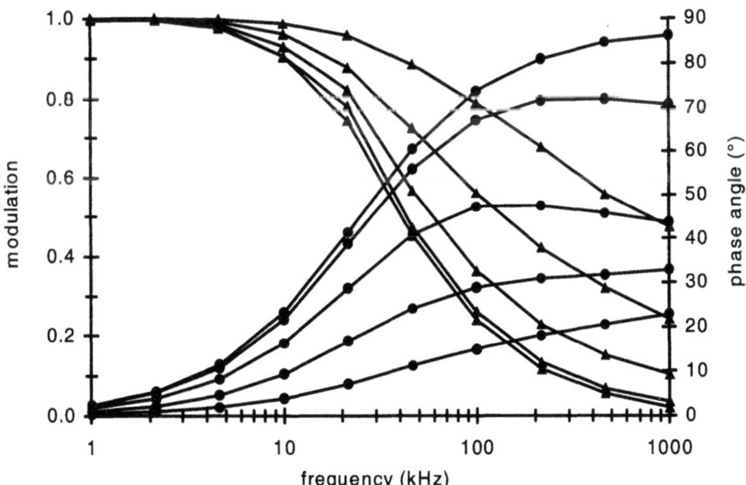

Fig. 1.26. Frequency spectra of beads dyed with Ru(dpp) and various concentrations of dye CY655 (2.43, 0.89, 0.32, 0.08, 0 mmol/kg beads). (•): phase angle; the largest phase shift (86°) refers to 0 mmol/kg; (▲): modulation; the smallest value (0.03) refers to 0 mmol/kg.

The apparent decay time in air consecutively decreases from 6.2 μs (no acceptor) to 0.4 μs (at the maximum acceptor concentration) depending on the applied cyanine and its concentration. Along with this, the quantum yield in air decreases from 0.38 to 0.07 (in case of 2.4 mmol dye per kg beads). The extent to which oxygen quenching occurs is very low throughout all arrays, typically a less than 4% change in τ on going from aerated to deaerated suspensions.

1.16
Beads Dyed with a Europium-based Label and Excitable with the 405-nm LED Diode Laser

Chelate complexes of the lanthanide ion europium(III) have been widely used in all kinds of binding assays [9, 10, 49] because of their long decay times which are in the range between several μs to even milliseconds. As a result, so-called gated measurements become possible. In such a measurement, the sample is excited with a short pulse of UV light, and the emission is gathered only after a delay of several 100 ns, during which the short-lived background luminescence is allowed to decay and therefore cannot interfere. After the delay time, the measurement gate is opened, typically for a period during which the intensity of the Eu label has dropped to 20 to 10% of its initial value.

All existing europium(III) (Eu) labels require UV excitation, typically between 330 and 370 nm. At present, this can be accomplished with rather expensive light sources only and causes strong luminescence background in the UV and near visible. Fortunately, the emission of the Eu labels is extremely red-shifted, with peaks between 610 and 620 nm. An Eu label that may be excited by the recently available blue (or UV) LEDs or the 405-nm diode laser is therefore highly desirable.

We have taken advantage of the fact that the europium-tetracycline complex (EuTc) has an absorption maximum at 405 nm and therefore exactly matches the line of the blue diode laser. The EuTc complex can be incorporated into polymers such as polystyrene or poly(acrylonitrile) (PAN), but also poly(acrylic acid) (PAA). The beads (and nanobeads) made from PAN or PAN/PAA display pink fluorescence if excited with light of wavelengths between 370 and 420 nm. Due to the presence of surface carboxy groups they can be surface-modified with biotin as shown for other nanobeads [37]. We consider these (nano)beads to be viable labels for various kinds of binding assays that make use of gated measurements.

1.17
A Europium(III)-based Probe for Use in Oxidase-Associated Reactions

1.17.1
Significance of Probes for Hydrogen Peroxide (HP)

All oxidases produce or consume hydrogen peroxide (HP). Conventional oxidases such as glucose oxidase, *produce* HP. Hence, by determining the quantity of HP formed by oxidases, a quantitative determination of the respective substrate is possible. This, in fact, forms the basis for practically all rapid diagnostic tests for blood glucose. Peroxidases (POx), in contrast to conventional oxidases, *consume* HP when oxidizing their respective substrates. Since POx are widely used markers in enzyme-linked immuno-sorbent assays, fluorescent probes for determination of POx (via measurement of the HP consumed) are of widest interest. We report on the use of the europium(III)-tetracycline complex (EuTc) as a new probe for HP, and on its use in oxidase-associated reactions.

1.17.2
A New Probe for Hydrogen Peroxide

The europium-tetracycline complex (EuTc) [50] undergoes a large increase in luminescence intensity if hydrogen peroxide (HP) is added even in low concentrations, and this has been exploited to increase the limits of detection of tetracycline in pharmaceutical analysis [51]. We find that the effect can be inversed in that it can be applied to the detection of HP if the concentration of added reagent (EuTc) is kept constant. Notably, the effect occurs at physiological pH's (6.9 is best). This is in contrast to chemiluminescence-based methods for HP where pHs between 8.5 and 9.0 have to be applied.

Fig. 1.27 shows the absorption and emission spectra of EuTc in the absence and in presence of HP. The emission of EuTc is characterized by the typical large Stokes' shift of europium(III)-ligand complexes. The absorption spectrum of EuTc has a peak at 405 nm and it is noted that this is the wavelength of the violet diode laser that recently has become available. The decay time of the luminescence of EuTc is in the order of 13 µs. All these factors make EuTc a promising luminescent probe for HP and, in particular, for oxidase-associated reactions. Its long decay time paves the way for gated measurements. In the following, several examples are given for new assays based on EuTc.

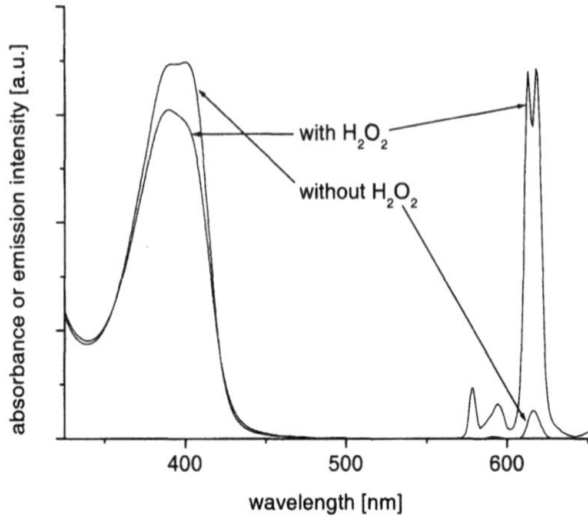

Fig. 1.27. Effect of hydrogen peroxide on the absorption and luminescence of EuTc

1.17.3
Glucose Assay Using the Europium Probe

Glucose oxidase (GOx) converts glucose to gluconolactone, and HP is a byproduct of this reaction. In order to study the usefulness of the EuTc reagent, we have added glucose solutions of various concentrations to solutions containing both the EuTc probe and GOx, and then monitored the enhancement of the luminescence of EuTc at 616 nm (under 405-nm excitation). It can be seen in Fig. 1.28 that luminescence intensity increases over time due to formation of hydrogen peroxide which in turn depends on the concentration of glucose.

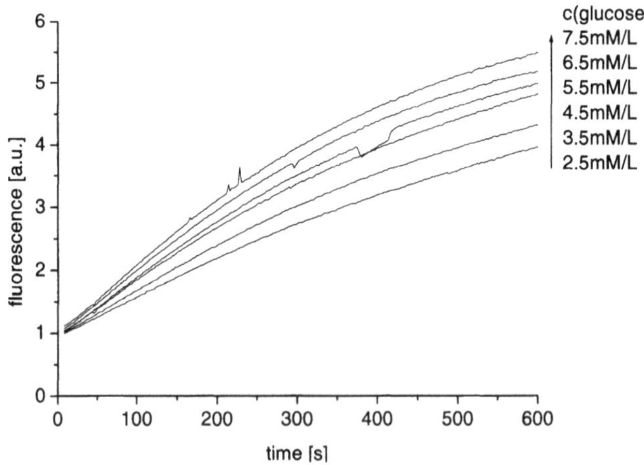

Fig. 1.28. Kinetic assay of glucose in the physiological concentration range via the increase in the luminescence of the EuTc complex following enzymatic formation of hydrogen peroxide

A more detailed study revealed that glucose can be determined in the range between 0.1 to 1.2 mM at a GOx concentration of 1.3 mg per mL. The detection limit (defined as 3 SD/slope) is 40 μmol L^{-1}. For the determination of glucose at human blood levels, i.e., between 2.5 and 30 (or even more) mmol L^{-1}, the activity of the enzyme was optimized (as a compromise between saturation by enzyme and saturation by substrate at any glucose concentration) to be 0.36 units per mL. The assay may also be run on a microplate reader using time resolution.

1.17.4
Peroxidase Assay Using the Europium Probe

Peroxidases (POx's) form a class of enzymes that use HP to oxidize their substrates [52]. We have studied horse raddish peroxidase (POx) for its capability of using the HP (contained in the EuTcHP complex) for oxidizing phenol. The kinetics of the scavenging of HP is shown in Fig. 1.29. It can be seen that with different concentrations of POx, the system displays different kinetic response. Since POx is a popular label in enzyme-linked immunoassay [53], this finding is of particular significance with respect to the detection of POx in ELISAs [54]. The assay envisioned here requires the reagent EuTcHP and phenol only. Different ratios of HP and phenol were applied to test whether the assay can be optimized for different POx activities. The results show that the widest linear range is obtained for a 1:1 ratio between phenol an HP, while at higher ratios (up to 100:1) the linear range becomes narrower.

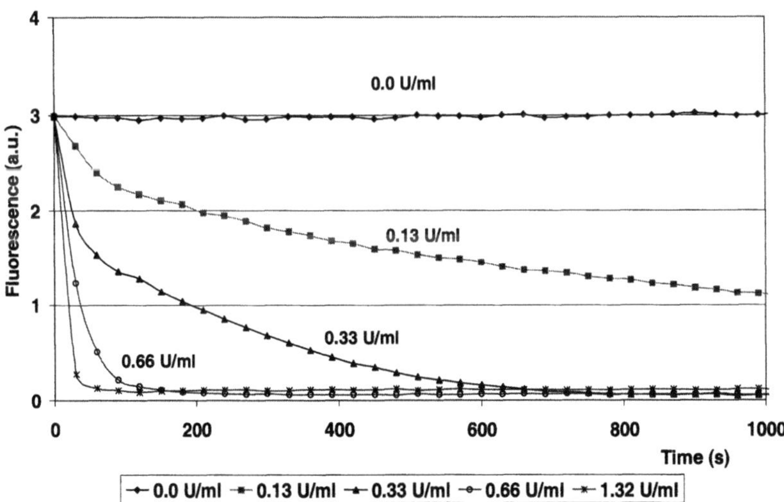

Fig. 1.29. Effect of increasing the activity of peroxidase (from horseraddish) on the fluorescence of the system EuTc-hydrogen peroxide. The upper graph reveals that the fluorescence of the system is constant over time in the absence of POx

We find, however, that not all peroxidases are compatible with this scheme. Gluthathione POx, for example, quenches the fluorescence of the EuTcHP system. Therefore – and in contrast to horseradish POx – it is not a viable POx.

1.17.5
Catalase

The catalases form a subgroup of peroxidases that catalyze the direct decomposition of hydrogen peroxide, i.e., without the aid of other substrates. It is known, however, that catalases also can exhibit a peroxidase-like activity [55]. We have studied catalase for with respect to the kinetics of the decomposition of HP to give water and oxygen. Fig. 1.30 shows the result of an experiment in which catalase was added to the EuTc-hydrogen peroxide (EuTcHP) complex. On increasing the activity of the catalase (EC 1.11.1.6; from bovine liver), EuTcHP is decomposed due to consumption of HP. Hence, fluorescence decreases to a level corresponding to that of EuTc only.

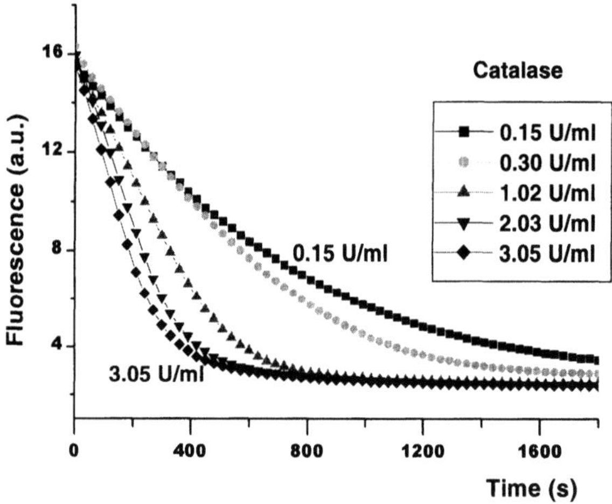

Fig. 1.30. Effect of increasing activities of catalase on the fluorescence of the system EuTc-hydrogen peroxide

1.17.6
A New Enzyme-linked Immunosorbent Assay (ELISA)

The ELISA has revolutionized immunoassay [53, 54, 56]. It is specific, sensitive, safe, and inexpensive. Since peroxidases (POx's) are quite stable under the conditions used for storage, cross-linking, and detection, it has become a popular label for ELISA. We have exploited the above discovery that the EuTc-hydrogen complex (EuTcHP) is a viable fluorescent probe for the acitivity of POx and have adapted this to an ELISA. In a typical experiment, the antigen (in our case IgG)

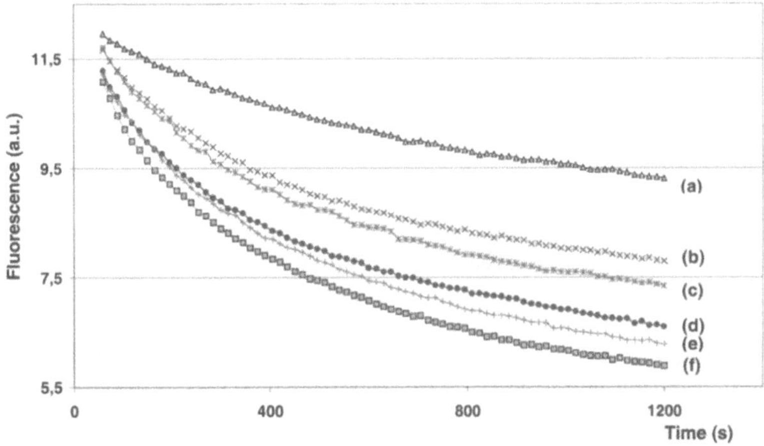

Fig. 1.31. Kinetics of the IgG sandwich ELISA in which the strongly fluorescent EuTc-hydrogen peroxide complex is decomposed by the POx-labeled anti-IgG, thus giving the weakly fluorescent EuTc complex. **a** blank (the EuTc-hydrogen peroxide complex only); **b** 0.3 ng/mL of POx-IgG; **c** 0.6 ng/mL; **d** 3.0 ng/mL; **e** 6.0 ng/mL; **f** 600 ng/mL

was attached to the bottom of a microtiterplate and reacted with POx-labeled antibody (anti-IgG). On addition of EuTcHP, the enzyme catalyzes the decomposition of hydrogen peroxide, and this results in a distinct decrease in fluorescence intensity. This is shown in Fig. 1.31.

The antibody-POx conjugates can be apllied both in a sandwich format and in a direct ELISA. The sandwich ELISA for IgG is almost as sensitive (in terms of limits of detection which is about 0.3 ng/mL) as for the conventional method [57] or by chemiluminescence [58]. Calibration graphs for the direct and sandwich assay, respectively, are given in Figure 1.32.

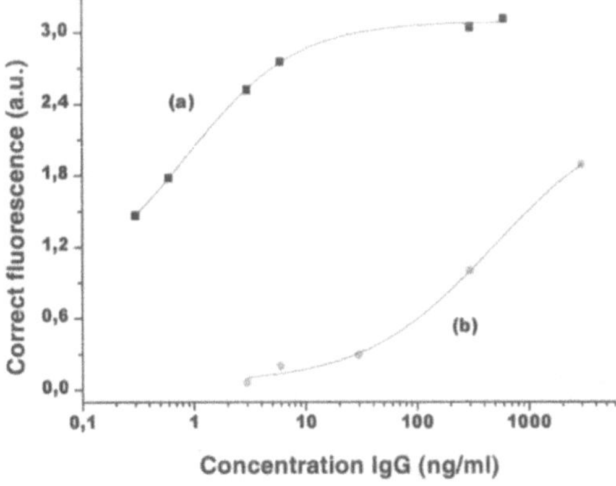

Fig. 1.32. IgG calibration graphs of the sandwich ELISA (curve **a**) and the direct ELISA (curve **b**) using POx-labeled anti-IgG and the fluorescent POx probe EuTcHP

Acknowledgement. This work was supported by the following grants: German DFG, grant LE-1203; EU grant CRAFT SMT4-CT98-5510. MW and ZL thank Fluorescent BioProbes GmbH (Regensburg) for a stipend.

References

1. Mason WT (1993) Fluorescent and luminescent probes for biological activity, Academic Press, London – San Diego
2. Brand L., Johnson ML (1997) Fluorescence spectroscopy, in: Methods in enzymology, vol 278, Academic Press, San Diego
3. Verga-Scheggi AM, Martelucci S, Chester AN, Pratesi R (1996) Biomedical optiocal instrumentation and laser-assisted biotechnology; NATO ASI Series E vol. 325, Kluwer Acad. Publ., Dordrecht (NL)
4. Szöllösi J, Damjanovich S, Matyus L (1998) Application of fluorescence resonance energy transfer in the clinical laboratory: routine and research. Cytometry 34:159
5. Wolfbeis OS, Leiner MJP (1985) Mapping of the total fluorescence of human blood serum as a new method for its characterization. Anal Chim Acta 167:203
6. Leiner MJP, Hubmann MR, Wolfbeis OS (1987) The total fluorescence of human urine. Anal Chim Acta 198:13
7. Selvin PR (1995) Fluorescence resonance energy transfer. In: Methods in enzymology, 246:300
8. Soini E, Lövgren T (1987) Crit Rev Anal Chem 18:105
9. Diamandis EP, Christopoulos TK (1990) Europium chelate labels in time-resolved fluorescence immunoassays and DNA hybridization assays. Anal Chem 62:1149A
10. Hemmillä I (1993) Progress in delayed fluorescence immunoassay, in: Wolfbeis OS (ed) Fluorescence spectroscopy: New methods and applications. Springer, Berlin-Heidelberg, pp. 259
11. Daehne S, Resch-Genger U, Wolfbeis OS (eds.) (1998) Near-infrared dyes for high technology applications. NATO ASI Ser. 3, vol. 53, Kluwer Acad. Publ., Dordrecht (NL)
12. Fabian J, Nakazami N, Matsuoka M (1992) Near infrared-absorbing dyes. Chem Rev 92:1197
13. Matsuoka M (ed) (1990) Infrared absorbing dyes. Plenum Press. New York.
14. Mujumdar RB, Ernst LA, Mujumdar SR, Lewis CJ, Waggoner AS (1993) Cyanine dye labeling reagents: sulfoindocyanine succinimidyl esters. Bioconjug Chem 4:105
15. Oswald B, Gruber M, Lehmann F, Probst M, Wolfbeis OS (2001) Novel diode laser-compatible fluorophores and their application to single molecule detection, protein labeling and fluorescence resonance energy transfer immunoassay. Photochem Photobiol 74:237
16. Oswald B, Lehmann F, Simon L, Terpetschnig E, Wolfbeis OS (2000) Red laser induced fluorescence energy transfer in an immunosystem. Anal Biochem 280:272
17. Oswald B, Patsenker L, Duschl J, Szmacinski H, Wolfbeis OS, Terpetschnig E (1999) Synthesis, spectral properties and detection limits of reactive squaraine dyes, a new class of diode laser compatible fluorescent protein labels. Bioconjugate Chem. 10:925

18. Böhmer M, Enderlein J (2002) Confocal laser scanning detection of molecules on surfaces, in: Zander C, Keller RR, Enderlein J (ed) Single-molecule detection in solution: Methods and applications., VCH-Wiley.

19. Böhmer M, Pampaloni F, Wahl M, Rahn H-J, Erdmann R, Enderlein J, Time-resolved confocal scanning device for ultrasensitive fluroescence detection. Rev. Sci. Instrum (in press).

20. Enderlein J, Goodwin PM, Van Orden A, Ambrose WP, Erdman R, Keller RA (1997) A maximum likelihood estimator to distinguish single molecules by their fluorescence decay. Chem. Phys. Lett. 270:464

21. Enderlein J, Sauer M (2001) J. Phys. Chem. A 105:48

22. Haugland RP (2001) Handbook of fluorescent probes and research chemicals, 8th ed, (see: www.probes.com) Molecular Probes, Eugene, OR (USA).

23. Lyttle HM, Carter CG, Dick DJ, Cook RM (2000) J Org Chem 65:9033

24. Probst M (2000) Diplomarbeit, University of Regensburg

25. De Marcos S, Wolfbeis OS (1996) Anal Chim Acta 334:149

26. Koncki R, Wolfbeis OS (1998) Anal Chem 70:2544

27. Koncki R, Wolfbeis OS (1999) Biosensors Bioelectron. 14:87

28. Parente AH, Marques ETA, Azevedo WM, Dinz FB, Melo EHM, Filo JLL (1992) Appl Biochem Biotech 37:262

29. Pringsheim E, Terpetschnig E, Wolfbeis OS (1997) Anal ChimActa 357:247

30. Grummt UW, Pron A, Zagorska M, Lefrant S (1997) Anal ChimActa 357:253

31. Pringsheim E, Zimin D, Wolfbeis OS (2001) Adv Mat 13:819

32. Ref. [22], chapter 6.

33. See: www.bang-labs.com

34. Bucheyska J. (1997) Modified polyacrylonitrile fibers. J Appl Polym Sci 65:1955

35. Korte S (1999) Physical constants of poly(acrylonitrile). In: Brandrup J, Immergut EH, Grulke EA (eds) Polymer Handbook V/59-V/66, Wiley & Sons, New York

36. Kürner JM, Klimant I, Krause C, Preu H, Kunz W, Wolfbeis OS (2001) Inert phosphorescent nanospheres as markers for optical assays. Bioconjug Chem (in press).

37. Kürner JM, Wolfbeis OS, Klimant I (2002) Homogeneous luminescence decay time-based assay using energy transfer from nanospheres. Anal Chem (submitted).

38. Liebsch G, Klimant I, Frank B, Holst G, Wolfbeis OS (2000) Luminescence lifetime imaging of oxygen pH, and carbon dioxide distribution using optical sensors. Appl Spectrosc 54:548

39. Klimant I, Huber Ch, Liebsch G, Neurauter G, Stangelmayer A, Wolfbeis OS (2001) New Trends in: Valeur B, Brochon JC (eds) Fluorescence spectroscopy: application to chemical and life sciences., Springer Verlag, Berlin (2001), chap. 13, pp 257–274

40. Liebsch G, Klimant I, Krause C, Wolfbeis OS (2001) Fluorescent imaging of pH with optical sensors using time domain dual lifetime referencing. Anal Chem 73:4354 and references cited therein.

41. Liebsch G, Klimant I, Wolfbeis OS (1999) Luminescence lifetime temperature sensing based on sol-gels and poly(acrylonitrile)s dyed with ruthenium metal ligand complexes. Adv Mat 11:1296

42. Walt, D. R. (2000) Bead-based fiber-optic arrays. Science 287:451

43. Fulton RJ, McDade RL, Smith PL, Kienker LJ, Kettman JR (1997) Advanced multiplexed analysis with the FlowMetrix[TM] System. Clinical Chem 43:1749–1756

44. Ferguson JA, Boles TC, Adams CP, Walt DR (1996) A fiber-optic DNA biosensor microarray for the analysis of gene expression. Nature Biotechnology 14:1681

45. Szurdoki F, Michael KL, Agrawal D, Taylor LC, Schultz SL, Walt DR (1999) Immunofluorescence detection methods using microspheres. Proc SPIE (The Int. Soc. for Optical Engin.) 3544:52

46. Méallet-Renault R, Denjean P, Pansu RB. (1999) Polymer beads as nano-sensors. Sensors and Actuators B 59:108

47. Fortin M, Hugo P (1999) Surface antigen detection with non-fluorescent, antibody-coated microbeads: an alternative method compatible with conventional fluorochrome-based labeling. Cytometry 36:27

48. Kürner JM, Klimant I, Krause C, Pringsheim E, Wolfbeis OS (2001) A new type of phosphorescent nanospheres for use in advanced time-resolved multiplexed bioassays. Anal Biochem 297:32

49. Alpha-Bazin B, Bazin H, Preaudat M, Trinquet E, Mathis G (2001) Rare earth cryptates and TRACE technology as tools for probing molecular interactions in biology. In: Valeur B, Brochon JC (eds) New trends in fluorescence spectroscopy, Springer, Berlin-Heidelberg, pp. 439–455

50. Hirschy LM, van Geel TF, Winefordner JD, Kelly RN, Schulman SG (1984) Anal Chim Acta 166:207

51. Rakicioglu Y, Perrin JH, Schulman SG (199) J Pharm Biomed Anal 20:397

52. Everse J, Everse KE, Grisham MB (eds) Peroxidase in chemistry and biology, vol. II, CRC Press, 991.

53. Miller JN, Niessner R, Knopp D (2001) Enzyme- and immunoassays, in: Guenzler H, Williams A (eds) Handbook of analytical techniques, Wiley, pp.162–166.

54. Kemeny a DM,Challacombe SJ (ed) ELISA and other solid phase immunoassays, John Wiley &Sons Ltd. 1988.

55. Schomburg D, Salzmann M, Stephan D (1994) GBF Enzyme Handbook, vol. 7; Springer-Verlag, Berlin, Heidelberg.

56. See: www.immunochemistry.com/2000services/2100immunoassaydevelopment/2150what-is-an-ia.htm

57. Tryphonas H, Karpinski K, O'Grady L, Hayward S (1991) Quantitation of serum immunoglobulins G, M, and A in the rhesus monkey using human monospecific antisera in the enzyme-linked immunosorbent assay: developmental aspects. J Med Primatol 20:58

58. Van Dyke K, Van Dyke R (1990) Luminescence immmunoassay and molecular applications, CRC Press, Boca Raton (Fla.), p. 35

CHAPTER 2

Fluorescence Spectral Engineering – Biophysical and Biomedical Applications

J. R. LAKOWICZ, I. GRYCZYNSKI, Y. SHEN, J. MALICKA, S. D'AURIA, AND Z. GRYCZYNSKI

Fluorescence spectroscopy is a central research tool in biology and has also become the dominant method enabling the revolution in biotechnology. At present, almost all fluorescence experiments are performed in a free space at condition, in which the oscillating dipole can radiate isotropically into space. We now describe a new opportunity, fluorescence spectral engineering (FSE), in which metallic surfaces or particles are used to modify the emission rates and spatial distribution of the emission. In this article we first summarize results from the physics literature which demonstrate the effects of metallic surfaces, colloids or islands to increase or decrease emissive rates, increase the quantum yields of low quantum yield chromophores, decrease the lifetimes, increase the distances for resonance energy transfer, and to direct the typically isotropic emission in specific directions. This latter effect is not the result of reflection, but rather as the result of the metal modifying the photon mode density surrounding the fluorophores and the fluorophore dipole interacting with free electrons in the metal. We then show experimental results, which demonstrate increased quantum yields, decreased lifetimes, increased photostability, and increased energy transfer distances near metal particles. Fluorescence spectral engineering provides opportunities for new types of fluorescence experiments and assays.

2.1
Introduction

In designing fluorescence experiments we take advantage of well understood principles which link molecular features of the sample with fluorescence spectral observables. These include quenching, environment effects, resonance energy transfer (RET) and rotational motions, and can be used to study the structure and dynamics of macromolecules and the interactions of macromolecules with each other. For example, quenching can reveal the exposure of fluorophores to the solvent or to other nearby quenching groups within the same macromolecule. RET reveals proximity between donors and acceptors located at regions of interest on the macromolecule(s), and anisotropy decays reveal the internal dynamics and segmental motions of proteins and membranes.

In most applications of fluorescence the spectral properties of the fluorophore are altered by rate processes which modify the non-radiative decay rates of the excited state population (Fig. 2.1). Both collisional quenching (k_q) and energy transfer (k_T) provide non-radiative pathways to the ground state. Quenching and RET have no significant effect on the rate of radiative decay (Γ), that is, the spontaneous rate at which fluorophores emit photons. The emissive rate is determined mostly by the excitation coefficient and absorption spectrum of the fluorophore [1], and the emission rate is nearly constant in most fluorescence experiments.

We now describe the unusual spectral properties and opportunities available if one can increase the radiative decay rates. This is possible using metallic surfaces and particles. While the theory of fluorophore–metal interactions is known, these interactions have not been considered within the framework of biochemical fluorescence. We then describe how fluorescence spectral engineering (FSE) can be applied to biochemical fluorescence and how FSE can be used in biomedical applications such as diagnostics and genomics.

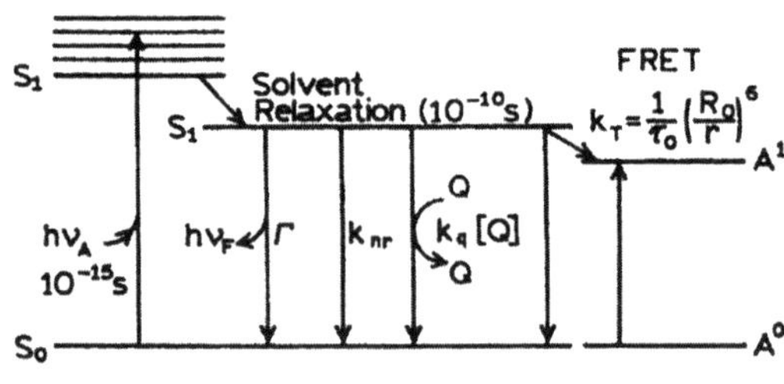

$$Q_0 = \Gamma/(\Gamma + k_{nr}) \qquad Q = \Gamma/(\Gamma + k_{nr} + k_q[Q])$$

$$\tau_0 = 1/(\Gamma + k_{nr}) \qquad \tau = 1/(\Gamma + k_{nr} + k_q[Q])$$

Fig. 2.1. Jablonski diagram for a typical fluorescence experiment, with the spectral properties changed by quenching or energy transfer

2.1.1
Fluorescence Spectral Engineering

Prior to describing the unusual effects of metal surfaces on fluorescence it is valuable to describe what we mean by FSE and spectral changes expected for increased radiative decay rates. We are accustomed to perform experiments using fluorophores in isotropic macroscopic solutions which are transparent to the emitted radiation. The fluorophores emit into free space and are observed in the far field.[1]

The emission properties are well described by Maxwell's equations for an oscillating dipole radiating into free space. For clarity we note that we are not considering reflection of the emitted photons from the metal surfaces, which occurs after emission has occurred. We are considering the effects of the nearby surface on altering the "free space" condition, and modifying Maxwell's equation from their free space counterparts [2, 3]. A fluorophore is an oscillating dipole, but one which oscillates at high frequency and radiates energy. Nearby metal surfaces can respond to the oscillating dipole and modify the rate of emission and the spatial distribution of the radiated energy. The electric field felt by a fluorophore is affected by interactions of the incident light with the nearby metal surface and also by interaction of the fluorophore's oscillating dipole with the metal surface. These interactions can increase or decrease the field incident on the fluorophore. The theory for such effects is complex. We will describe these effects in an intuitive manner with minimal use of theory.

Depending upon the distance and geometry, metal surfaces or particles can result in quenching of fluorescence or enhancement of fluorescence by factors of up to 1000 [4–6]. The effects of surfaces on fluorophores are due to at least three mechanisms. One is energy transfer quenching to these metals with a d^{-3} dependence [5]. The quenching can be understood by damping of the dipole oscillators by the nearby metal. A second mechanism is an increase in intensity due to the metal concentrating the incident field, which has been seen for metal colloids [6–8]. The enhancement of the local excitation intensity is sometimes called the "lightening rod effect". These effects of quenching and enhancing the local fields are important. However, another more important effect of metal surfaces and particles is possible.

In our opinion a remarkable opportunity for new fluorescence methods is available from a less known fluorophore–metal interaction. This effect is an increase in the intrinsic radiative decay rate of the fluorophore. This is a highly unusual effect. We usually have no significant control over the radiative rate (Γ). The spectral observables of quantum yields and lifetimes are governed by the magnitudes of the radiative rate Γ and the sum of the non-radiative decay rates (k_{nr}). To under-

---

[1] Abbreviations: FSE: Fluorescence Spectral Engineering; AO: Acridine orange; BF: Basic Fucsin; DAPI: 4',6-diamidino-2-phenylindole; dppz: Dipyrido[3,2-a:2',3'-c]phenazine; DMF: Dimethylformamide; RhB: Rhodamine B; RB: Rose Bengal; RET: Resonance energy transfer; SERS: Surface-enhanced Raman scattering; SEF: Surface-enhanced Fluorescence; SPR: Surface Plasmon Resonance; TIRF: total internal reflection fluorescence.

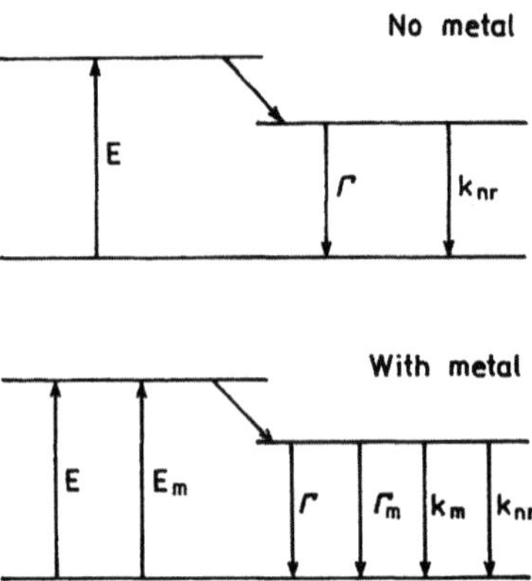

Fig. 2.2. Jablonski diagram without (top) and with (bottom) the effects of near metal surfaces

stand the value of controlling the radiative decay rate (Γ) it is informative to consider how this rate affects the quantum yield Q_0 and lifetime τ_0 of a fluorophore in the absence of a metal surface. Consider a simplified Jablonski diagram in Fig. 2.2 (top). The quantum yield of the fluorophore in the absence of other quenching interactions is given by

$$Q_0 = \Gamma / (\Gamma + k_{nr}) \tag{2.1}$$

The natural lifetime of a fluorophore (τ_N) is the inverse of the radiative decay rate ($\tau_N = \Gamma^{-1}$) or the lifetime which would be observed if their quantum yield were unity or equivalently if $k_{nr} = 0$. The quantum yield is the fraction of the excited fluorophores which decay by emission (Γ) relative to the total decay ($\Gamma + k_{nr}$), and thus by the ratio in Eq. 2.1. Fluorophores with high radiative rates have high quantum yields and short lifetimes. The radiative decay rate is essentially constant for any given fluorophore. Hence, the quantum yield can only be increased by decreasing the non-radiative rate k_{nr}, which usually occurs at lower temperatures. The lifetime of a fluorophore is determined by the sum of the rates, which depopulate the excited state. In the absence of other quenching interactions the lifetime is given by

$$\tau_0 = (\Gamma + k_{nr})^{-1} \tag{2.2}$$

The lifetime of a fluorophore can be increased or decreased by changing the value of k_{nr}. Almost invariably, the lifetimes and quantum yields increase or decrease together.

The brightness of a fluorophore is dependent on the values of the radiative and non-radiative decay rates. For instance, the radiative decay rates of the nucleic acid bases are comparable to those of typical fluorescent probes. However, the emission is very weak ($Q_0 < 10^{-4}$) and the lifetimes very short (≈ 10 ps) because of the high rates of non-radiative decay ($\approx 10^{11}$ s^{-1}) (Eq. 2.1) [9, 10]. Similarly, the quantum yields for phosphorescence are usually very low because the radiative decay rates are slow compared to typical non-radiative rates and quenching processes [11, 12]. In general there is no way to control the lifetime or quantum yield of a fluorophore other than by altering the values of k_{nr} or k_T.

It is informative to consider the novel spectral effects expected by increasing the radiative rate. Assume the presence of a nearby metal (m) surface increases the radiative rate by addition of a new rate Γ_m (Fig. 2.2, bottom). For the moment we are assuming the metal does not cause significant quenching ($k_m = 0$). In this case the quantum yield and lifetime of the fluorophore near the metal surface are given by

$$Q_m = \frac{\Gamma + \Gamma_m}{\Gamma + \Gamma_m + k_{nr}} \tag{2.3}$$

$$\tau_m = (\Gamma + \Gamma_m + k_{nr})^{-1} \tag{2.4}$$

The new radiative rate Γ_m results in unusual predictions for a fluorophore near a metal surface. As the value of Γ_m increases the quantum yield increases and the lifetime decreases. To illustrate this point we calculated the lifetime and quantum yield for fluorophores with an assumed natural lifetime $\tau_N = 10$ ns and various values for the non-radiative decay rates and quantum yields from 1.0 to 0.01. Since Γ_m is a rate process returning the fluorophore to the ground state, the lifetime decreases as Γ_m becomes comparable and larger than Γ (Fig. 2.3, top). This is a typical result, similar to that which occurs for increasing amounts of collisional quenching. However, an unusual effect is expected for the quantum yield. As Γ_m increases the quantum yield increases (Fig. 2.3, bottom), the quantum yield increases as the lifetime decreases. The most dramatic relative changes are found for fluorophores with the lowest quantum yields. If $Q = 1.0$, then changing Γ_m has no effect. If Q is low, such as 0.1 in Fig. 2.3, the metal-induced rate Γ_m increases the quantum yield. At sufficiently high values of Γ_m, the quantum yields of all fluorophores approaches 1.0. The increases in quantum yields which can occur near metal surfaces are different from the typical increase in quantum yield which occurs when a solvent sensitive fluorophore like ANS is dissolved in a low polarity solvent or binds to a protein [13]. In these cases the quantum yield increases because of a decrease in the non-radiative decay rate k_{nr} (Eq. 2.1).

2.1.2
Overview of Metallic Surface Effects on Fluorescence

The effects of metals on fluorophores have been described in the physics literature. We now summarize those results which reveal the potential of using these interactions in fluorescence spectroscopy. Many of these studies use metal islands, metal colloids, metal surfaces or mirrors. More dramatic effects have been found for islands and colloids rather than for continuous metallic surfaces.

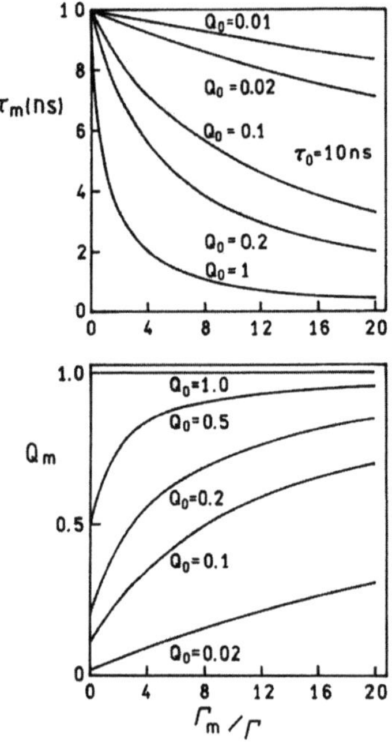

Fig. 2.3. Effect of an increase in the metal-induced radiative rate on the lifetime and quantum yields of fluorophores

The possibility of altering the radiative decay rates was demonstrated by measurements of the decay times of europium (Eu^{3+}) positioned at various distances from a planar silver mirror [14–16]. In a mirror the metal layer is thicker than the optical wavelength. The lifetimes oscillate with distance but remain a single exponential at each distance (Fig. 2.4). This effect can be explained by changes in the phase of the reflected field with distance and the effects of this reflected field on the fluorophore. A decrease in lifetime is found when the reflected field is in-phase with the fluorophore. An increase in the lifetime is found if the reflected field is out-of-phase with the fluorophore. As the distance increases the amplitude of the oscillations decreases. The effects of a plane mirror occur over distances comparable to the excitation and emission wavelengths. At short distances below 20 nm the emission is quenched. This effect is due to coupling of the dipole to oscillating surface changes on the surface of the metal.

The effects of metallic surfaces on optical spectra are strongly dependent on the nature of the metal surface and/or metal particles. In most cases more dramatic effects are observed for metal colloids than planar mirrored surfaces. For example, Raman signals are remarkably enhanced by metal colloids or islands [17, 18], which has resulted in the field of surface enhanced Raman spectroscopy (SERS). One frequently used system is silver island films, which are made by depositing silver on a glass substrate. Under suitable conditions the glass becomes covered with approximate circular islands about 200 Å in diameter. About 40% of the surface is covered by the silver. If the silver were uniformly distributed the mass

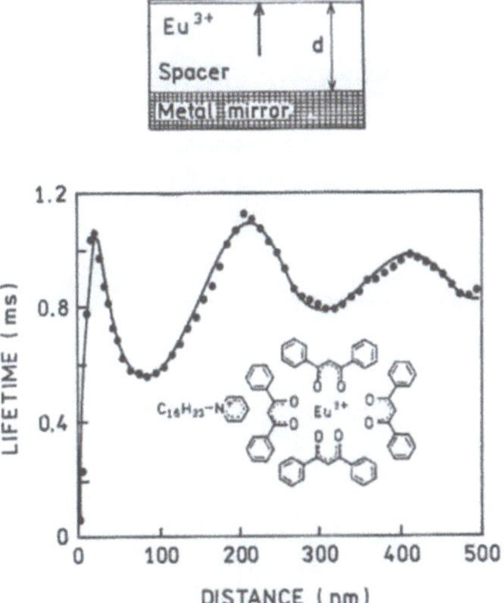

Fig. 2.4. Lifetime of Eu (ETA)$_3$ ions in front of a Ag mirror as a function of separation between the Eu^{3+} ions and the mirror. The solid curve is a theoretical fit

thickness would be near 40 Å. Silver island films are distinct from semi-transparent metallic surfaces, which are obtained at higher mass thicknesses.

A dramatic effect of silver islands on fluorescence is shown in Fig. 2.5. Silver islands were coated with a thin film of Eu (ETA)$_3$, where ETA is a ligand which chelates europium. This chelate displayed a quantum yield near 0.4. The sample contained an inert coating between the islands so Eu^{3+} chelates positioned between the islands were not emissive. When the Eu^{3+} chelate was deposited on the silica substrate, without the silver islands, it displayed a single exponential decay time of 280 μs and a quantum yield near 0.4. However, when deposited on silver island films the intensity increases about 5-fold and the lifetime decreases by about 100-fold to near 2 μs (Fig. 2.5). Also, the decay is no longer a single exponential [19] on the silver island films. The silver islands had the remarkable effect of increasing the intensity 5-fold while decreasing the lifetime 100-fold. Such an effect can only be explained by a increase in the radiative rate.

The 5-fold increase in the quantum yield of Eu(ETA)$_3$ results in an apparent quantum yield of 2.0, which is obviously impossible. This result is due to an increase in the local excitation field near the metal particle. For this reason, it is important to recognize the intensities measured on surfaces represent "apparent" quantum yields, which can include an unknown factor due to incident field enhancement. This increase in the local intensity of the incident light cannot explain the decreased lifetime because an unperturbed Eu^{3+} chelate, excited by this enhanced field, would still decay with a 280 μs lifetime. According to the authors [19] the decreased lifetime is due to electromagnetic coupling between the Eu^{3+} and the silver islands.

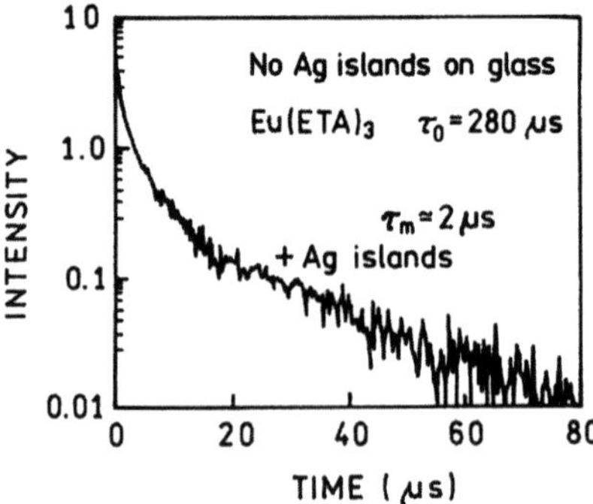

Fig. 2.5. Fluorescence decay of Eu^{3+} on silver-island films. Eu^{3+} was complexed with thenolytrifluoro-acetonate (ETA)

2.1.3
Theory for Fluorophore Metal Interactions

Prior to describing the effects of metals, it is useful to review the optical properties of metal colloids and islands. Metal colloids have been used for centuries to make some colored glasses [20]. The origin of the color as due to metallic colloids was first recognized by Faraday [21]. A typical absorption spectrum of gold colloid is shown in Fig. 2.6. The long wavelength absorption is called the surface plasmon absorption, which is due to electron oscillations on the metal surface. These spectra can be calculated for the small particle limit ($r << \lambda$) from the properties of the metal [22, 23]. Larger particles display longer wavelength absorption. The absorption spectra are also dependent on the shape of the particles, with prolate spheroids displaying longer absorption wavelengths. Most studies of surface effects on fluorescence have been performed using silver particles to avoid the longer wavelength absorption of gold.

The physics of the interactions between fluorophores and metallic surface or particles is understood in moderate detail [24–28]. Several effects are present, and are shown schematically in Fig. 2.7. Quenching occurs by coupling to the surface plasmon absorption of the metal. This effect decreases with the cube of the distance (d) between the metal surface and the fluorophore (d^{-3}) [5].

In addition to quenching there are two effects which determine the apparent quantum yield Y:

$$Y = \left|L(\omega_{ex})\right|^2 Z(\omega_{em})$$

(2.5)

The first term describes the local intensity. The local field is proportional to $L(\omega_{ex})E_0$ where E_0 is the incident field. The first term in Eq. 2.5 is proportional to the product of the quantum yield in free space and the amplification of the incident field. The term "apparent quantum yield" refers to the intensity of the samples,

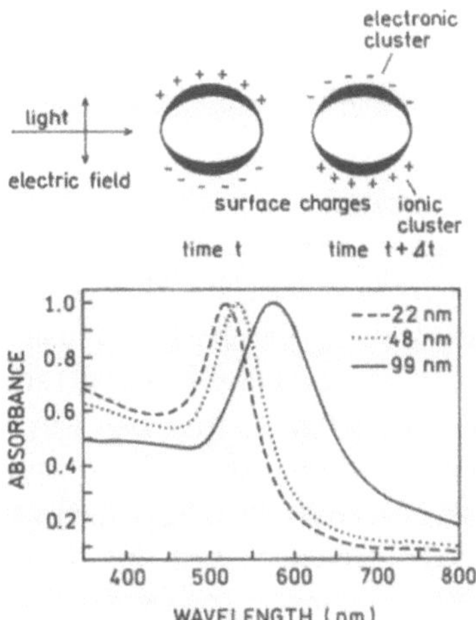

Fig. 2.6. Absorption spectra of gold colloidal spheres

relative to the control sample, measured with the same intensity of the incident light. In studies with metal particles it is necessary to distinguish the actual quantum yield from the apparent quantum yield Y because of amplification of the incident field by the particles. This distinction of Y from the true quantum yield (Q) is not needed in the absence of metals because the field felt by the fluorophore is always the same for the sample and the reference. The field concentration has been modeled for ellipsoidal particles and the maximum enhancement in the magnitude of the local field is about 140 [26].

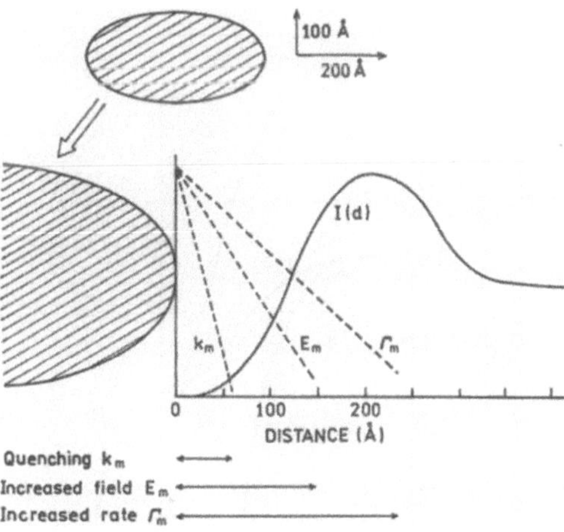

Fig. 2.7. Effects of a metallic particle on transitions of a fluorophore

The second term $Z(\omega_{em})$ describes the partition of energy into the radiative and non-radiative decay pathways, as modified by the metal particles. In the absence of metal the quantum yield is given by Eq. 2.1. The quantum yield in the presence of the metal is given by Eq. 2.3. The larger overall radiative decay rate results in a larger quantum yield. The enhanced field and increased radiative rates occur at longer distances from the metal than quenching. Hence there exists a region 50–200 Å from the metal surfaces were the emission is enhanced (Fig. 2.7).

It is interesting to consider how a metal surface affects fluorophores with high and low intrinsic quantum yields (Q_0). If the dye has a high quantum yield ($Q_0 \to 1$) the additional radiative decay rate cannot substantially increase the quantum yield. In this case energy transfer quenching to the metal will dominate and $Z(\omega_{em})$ will be less than one. The more interesting case is for low quantum yield chromophores. In this case $Z(\omega_{em})$ can be as large as $1/Q_0$ [28]. For this reason it is of interest to study fluorophore-metal interactions with low quantum yield fluorophores. While the actual mechanism is complex, one can imagine the particles serve as an antenna, which in combination with the chromophore, radiate faster than k_{nr}. This suggests the emission from weakly fluorescent substances can be increased if they are positioned at an appropriate distance from a metal surface or colloid.

The value of chromophore-metal interactions in optical spectroscopy was demonstrated by Surface Enhanced Raman Spectroscopy (SERS), in which the Raman lines are dramatically increased by the metal surface. The value of this enhancement is seen from recent reports of DNA detection using SERS, without the resonance effect [29–31]. The enhancement is so significant that single fluorophores and single nucleotides have been detected by SERS [32–33]. Additionally, the enhancement may be due to a small fraction of the particles, which display enhancements of up to 10^{15} [34, 35]. It seems probable that such "hot" particles will also be found for surface-enhanced fluorescence.

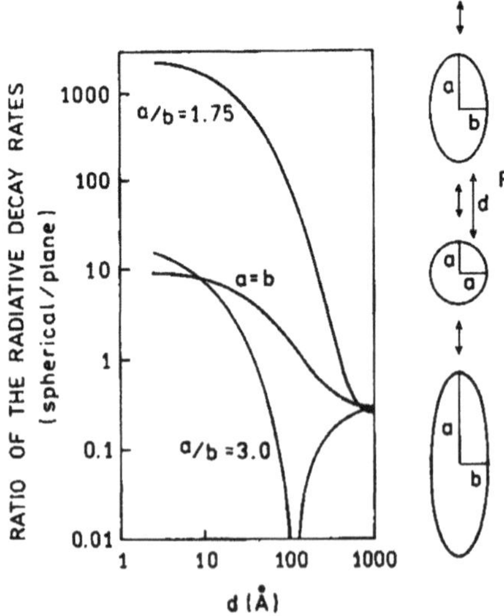

Fig. 2.8. Effect of a metallic spheroid on the radiative decay rate of a fluorophore

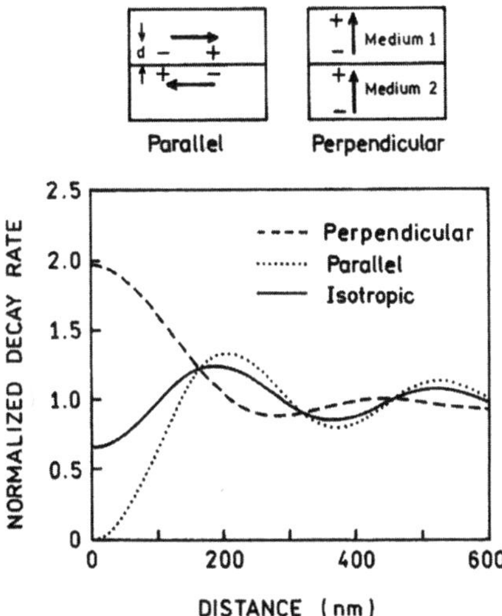

Fig. 2.9. Effect of fluorophore orientation on the decay rate of a fluorophore near a metallic surface

Several groups have considered the effects of metallic spheroids on the spectral properties of nearby fluorophores [24–27]. A typical model is shown in Fig. 2.8, for a prolate spheroid with an aspect ratio of a/b. The particle is assumed to be a metallic ellipsoid with a fluorophore positioned near the particle. The fluorophore is located outside the particle at a distance d from the surface. The fluorophore is located on the major axis and can be oriented parallel or perpendicular to the metallic surface.

The presence of a metallic particle can have dramatic effects on the radiative decay rate of a nearby fluorophore. Fig. 2.8 shows the radiative rates expected for a fluorophore at various distances from the surface of a silver particle and for different orientations of the fluorophore transition moment. The most remarkable effect is for a fluorophore perpendicular to the surface of a spheroid with $a/b = 1.75$. In this case the radiative rate can be enhanced by a factor of 1000-fold or greater. The effect is much smaller for a sphere ($a/b = 1.0$) and much smaller a more elongated spheroid ($a/b = 3.0$) when the optical transition is not in resonance with the particle. In this case the radiative decay rate can be decreased by over 100-fold. This effect could result in 100-fold longer lifetimes. The magnitude of these effects depends on the location of the fluorophore around the particle and the orientation of its dipole moment relative to the metallic surface. The dominant effect of the perpendicular orientation is thought to be due to an enhancement of the local field along the long axis of the particle.

The transition dipole-surface orientation can result in unusual effects. This is illustrated in Fig. 2.9, which shows the expected decay rates of a fluorophore near a solid silver surface [16]. The decay depends on orientation relative to the surface. In the parallel orientation the dipole in the metal cancels the dipole in the fluorophore, which slows the decay. In the perpendicular orientation the fluorophore's dipole and the dipole in the metal are synergistic and increase the decay rate. This effect of orientation results in an unusual possibility. In almost all known cases the anisotropy decay of a fluorophore represents motions of the fluorophore and is independent of the intensity decay. For a fluorophore near a plane metal surface (Fig. 2.9) or a metal particle (Fig. 2.8) the intensity decay will be coupled to the anisotropy decay. For instance, a fluorophore oriented perpendicular to the surface of a particle will decay 1000-fold faster if it can rotate rather than remain stationary.

2.1.4
Spatial Distribution of Emission Near Metal Surfaces

In a typical cuvette experiment the fluorophores radiate into free space. The emission from a randomly oriented population is isotropic in all directions. Even for oriented fluorophores, or for an isotropic solution excited with polarized light, the emission is nearly isotropic. As a result the sensitivity of fluorescence detection is limited by the light collection efficiency possible with detectors of practical size. Radiative decay engineering offers the possibility of directing the emission in specific directions, which may be towards the detector. This can occur when fluorophores are deposited directly, with no inert spacer, onto silver grating. In this case energy from the fluorophores is transferred into the metal and radiated by surface plasmon polaritrons [36–38]. This remarkable result suggests the design of high sensitivity nanoscale sample chambers, which direct the emission into the detector.

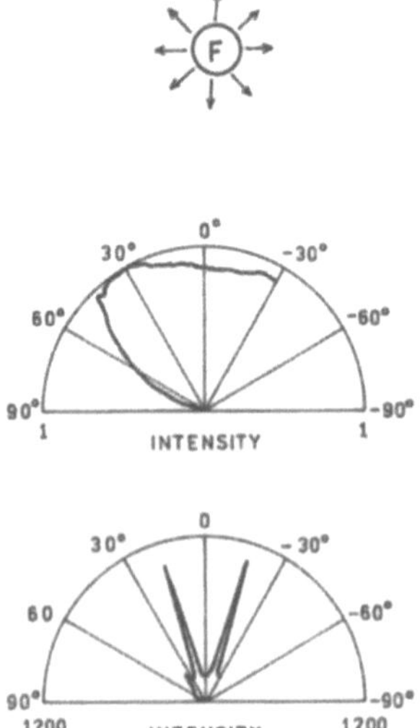

Fig. 2.10. Directional emission from fluorophores near a plane (middle) or periodic metallic surface (bottom). The lower panel is for the TE-polarized emission. In the absence of a metal surface the emission is essentially isotropic (top)

2.1.5
Resonance Energy Transfer

Nearby metallic surfaces can have dramatic effects on resonance energy transfer (RET) [39, 40]. Suppose the donor and acceptor are located along the long axis of an ellipsoid with the dipoles also oriented along this axis. Fig. 2.11 shows the enhancement of the rate of energy transfer due to the metal particle, that is the ratio of the rates of transfer in the presence (k_T^m) and absence (k_T^0) of the metal. Enhancements of 104 are possible. The enhancement depends on the transition energy being in resonance with the particle. A smaller but still significant enhancement is found for a less resonant particle. The enhanced rate of energy transfer persists for distances much larger than typical Forster distances. While these simulations are for dipoles on the long axis and oriented along this axis, the enhancements are still large when the donors and acceptors have different orientations and locations around the particle [40]. The simulation in Fig. 2.11 do not consider the increased emission rates of fluorophores near surfaces. Since emission and RET are competitive processes, the RET efficiency depends on the relative values of the emission and energy transfer rates. Examination of the available simulations [39, 40] suggests that the enhancements in RET are larger than the enhancements in the emissive rates. This indicates that RET will occur over longer distances near metal particles even if the donor lifetimes are decreased by more rapid emission.

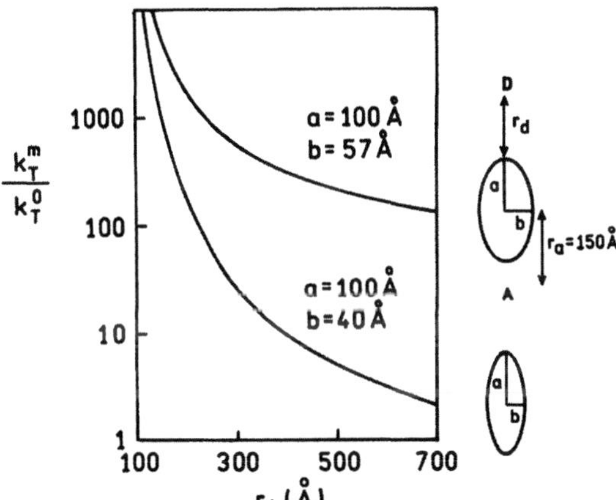

Fig. 2.11. Enhancements in the rate of energy transfer in the presence of a silver particle

2.2
Experimental Results on Fluorophore–metal Interactions

2.2.1
Silver Island Films and Experimental Geometry

We performed experiments with fluorophores between silver island films to confirm the predicted effects of metallic surfaces. Silver islands were formed by chemical reduction of silver nitrate. Fig. 2.12 shows the absorption spectra of our silver island films. This spectrum indicates that the particles are sub-wavelength in size. The shape and size distribution of the particles is almost certainly heterogeneous. To determine the effects of silver islands on fluorescence the samples were placed between two such silver island plates. From the absorption spectra of rose bengal between two quartz plates or two silver island coated plates we estimate the distance between the plates to be 1 to 1.5 μm.

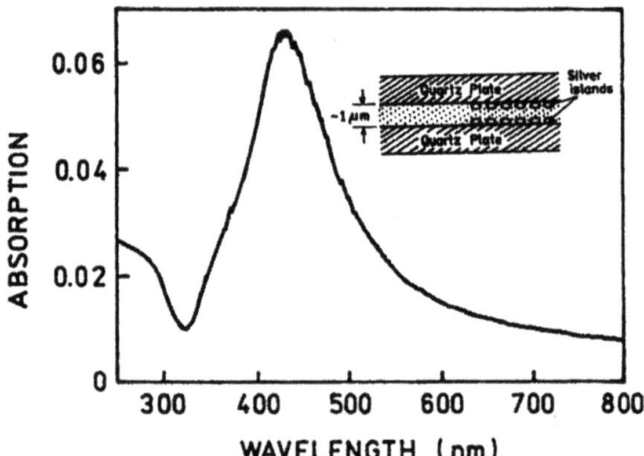

Fig. 2.12. Absorption spectrum of silver islands deposited on a quartz plate

2.2.2
Effects of Silver Island Films on Emission Spectra of Rhodamine B and Rose Bengal

We examined the emission spectra of rhodamine B (RhB) and rose bengal (RB) between uncoated quartz plates (Q) or silver island films (S). We selected these two fluorophores because of their similar absorption and emission spectra but different quantum yields of 0.48 and 0.02 for RhB and RB, respectively. In the case of RhB the intensities are similar in the absence and presence of these silver islands (Fig. 2.13, top). There may be a small decrease in the RhB intensity due to the silver islands, which may be due to the quenching effects of metals at short distances. Since the quantum yield of RhB is high, its quantum yield cannot be substantially increased by the silver islands.

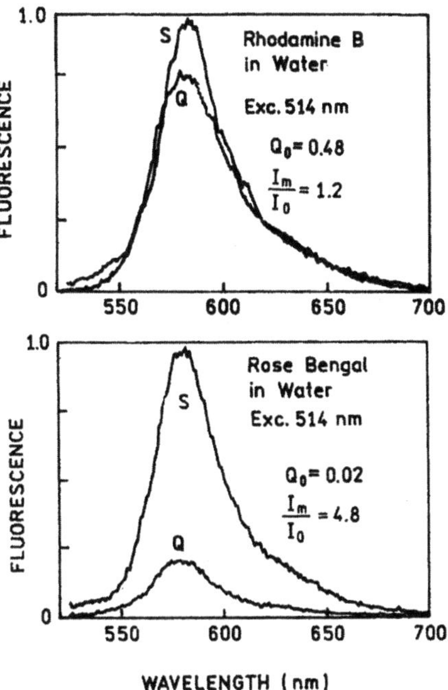

Fig. 2.13. Emission spectra of rhodamine B (top) and rose bengal (bottom) between silver island films (S) or unsilvered quartz plates (Q)

Contrasting results were obtained for rose bengal (Fig. 2.13, bottom). In this case the intensity increased about 5-fold in the presence of silver islands. It is important to recognize that the increased intensity observed for RB represents an underestimation of the quantum yield of RB near the silver islands. This is because only a small fraction of the RB molecules are within the distance over which metallic surfaces can exert effects. The region of enhanced fluorescence is expected to extend about 200 Å into the solution. Hence only about 4% of the liquid volume between the plates is within the active volume. This suggests that the quantum yield of RB within 200 Å of the islands is increased 125-fold. Of course this is larger than possible if the quantum yield of 0.02 is correct. Nonetheless, the spectra for RB in Fig. 2.13 indicates a substantial increased in quantum yield for the molecule within 200 Å of the silver islands.

It is known that metallic particles can concentrate the electric field of the incident light by a factor of up to 140 [41], which in turn can result in increased excitation as compared to a sample without metallic particles. The effect of a concentrated electric field will be the same for low and high quantum yields fluorophores. Hence, the lack of an increase in the intensity of RhB suggests this effect is not the dominant cause of the intensity increase for rose bengal in Fig. 2.13. For RhB the emission occurs for RhB molecules both near to and distant from the silver islands, so that the field concentration effects could be marked by a dominant emission from the RhB molecules distant from the silver islands.

2.2.3
Effect of Silver Island Films on Photostability

The photostability of fluorophores is an important property in fluorescence microscopy and in single molecule detection [42–44]. It is known that this extent of fluorophore photo-decomposition is roughly proportional to the time a fluorophore remains in the excited state, as has been shown by increased donor photostability in the presence of a RET acceptor [45]. This suggests that the increased radiative decay rates of fluorophores near metal islands or colloids could result in increased photostability and easier single molecule detection.

We examined the effects of silver island films on the photostability of RhB and rose bengal (RB). For RhB the rate of photobleaching was unchanged by the silver island films (Fig. 2.14, right), which is consistent with most of the emission being due to RhB molecules distant from the metal-island films. For RB there was a dramatic increase in photostability in the presence of the silver island films (Fig. 2.14, left). This result is consistent with an increase in the radiative decay rate of RB in the fraction of molecules close to the films, and also with the decreased lifetimes (below). These results indicate that the increased intensity seen for RB (Fig. 2.13) is not due to an increased local field or rate of excitation because such effects would increase the rate of photobleaching or leave the rate unchanged. These results suggest that fluorophores on metal colloids may provide photostable probes for microscopy, and that the number of photons detectable from a single fluorophore can be increased near metallic particles.

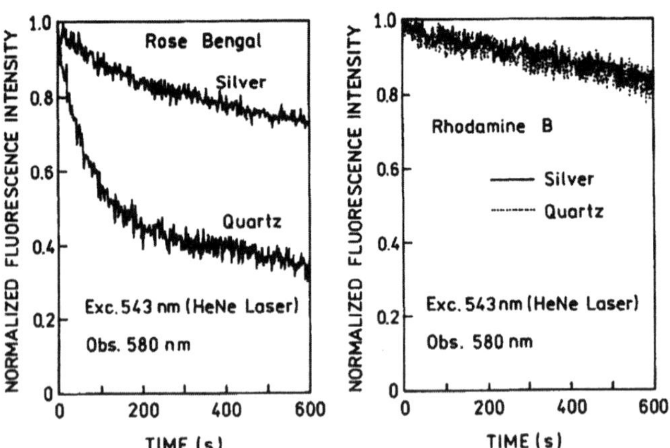

Fig. 2.14. Photostability of rose bengal and rhodamine B aqueous solutions on quartz and on silver island films. The excitation source was 543 nm HeNe laser (Melles Griot, 0.8 mW) focused into a 0.5 mm spot

2.2.4
Effects of Silver Island Films on the Lifetime of Rhodamine B and Rose Bengal

The effects of an increased radiative rate and concentrated electric field can be distinguished by lifetime measurements. An increase in the radiative rate will decrease the lifetime whereas an increased rate of excitation will not change the lifetime. We measured the intensity decays of RhB and rose bengal in the absence and presence of silver islands. In an standard cuvette the intensity decay of RhB was found to be a single exponential with a lifetime $\tau = 1.56$ ns (Fig. 2.15). In the presence of silver islands the intensity decay becomes heterogeneous. The data could be fit to two decay times with the long lifetime of 1.81 ns being comparable to that found in a cuvette. A short lifetime of 0.14 ns appeared for RhB between the silver islands, which we attribute to RhB molecules in close proximity to the silver islands. The fractional steady state intensity of this short component is about 10%. Control measurements showed that this component was not due to scattered light. Measurements were also performed for RhB between quartz plates without silver islands. In this case the decay was also double exponential, but less heterogeneous than in the presence of islands.

Frequency-domain intensity decays for rose bengal are shown in Fig. 2.16. In a cuvette the decay is a single exponential with $\tau = 94$ ps. The decay becomes

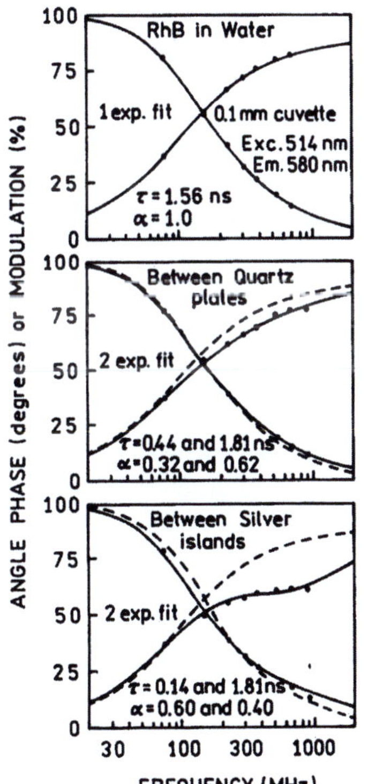

Fig. 2.15. Frequency-domain intensity decays of rhodamine B in water when placed in a 0.1 mm cuvette (top), between quartz plates (middle) and between silver islands (bottom). In the lower two panels the dashed lines represent the FD data for RhB in the 0.1 mm cuvette (from the top panel)

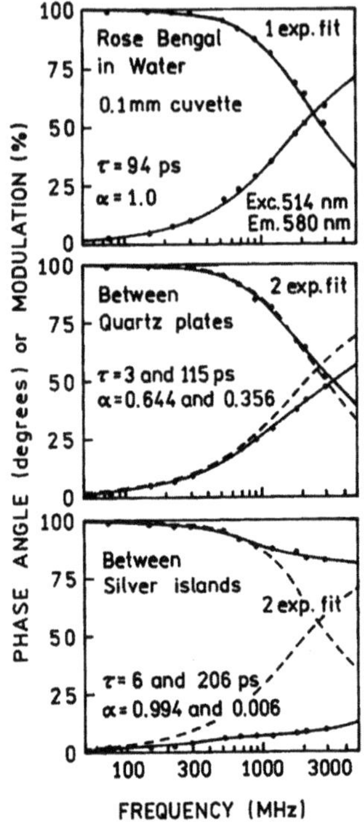

Fig. 2.16. Frequency-domain intensity decays of rose bengal in water when placed in 0.1 mm cuvette (top), between quartz plates (middle) and between silver islands (bottom). In the lower two panels the dashed lines represent the FD data for rose bengal in the 0.1 mm cuvette (from the top panel)

slightly heterogeneous for RB between uncoated quartz plates. However, the intensity decay of RB changed dramatically when between silver islands. In this case the dominant lifetime became a 6 ps component, which we assign to rose bengal molecules adjacent to the silver islands. This dramatic decrease in the lifetime of RB is consistent with the increased photostability seen near the silver islands (Fig. 2.14).

2.2.5
Effect of Quantum Yield on Silver Island Enhancements

We examined a number of additional fluorophores between uncoated quartz plates and between silver island films. The enhancements for 10 different fluorophores solutions are shown in Fig. 2.17. In all cases lower bulk-phase quantum yields result in larger enhancements for samples between silver island films. These results (Fig. 2.17) provide strong support for our assertion that proximity of the fluorophore to the metal islands resulted in increased quantum yields. It is unlikely that these diverse fluorophores would all bind to the silver islands or display other unknown effects which would result in enhancements which increased monotonically with decreased quantum yields.

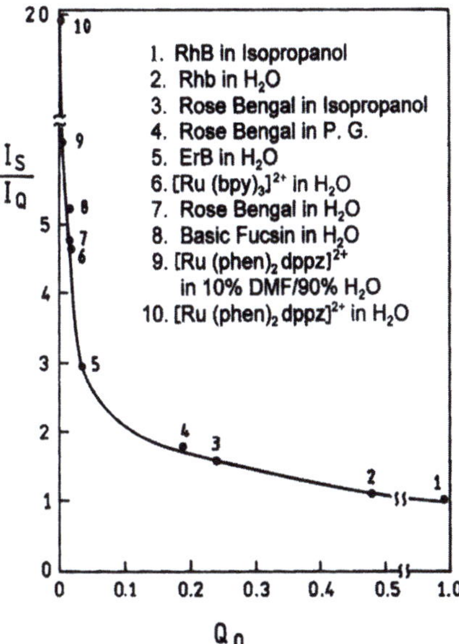

Fig. 2.17. Enhancement of the emission of fluorophores with different quantum yields

2.2.6
Effects of Silver Islands on Intrinsic Protein Fluorescence

We examined the emission spectra of two proteins in the presence and absence of silver islands (Fig. 2.18). The proteins β-galactosidase and glyoxalase were selected for their modest and low quantum yields. The quantum yield of *E. coli* β-galactosidase was found to be 0.18 relative to *N*-acetyl-L-tryptophamide (NATA) [46] which is reported to be 0.13 [47]. The quantum yield of human glyoxalase was found to be about 10-fold less, and thus near 0.013. β-galactosidase is a tetrameric protein, 480,000 molecular mass, which contains 26 tryptophan residues in each 120,000 dalton subunit [48]. Human glyoxalase is a 66,000 dalton monomer which contains two tryptophan residues [49]. For the higher quantum yield β-galactosidase there was no significant effect of the silver islands on the emission spectra. For the lower quantum yield human glyoxalase we observed both a blue shift and an increase in emission intensity. We attribute the spectra changes in Fig. 2.18 (bottom) as due to increased emission from a highly quenched tryptophan residue in glyoxyalase. Because the enhanced emission spectrum is blue shifted we conclude that the low quantum yield tryptophan residue is shielded from the solvent. The absence of a spectral shift or enhancement in β-galactosidase is understandable given its large number of tryptophan residues, since it is unlikely that a significant fraction of these are highly quenched. These results suggest that silver islands can result in increased emission from quenched aromatic amino acid residues in proteins.

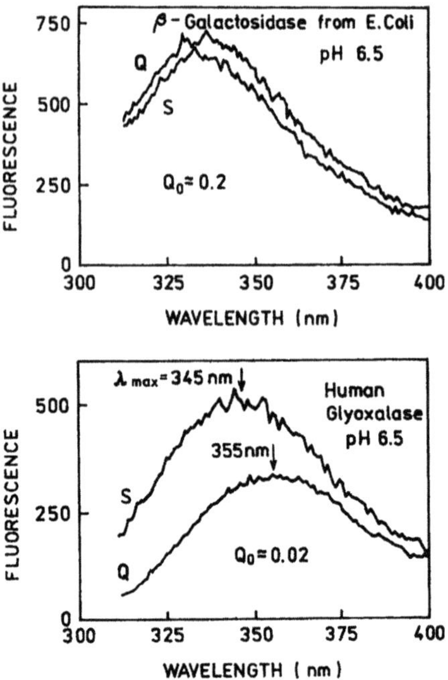

Fig. 2.18. Emission spectra of a higher quantum yield protein β-galactosidase from *E. Coli* (0.05 mg/mL) and a lower quantum yield protein human glyoxalase (0.15 mg/mL) between quartz plates (Q) and silver island films (S). The excitation wavelength was 295 nm

2.2.7
Effects of Silver Islands on Nucleic Acid Bases and DNA

The intrinsic emission from DNA, nucleotides and nucleic acid bases is very weak [50] and is difficult to observe even with modern instrumentation [9, 10, 51]. We questioned whether silver islands could enhance this weak intrinsic fluorescence. Emission spectra of the single stranded oligonucleotides poly T and poly C (Fig. 2.19). The long wavelength emission maximum of poly C is in agreement with that reported previously [54]. Additionally, we have observed enhanced emission from double helical DNA near silver islands and the increased emission was accompanied by a decreased lifetime [52]. These results could be of importance for the growing use of DNA arrays or gene chips or attempts to sequence DNA using a single strand of DNA.

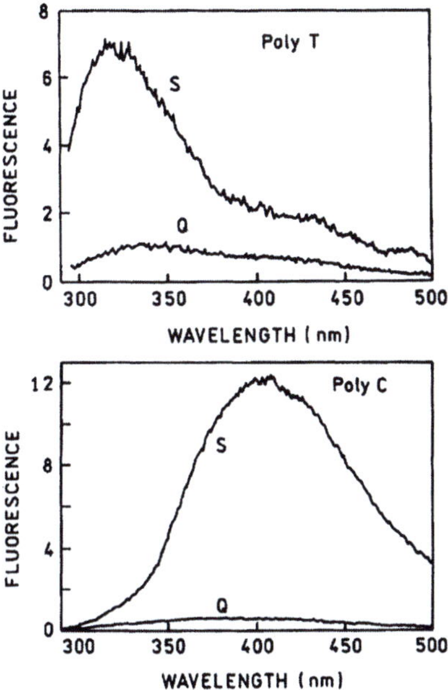

Fig. 2.19. Effects of silver island films on nucleic acid bases. The concentrations of adenine and thymine were 1.2 and 1.3 mM, respectively in 100 mM phosphate buffer, pH 7

2.2.8
Effects of Silver Islands on Resonance Energy Transfer

Resonance energy transfer (RET) is widely used in biochemical and biomedical research. RET occurs whenever fluorophores with suitable spectral properties come within the Förster distance R_0. Förster distances range from 20 to 40 Å, and are rarely larger than 50 Å. Theoretical studies of donors and acceptors at appropriate locations near metal particles have been predicted to increase the rates of energy transfer by a factor of 100-fold or larger at distances as large as 700 Å [39, 40]. To the best of our knowledge there have been no experimental demonstrations of increased energy transfer near metal surfaces.

We examined resonance energy transfer from DAPI to acridine orange (AO) when bound to double helical calf thymus DNA (Fig. 2.20). There is a dramatic increase in the acceptor emission near 520 nm, which we believe is due to a metal-enhanced increase in the extent of energy transfer. This interpretation is supported by the frequency-domain intensity decays of the DAPI donor (Fig. 2.21). The mean decay time of the donor alone (D) decreases for 2.80 ns between the quartz plates to 2.39 ns between the silver islands. In contrast, for DAPI in the presence of acceptor the mean decay time (DA) decreases nearly two-fold, from 2.31 to 1.37 ns between quartz and silver islands, respectively. These results indicate a significant increase in energy transfer near the silver islands.

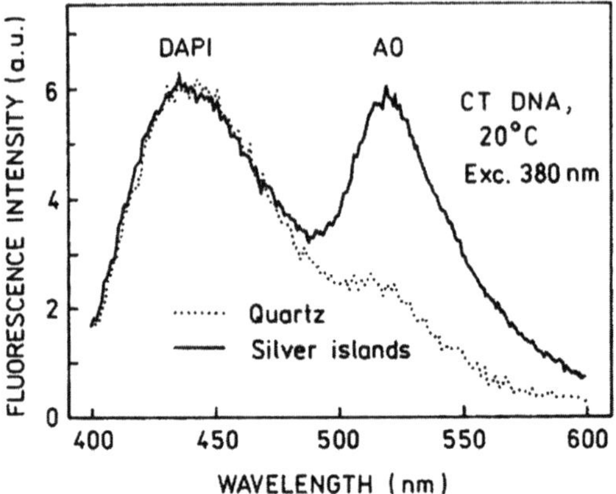

Fig. 2.20. Emission spectra of DNA labeled with DAPI (donor) and acridine orange (acceptor) between quartz plates (Q) and silver island films (S). The spectra are normalized to the donor emission. The concentration of DNA in 100 mM phosphate buffer, pH 7, was 2 mM as base pairs. The concentration of DAPI and acridine orange were 1 per 100 base pairs and 1 per 7000 pairs, respectively

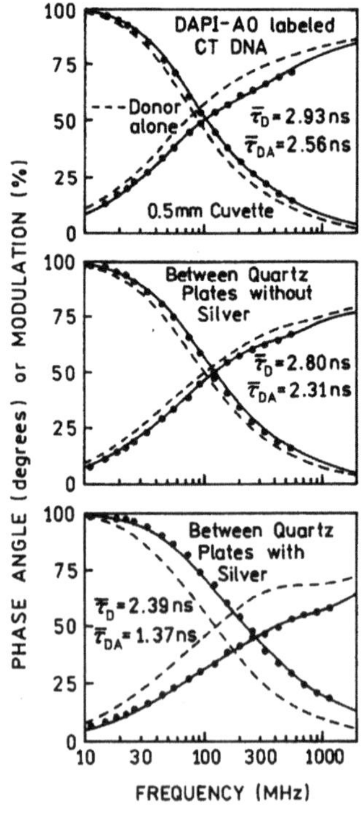

Fig. 2.21. Frequency-domain DAPI donor decays when bound to DNA

2.3
Conclusion

Our preliminary experiments show that metallic particles can have dramatic and useful effects on biochemical fluorophores, and importantly, increased quantum yields from weakly fluorescent molecules. One can imagine numerous opportunities when using these effects of metallic surfaces, including detection of DNA based on its intrinsic fluorescence, observation of weakly fluorescent residues in proteins, detection of membranes adjacent to metal surfaces based on spectral shifts of polarity-sensitive fluorophores and detection of binding events using weakly fluorescent probes which become localized near the metal surfaces. Additionally, the metal-induced increase in energy transfer may allow immunoassays and DNA hybridization measurements with widely spaced donors and acceptors.

Acknowledgments. This work was supported by the NIH National Center for Research Resources, RR-08119, with additional support from the Juvenile Diabetes Foundation International 1-2000-5 and the American Diabetes Foundation. The authors would like to thank Dr. Chris Geddes for presenting this talk on their behalf. The crisis of September 11, 2001 prevented our attendance at this valuable conference.

References

1. Strickler SJ, Berg RA (1962) Relationship between absorption intensity and fluorescence lifetimes of molecule. J Chem Phys 37:814–822
2. Ford GW, Weber WH (1984). Electromagnetic interactions of molecules with metal surfaces. Phys Rep 113:195–287
3. Chance RR, Prock A, Silbey R (1978) Molecular fluorescence and energy transfer near interfaces. Adv Chem Phys 37:1–65
4. Glass AM, Liao PF, Bergman JG, Olson DH (1980) Interaction of metal particles with adsorbed dye molecules: absorption and luminescence. Optics Lett 5(9):368–370
5. Campion A, Gallo AR, Harris CB, Robota HJ, Whitmore PM (1980) Electronic energy transfer to metal surfaces: A test of classical image dipole theory at short distances. Chem Phys Letts 73(3):447–450
6. Sokolov K, Chumanov G, Cotton TM (1998) Enhancement of molecular fluorescence near the surface of colloidal metal films. Anal Chem 70:3898–3905
7. Hayakawa T, Selvan ST, Nogami M (1999) Field enhancement effect of small Ag particles on the fluorescence from Eu^{3+}-doped SiO_2 glass. Appl Phys Lett 74(11):1513–1515
8. Selvan ST, Hayakawa T, Nogami M (1999) Remarkable influence of silver islands on the enhancement of fluorescence from Eu^{3+} ion-doped silica gels. J Phys Chem B 103:7064–7067
9. Ballini JP, Vigny P, Daniels M (1983) Synchrotron excitation of DNA fluorescence decay time – evidence for excimer emission at room temperature. Biophys Chem 18:61–65

10. Georghiou S, Nordlund, Thomas M, Saim AM (1985) Picosecond fluorescence decay time measurements of nucleic acids at room temperature in aqueous solution. Photochem Photobiol 41(2):209–212

11. Hurtubise RJ (1990) Phosphorimetry: Theory, instrumentation, and applications. VCH Publishers, New York, p 370

12. Subramaniam V, Steel DG, Gafni A (2000) Room temperature tryptophan phosphorescence as a probe of structural and dynamic properties of proteins. In: Lakowicz JR (ed) Topics in Fluorescence Spectroscopy, Vol. 6: Protein Fluorescence, Kluwer Academic/Plenum Publishers, New York, pp 43–65

13. Slavik J (1994) Fluorescent probes in cellular and molecular biology. CRC Press, Boca Raton, Chapter 5, pp 125–137

14. Drexhage KH (1974) Interaction of light with monomolecular dye lasers. In: Wolfe E (ed) Progress in optics, North-Holland Publishing Company, Amsterdam, pp 161–232

15. Amos RM, Barnes WL (1997) Modification of the spontaneous emission rate of Eu^{3+} ions close to a thin metal mirror. Phys Rev B 55(11):7249–7254

16. Barnes WL (1998) Fluorescence near interfaces: the role of photonic mode density. J Modern Optics 45(4):661–699

17. Michaels AM, Jiang J, Brus L (2000) Ag nanocrystal junctions as the site for surface-enhanced Raman scattering of single rhodamine 6G molecules. J Phys Chem B 104:11965–11971

18. Freeman RG, Grabar KC, Allison KJ, Bright RM, Davis JA, Guthrie AP, Hommer MB, Jackson MA, Smith PC, Walter DG, Natan MJ (1995) Self-assembled metal colloid monolayers: An approach to SERS substrates. Science 267:1629–1632

19. Weitz DA, Garoff S, Hanson CD, Gramila TJ (1982) Fluorescent lifetimes of molecules on silver-island films. Optics Lett 7(2):89–91

20. Kerker M (1985) The optics of colloidal silver: Something old and something new. J Coll Interface Science 105:297–314

21. Faraday M (1857) The bakerian lecture – experimental relations of gold (and other metals) to light. Philos Trans 147:145–181

22. Link S, El-Sayed MA (2000) Shape and size dependence of radiative, non-radiative and photothermal properties of gold nanocrystals. Int Rev Phys Chem 19:409–453

23. Kreibig U, Vollmer M (1995) Optical properties of metal clusters. Springer Series in Materials Science, p 532

24. Philpott MR (1975) Effect of surface plasmons on transitions in molecules. J Chem Phys 62:1812–1817

25. Chance RR, Prock A, Silbey R (1978) Molecular fluorescence and energy transfer near interfaces. Adv Chem Phys 37:1–65

26. Gersten J, Nitzan A (1981) Spectroscopic properties of molecules interacting with small dielectric particles. J Chem Phys, 75(3):1139–1152

27. Weitz DA, Garoff S, Gersten JI, Nitzan A (1983) The enhancement of Raman scattering, resonance Raman scattering, and fluorescence from molecules absorbed on a rough silver surface. J Chem Phys 78(9):5324–5338

28. Kummerlen J, Leitner A, Brunner H, Aussenegg FR, Wokaun A (1993) Enhanced dye fluorescence over silver island films: analysis of the distance dependence. Mol Phys 80(5):1031–1046

29. Garcia-Ramos JV, Sanchez-Cortes S (1997) Metal colloids employed in the SERS of biomolecules: activation when exciting in the visible and near-infrared regions. J Mol Structure 405:13–28

30. Graham D, Mallinder BJ, Smith WE (2000) Surface-enhanced resonance Raman scattering as a novel method of DNA discrimination. Angew Chem Int Ed 39(6):1061–1063

31. Graham D, Mallinder BJ, Smith WE (2000) Detection and identification of labeled DNA by surface enhanced resonance Raman scattering. Biopolymers 57:85–91

32. Nie S, Emory SR (1997) Probing single molecules and single nanoparticles by surface-enhanced Raman scattering. Science 275:1102–1106

33. Kneipp K, Kneipp H, Bhaskaran Kartha V, Manoharan R, Deinum G, Itzkan I, Dasari RR, Feld MS (1998) Detection and identification of a single DNA base molecule using surface-enhanced Raman scattering (SERS). Phys Rev E 57(6):R6281–R6284

34. Emory SR, Nie S (1998) Screening and enrichment of metal nanoparticles with novel optical properties. J Phys Chem B 102:493–497

35. Michaels AM, Nirmal M, Brus LE (1999) Surface enhanced raman spectroscopy of individual Rhodamine 6G molecules on large Ag nanocrystals. J Am Chem Soc 121:9932–9939

36. Amos RM, Barnes WL (1999) Modification of spontaneous emission lifetimes in the presence of corrugated metallic surfaces. Phys Rev B 59(11):7708–7714

37. Kitson SC, Barnes WL, Sambles JR (1996) Photoluminscence from dye molecules on silver gratings. Optics Commun. 122:147–154

38. Kitson SC, Barnes WL, Sambles JR, Cotter NPK (1996) Excitation of molecular fluorescence via surface plasmon polaritons. J Mod Optics 43(3):573–582

39. Chance RR, Prock A, Silbey R (1978) Molecular fluorescence and energy transfer near interfaces. Adv Chem Phys 37:1–65

40. Weitz DA, Garoff S, Gersten JI, Nitzan A (1983) The enhancement of Raman scattering, resonance Raman scattering, and fluorescence from molecules adsorbed on a rough silver surface. J Chem Phys 78(9):5324–5338

41. Gersten JI, Nitzan A (1984) Accelerated energy transfer between molecules near a solid particle. Chem Phys Letts 104(1):31–37

42. Soper SA, Nutter HL, Keller RA, Davis LM, Shera EB (1993) The photophysical constants of several fluorescent dyes pertaining to ultrasensitive fluorescence spectroscopy. Photochem Photobiol 57(6):972–977

43. Van Orden A, Machara NP, Goodwin PM, Keller RA (1998) Single-molecule identification in flowing sample streams by fluorescence burst size and intraburst fluorescence decay rate. Anal Chem 70:1444–1451

44. Ambrose WP, Goodwin PM, Jett JH, Van Orden A, Wemer JH, Keller RA (1999) Single molecule fluorescence spectroscopy at ambient temperature. Chem Rev 99:2929–2956

45. Jovin TM, Arndt-Jovin DJ (1989) FRET microscopy: digital imaging of fluorescence resonance energy transfer. Application in cell biology. In: Kohen E, Hirschberg JG, Ploem JS (eds) Cell structure and function by microspectro-fluorometry Academic Press, London, pp 99–117

46. D'Auria S, DiCesare N, Gryczynski I, Rossi M, Lakowicz JR (2001) On the effect of SDS on the structure of β-galactosidase from E. coli. J Biochem (in press)

47. Demchenko AP (1981) Ultraviolet spectroscopy of proteins. Springer-Verlag, New York

48. Jacobson RH, Zhang XJ, DuBose RF, Matthews BW (1994) Three-dimensional structure of β-galactosidase from E. coli. Nature 369:761–766

49. D'Auria S (unpublished results)

50. Daniels M, Hauswirth W (1971) Fluorescence of the purine and pyrimidine bases of the nucleic acids in neutral aqueous solution at 300 K. Science 171:675–677
51. Georghiou S, Braddick TD, Philippetis A, Beechem JM (1996) Large-amplitude picosecond anisotropy decay of the intrinsic fluorescence of double-stranded DNA. Biophys J 70:1909–1922
52. Lakowicz JR, Shen Y, Gryczynski Z, D'Auria S, Gryczynski I (2001) Intrinsic fluorescence from DNA can be enhanced by metallic particles. Biochem Biophys Res Commun. (in press)

CHAPTER 3

Fluorescence Nanometrology in Sol-Gels

D.J.S. Birch, C.D. Geddes, J. Karolin, R. Leishman, and O.J. Rolinski

We describe recent fluorescence studies of the formation dynamics and structure of sol-gel glasses from nanometre particles composed of silica clusters in sols to nanometre pores in silica gels. The "kinetic life-history" of silica produced under both acidic and alkaline conditions from sodium silicate in a hydrogel and from an alkoxide in an alcogel is now starting to be revealed by fluorescence techniques and the influence of key parameters such as pH and silica concentration quantified at the molecular level. Through careful choice of fluoro-probe, anisotropy decay has been shown to provide particle size as well as viscosity information and offer advantages over traditional techniques for silica particle sizing based on small angle neutron, X-ray or light scattering. Fluorescence resonance energy transfer (FRET) can now be used to determine the donor-acceptor spatial distribution function without making any *a-priori* assumptions as to its form. This in turn promises to make FRET a better means of monitoring pore morphology in the wet gel during drying and ageing, offering distinct advantages over dry gel techniques such as mercury porosimetry and nitrogen adsorption. The insight into sol-gel processes provided by these new interpretations of fluorescence decay data promises to have implications for both our fundamental understanding and the production of sol-gel systems.

3.1
Introduction

Fluorescence resonance energy transfer (FRET) is often cited as being "an Å ruler" or "spectroscopic ruler" and these phrases highlight one of the prime features of modern day fluorescence spectroscopy, namely the resolution for reporting on molecular structure and dynamics at a functional level. FRET works fine over ~10 to 100 Å, but of course fluorescence anisotropy decay can also provide a distance measurement via the hydrodynamic radius. In this chapter we consider recent work combining these two powerful fluorescence techniques in an area of metrology aimed at unravelling the molecular mysteries leading up to the silica sol to gel transition and beyond.

The silica sol-gel process is a room temperature polymerisation whereby the precursor solution "*the sol*" forms a rigid network spanning the containing vessel, after a time t_g [1]. This point marks the onset of "*the gel*" and t_g is strongly pH, temperature and silica concentration dependent. Prior to t_g ramified nanometre size clusters of silica form and diffuse. Eventually, this process leads to the rigid network appearing at t_g and other processes start to dominate as the solvent evaporates, pores form as the particles aggregate, condensation occurs and the gel shrinks (syneresis) and ages. Although the chemistry is in general well understood the complexity of physics can be gauged from the fact that many of the chemical processes occur simultaneously, not sequentially, although at different times different reaction rates may dominate. For example, for a while even after t_g most of the volume is still a liquid. Control of the polymerisation produces a wide range of materials. These include stable colloids of well defined nanoparticles, e.g., Dupont's Ludox (which has many uses though known in fluorescence as a light scattering medium for recording excitation pulse profiles), optical quality components used in photonics, porous glasses used in sensors and the ubiquitous uses of silica gel powder.

A clearer picture of how controlling the competitive rates in the sol in the very early stages translate into the final gel is not only important to our fundamental understanding, but also in optimising manufacturing processes. The link between particle size in the sol and pore size in the gel is a classic example, which spans the whole "life-history" of silica gel. Of course any technique for monitoring such changes should ideally be capable of monitoring *in-situ* the whole process through to completion. Fluorescence is *par excellence* a method of determining reaction rates and, although ideally suited to the task in sol-gels, its capabilities have perhaps been hitherto under-appreciated in this context.

Traditionally, small angle scattering of laser light, X-rays or neutrons have been used to study silica particle growth. For example, light scattering measurements found a primary particle hydrodynamic diameter of 1.0 nm increasing to 2.4 nm prior to gelation [2]. Small angle X-ray scattering studies of silica gel have revealed evidence for 1 nm particles [3] and similar studies on a silica sol indicate that 2 nm diameter primary particles aggregate to form secondary particles of 6 nm diameter prior to gelation [4]. Small angle neutron scattering has found comparable primary particle dimensions [5]. However, scattering methods have a number of drawbacks. For example, they need low silicate concentrations to avoid multiple scattering and dilution is not the answer, as this can cause depolymerisation. Moreover, scattering by the gel matrix after t_g corrupts the particle scattering measurement. X-ray and neutron scattering in particular are also very expensive and unsuitable for on-line use. Light scattering is limited in resolution by the wavelength of light and electron microscopy can only be used on dry colloids.

Fluorescence correlation spectroscopy (FCS) [6] and fluorescence recovery after photobleaching (FRAP) [7] possess sufficient resolution for silica particle metrology, but also suffer from the need to dilute the sols and have the added complication of requiring a microscope.

We include in this chapter the use of fluorescence anisotropy decay to determine the growth in hydrodynamic radius of silica particles [8, 9]. Although fluorescence anisotropy decay has been widely used in biochemistry to determine structure and dynamics in membranes and proteins [10] it has hitherto found little if any application where the hydrodynamic radius of the fluorescing rotor species changes continuously with time. And yet fluorescence anisotropy decay is ideally suited to particle metrology during sol-gel polymerisation, overcoming many of the drawbacks of scattering methods. For example, because in the absence of energy migration fluorescence anisotropy will decay only by rotating particles, growth can be studied at higher silicate concentrations than other techniques and even after gelation has occurred. Interestingly, although fluorescence has been widely used to study the sol to gel transition the interpretation was for a long time confined to viscosity [11–15]. The realisation that the observed second and longer rotational correlation time corresponded to dye bound to particles [8, 9] has bridged the gap between scattering and fluorescence techniques and is providing new insight into silica growth mechanisms.

Fluorescence also has much to offer in gel studies after t_g. Traditional techniques for pore size and surface area measurement such as mercury porosimetry, nitrogen adsorption, and BET analysis can only be used on dry gels [1]. FRET has demonstrable capabilities for studying wet gels *in-situ*. The dual assumptions of a random donor-acceptor distance distribution function for $\rho(r)$ in a specific geometry such as a cylinder, sphere or fractal enables the donor fluorescence decay to be analysed and pore size estimates determined [16–19]. Recently, we have started to apply a new approach to the problem whereby $\rho(r)$ is determined from fluorescence decay measurements without making any *a-priori* assumptions as to its form and changes in pore morphology are then analysed by applying the single assumption of a specific geometrical model. The theoretical basis of finding $\rho(r)$ [20] has been shown to work well in describing the discrete acceptor binding sites found in a protein [21–23] and the porous polymer Nafion used in FRET based metal ion sensing [24], which is in some ways analogous to sol-gel pores. In this chapter we show that in a sol-gel not only pore diameters, but also wall thickness between pores can be obtained by determining (i.e., not assuming) a form for $\rho(r)$.

The combination of fluorescence anisotropy and FRET enables the "kinetic life-history" of silica gel to be tracked *in-situ* from particle to pore. Fig. 3.1 de-

Fig. 3.1. Generalised depiction of the sol to gel transition in silica gel. **A** Nanometre scale particles composed of clusters of silica form and join together to form a growing network (**B**), which spans the containing vessel at a time t_g and then shrinks forming pores (**C**)

picts the problem, which we will expand upon in terms of critical parameters in the sol, such as pH and silica concentration through illustrative examples.

3.2
Sol-gel Chemistry

The sol-gel process involves the transformation of a liquid like solution, *the sol*, to *the gel*, a highly porous matrix filled with solvent, through a series of hydrolysis and polycondensation steps.

Simplified the gel-forming step can be given by:

$$n \, Si(OR)_4 + 2 \, n \, H_2O \rightarrow n \, SiO_2 \text{ (a growing gel network)} + 4 \, n \, ROH \qquad (3.1)$$

where R is hydrogen for the case of hydrogels (an inorganic polymerisation) and methyl or ethyl or propyl etc (an organic polymerisation) for the case of alcogels, both so named in accordance with the solvents used and condensed. The rates of formation and properties of the final gels derived from these organic and inorganic polymerisations are strongly pH, temperature, solvent and SiO_2 concentration dependent [1, 25]. There are many similarities between both processes and similar end products are obtained. The alkoxide "alcogel" route is the one that has been mostly used for research into sol-gel processes because it has better defined reactants and is typically simpler to prepare than a hydrogel; the lower cost of the latter making it ideal for industrial applications requiring mass production.

For hydrogels at pH < 2 (acid catalysed), gelation is thought to occur by means of rapid monomer additions initially to form primary clusters followed by inter-cluster condensation reactions between silanol (Si-OH) bonds to form ramified siloxane type (Si-O-Si) secondary clusters [1]. For hydrogels polymerised from sodium silicate solution (water glass), a ternary system of SiO_2, Na_2O and H_2O, then the gel-forming step can be crudely described by:

$$Na_2O \cdot x\, SiO2 \cdot Y\, H_2O \text{ (water glass)} + H_2SO_4 \rightarrow$$
$$SiO_2 + Na_2SO_4 + (Y+1)H_2O \tag{3.2}$$

where x denotes the weight ratio (w/w) of the glass, i.e., SiO_2:Na_2O, which for the work described here is ≈ 3.3. At a sol pH < 2, the gel times are typically quite long (Fig. 3.2) where the polymerisation rate is thought to be proportional to [H^+] and the silicate species are thought to be positively charged but not highly ionised [25]. Also, in the absence of fluoride ions, the solubility of silica below pH 2 is quite low. It is therefore likely [1] that the formation and aggregation of primary particles occurs rapidly together and that Ostwald ripening (a process whereby particles grow in size but decrease in number as highly soluble small particles dissolve and reprecipitate on larger less soluble nuclei) contributes little to particle growth after the particles exceed 1 nm radius. Therefore, gels polymerised at pH < 2 are thought to be comprised of the joining together of very small particles indeed.

At intermediate sol pH of 2–6 (Fig. 3.2), the gel times steadily decrease and it is thought that above the isoelectric point at around pH 2 the condensation rate is proportional to [OH^-] as shown below [1].

$$\equiv\!Si\text{-}OH + OH^- \rightarrow\, \equiv\!Si\text{-}O^- + H_2O \qquad \text{(Fast process)} \tag{3.3}$$

$$\equiv\!Si\text{-}O^- + HO\text{-}Si \rightarrow\, \equiv\!Si\text{-}O\text{-}Si\!\equiv + OH^- \qquad \text{(Slower process)} \tag{3.4}$$

At a pH above 7, hydrogel sol polymerisation occurs by the same nucleophilic mechanism as for sols in the pH range 2–7, Eqs. 3.3 and 3.4. However, because all condensed species are likely to be negatively charged (highly ionised) and therefore mutually repulsive, growth occurs primarily by the addition of monomers to more highly condensed particles *rather* than by particle aggregation [1]. Particles

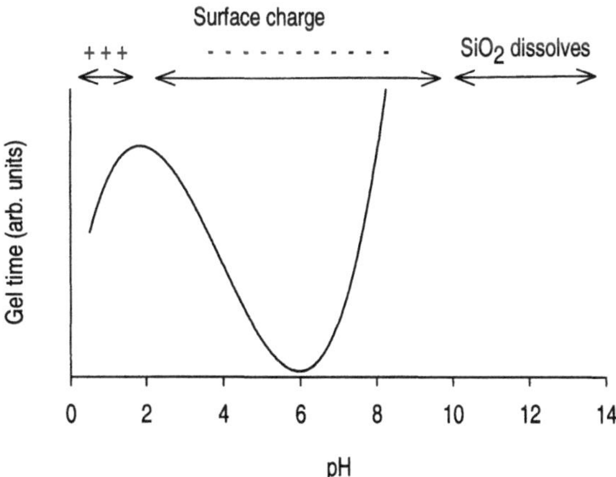

Fig. 3.2. Effect of pH on gel time and sol stability (adapted from [25])

of ≈ 1 nm radii are typically formed within a few minutes above pH 7. Due to the greater solubility of silica and the greater size dependence of solubility above pH 7, the growth of primary particles continues by Oswald ripening. At a given pH, particles grow to a size that depends mainly on the temperature and SiO_2 concentration, with the *growth rate* also depending on the particle size distribution. Due to lack of salt and the mutually repulsive nature of the silica particles, stable, fairly high pH, non-gelling sols can be readily prepared [25] (Fig. 3.2), e.g., DuPont's colloidal silica range. At a much higher sol pH, ≥ 12, most silanol groups are deprotonated and the primary building blocks are composed primarily of cyclic trimers and tetramers.

The alkoxide "alcogel" route, which is similar in many respects to the polymerisation of aqueous silicates, can be described at the functional group level by three simple reactions [1]:

$$Si(OR)_4 + n\,H_2O \leftrightarrow Si(OR)_{4-n}(OH)_n + n\,ROH \quad \text{Hydrolysis} \tag{3.5}$$

$$\equiv Si\text{-}OR + HO\text{-}Si\equiv \;\leftrightarrow\; \equiv Si\text{-}O\text{-}Si\equiv + ROH \qquad \text{Alcohol Condensation} \tag{3.6}$$

$$\equiv Si\text{-}OH + HO\text{-}Si\equiv \;\leftrightarrow\; \equiv Si\text{-}O\text{-}Si\equiv + H_2O \qquad \text{Water Condensation} \tag{3.7}$$

and overall in the case of tetramethylorthosilicate (TMOS) reported here by:

$$n\,Si(OCH_3)_4 + 2\,n\,H_2O \rightarrow n\,SiO_2 + 4\,n\,CH_3OH \quad \text{Net Reaction} \tag{3.8}$$

The hydrolysis reaction, Eq. 3.5, effectively replaces alkoxide groups with hydroxyl groups, which can then readily condense to produce either water, Eq. 3.7, or alcohol, Eq. 3.6, where both reactions result in siloxane (Si-O-Si) bonds. Due to the fact that water and alkoxides are immiscible, a mutual solvent is typically used as a homogenising agent, e.g., an alcohol. As indicated by Eqs. 3.5 and 3.6, alcohol is not just a solvent but can participate in the reverse esterification and alcoholysis reactions, respectively. The hydrolysis scheme, Eq. 3.5, is generally acid or base catalysed where the rate and nature of polymerisation is pH dependent [1]. Other parameters such as temperature, pressure, the type of alcohol used and the molar H_2O:Si ratio (sometimes denoted R:1) significantly influence the properties of the final gel [1].

Even after t_g particles continue to play a role but new processes now start to occur as well. These include ageing involving further condensation, dissolution or reprecipitation of monomers and oligomers depending on pH, etc., and syneresis as the gel shrinks due to condensation, drying and associated expulsion of liquid from pores (see Fig. 3.1).

3.3
Anisotropy Theory

Vertically and horizontally polarised fluorescence decay curves, $F_V(t)$ and $F_H(t)$, orthogonal to pulsed and vertically polarised excitation, recorded at different delay times following initial mixing of the sol, lead to an anisotropy function $R(t)$ [10] describing the rotational correlation function where:

$$R(t) = \frac{F_V(t) - F_H(t)}{F_V(t) + 2\,F_H(t)} \tag{3.9}$$

If continuous excitation is used the time-dependencies remain unresolved and a weighted average of the individual time-resolved anisotropies is observed in a multi-component system such as a sol-gel.

Our previous analysis of $R(t)$ [8, 9] for silica hydrogels and that of Narang and coworkers [14] on TMOS showed that the best description was provided by two rotational correlation times τ_{r1} and τ_{r2} in the form:

$$R(t) = (1-f)R_0\exp(-t/\tau_{r1}) + f\,R_0\exp(-t/\tau_{r2}) \tag{3.10}$$

where R_0 is the initial anisotropy. We interpret 'f' as the fraction of fluorescence due to probe molecules bound to silica particles and hence $1-f$ the fraction due to free dye in the sol. From the Stokes-Einstein relation, τ_{r1} gives the sol microviscosity $\eta_1 = 3\tau_{r1}kT/4\pi r^3$, where r is the hydrodynamic radius of the dye and likewise using η_1 and τ_{r2} gives the average silica particle hydrodynamic radius.

By expanding $\exp(-t/\tau_{r2})$ and putting $\tau_{r1} \ll \tau_{r2}$, to reflect the unbound probe molecules rotating much faster than those which are bound to silica particles, then in the case where the fluorescence lifetime $\tau_f \ll \tau_{r2}$ a similar expression to that encountered for the hindered rotation of a fluorophore in a membrane or protein [10] can be expected to hold in a sol-gel, i.e., a residual anisotropy is observed:

$$R(t) = (1-f)R_0\exp(-t/\tau_{r1}) + f\,R_0 \tag{3.11}$$

If a fraction of the fluorescence 'g' is attributed to dye bound rigidly within the gel after t_g as well as both free solvated dye and dye bound to silica particles then, if appropriate, Eq. 3.10 could be further extended to

$$R(t) = (1-f-g)R_0\exp(-t/\tau_{r1}) + f\,R_0\exp(-t/\tau_{r2}) + gR_0 \tag{3.12}$$

Given the potential complexity of the molecular rotations likely to be observed, multiphoton excitation as well as one-photon excitation is useful [26]. Multiphoton excitation increases the initial value of the fluorescence anisotropy, R_0, and hence the dynamic range over which rotations are observed. This is particularly useful when the fluorescence lifetime is significantly less than the rotational correlation time, as is the case here. For i-photon excitation R_0 can be expressed as [26]:

$$R_{0i} = \frac{2i}{2i+3}\left[\frac{3}{2}cos^2\beta_i - \frac{1}{2}\right]$$ (3.13)

where β_i is the intramolecular angle between the dominant absorption and emission transition moments. In the collinear ($\beta_i = 0$) case, $R_{0i} = 0.4$ for $i = 1$ and 0.57 for $i = 2$.

3.4
FRET Donor-acceptor Distribution Theory

Among several molecular mechanisms which can be applied to the detection of a structure of a complex media, FRET from the excited donor molecule D to the acceptor molecule A provides quite specific site (and hence structural) information. This is because the rate of FRET, $w(r)$, is

$$w(r) = (1/\tau_0)(D_0/r)^6$$ (3.14)

and depends strongly on the donor/acceptor spectral characteristics (D_0 being a measure of donor fluorescence and acceptor absorption spectral overlap), providing potential for high selectivity, but also on the donor-acceptor distance r. τ_0 is the donor fluorescence lifetime in the absence of FRET, and D_0 is defined as the critical transfer distance at which the probability of FRET is ½ and is specific for each donor-acceptor pair.

In the presence of an acceptor the donor fluorescence impulse response function $I_D(t)$ is modified from a monoexponential function $exp(-t/\tau_0)$ to the form [20]:

$$I_D(t) = \exp\left[-\frac{t}{\tau_0} - \int_0^\infty dr\, \rho(r)(1-\exp[-tw(r)])\right]$$ (3.15)

Here $\rho(r)$ is the donor-acceptor distribution function, where $\int_0^R \rho(r)dr$ is the number of acceptor molecules in a volume of a sphere of radius R.

Equation 3.15 constitutes an inverse problem in which the donor-acceptor distribution function $\rho(r)$ is an input information which is to be recovered from the output signal $I_D(t)$. The transmitted information might be modified during measurements by the detection method applied, autofluorescence and scatter.

Problems similar to that posed by Eq. 3.15 arise in a wide variety of biochemical and porous solid applications, and consist of using mathematical models in order to determine unknown system inputs, sources or parameters from observed system outputs and responses. A common approach is to use a mathematical model to fit, often using least-squares error analysis, predictions of the model to the measured system outputs by adjusting the unknown model parameters. For example, the integral in Eq. 3.15 is usually solved for the assumed random 3 or 2-dimensional distribution $\rho(r)$ and the parameter γ, indicating the acceptor concentration, is determined from the experimental data. This simple approach is not appropriate to systems of complex structures.

Mathematically, Eq. 3.15 belongs to a class of Fredholm integral equations of the first kind, which are known to be ill-posed. This means that the solution $\rho(r)$ may not be unique, may not exist and may not depend continuously on the data. The extent of such problems mainly depends on the property of the integral kernel, which, together with the accuracy of the fluorescence impulse response function measurement $I_D(t)$ and smoothing techniques, will determine how much information about $\rho(r)$ can be extracted from the measurements. In this approach the donor-acceptor distribution function $\rho(r)>0$ is expressed as the infinite series of the orthonormal Laguerre polynomials $L_k^s(r)$ with the coefficients $a_k^{(s)}$ without sacrificing generality of the function $\rho(r)$:

$$\rho(r) = r^5 \sum_{k=0}^{\infty} a_k^{(s)} L_k^s(r) \tag{3.16}$$

where s is an arbitrary chosen parameter. Combining Eq. 3.16 and Eq. 3.15 and then applying to both sides of the resulting expression the operator:

$$\int_0^{\infty} dt\, e^{-t} L_V(t)* \tag{3.17}$$

defined in time space, we obtain:

$$f_{\alpha,p} = \sum_{k=0}^{m} a_k^{(s)} b_{kp}^s(\alpha) \tag{3.18}$$

Equation 3.18 establishes fundamental relationship between the experimentally available vector $f_\alpha = \{f_{\alpha,0}, f_{\alpha,1}, f_{\alpha,2},...,f_{\alpha,p}\}$ defined in the time domain, the sought for vector $a^{(s)} = \{a_0^s, a_1^s, a_2^s,...,a_M^s\}$ defined in the distance domain, and the matrix $b^{(s)}(\alpha)$, which depends on $\alpha = D_0^6/\tau_0$ only. The basic procedure for determining of $\rho(r)$ in this approach assumes several steps. In the first step the parameter α is estimated from the steady-state (D_0) and time-resolved (τ_0) measurements. Next, the donor-acceptor system for which $\rho(r)$ is to be determined is prepared, the donor fluorescence decay is measured and the f_α vector calculated. In the next step the mode of representation of the donor-acceptor distribution function (s value) has to be chosen and the matrix $b^{(s)}(\alpha)$ calculated. Finally, vector $a^{(s)}$ is found from solving the set of Eq. 3.18 and $\rho(r)$ is calculated on the basis of Eq. 3.16. The donor/acceptor pair of particular D_0 value reports only the value of $\rho(r)$ in the narrow region ($D_0 - dD_0$, $D_0 + dD_0$) [20]. Thus, improved accuracy of $\rho(r)$ requires the measurements to be made over a range of D_0 values. Here we will report only $\rho(D_0)$ determination on a shrinking gel at a single D_0 value. However, recently [23] we have shown how $\rho(r)$ can be determined in the range for $0 < r < 2\,D_0$ for a glucose-binding protein and work is currently underway to apply this approach to sol-gels.

3.5
Acidic Hydrogels

Fig. 3.3 shows a typical silica particle growth and microviscosity at pH < 1 for a hydrogel using an analysis of anisotropy decays according to Eq. 3.10. Under such strongly acidic conditions (c.f. Eq. 3.2) many dyes are unstable but the cationic dye JA120 [27] was used successfully as a probe (radius 0.75 nm) and excited at 650 nm using a diode laser of 50 ps pulses at 1 MHz repetition rate and time-correlated single-photon timing detection [28]. Working in the near-infrared overcomes the sol auto-fluorescence but places a further restriction on the choice of dye. Polarized fluorescence decays were accumulated for 1 minute durations during polymerisation. At pH < 1 the particles are slightly positively charged and hence a cationic dye is only slowly taken up by the silica particles such that both the microviscosity and the particle radius can be determined together experimentally using Eq. 3.10. Notable features include the constancy of the microviscosity in the range 1–2 cP, the reducing growth rate as the particle number density decrease as particles bind and grow via inter-particle condensation reactions (e.g., Eq. 3.7) and particle syneresis due to intra-particle condensation. The initial growth kinetics are well described for a particle radius r by:

$$r = r_0 + (r_{max} - r_0)\,(1 - e^{-kt}) \tag{3.19}$$

Modelling such as Fig. 3.3 quantifies a number of key parameters in sol-gel kinetics. A growth rate $k \sim 8.6 \times 10^{-5}$ s^{-1} at early times is seen to dominate over a syneresis rate $\sim 6 \times 10^{-6}$ s^{-1} until later times. The fluorescence evidence is consistent with other evidence [1, 25]. At very early times silicic acid monomer units rapidly combine under diffusion control (i.e., the probability of collisions producing binding approaches unity) to form primary particles of mean hydrodynamic radius 1.5–2 nm, which then aggregate under chemical reaction control (i.e., the collisional binding probability is significantly less than unity) and grow ~ 3 fold to

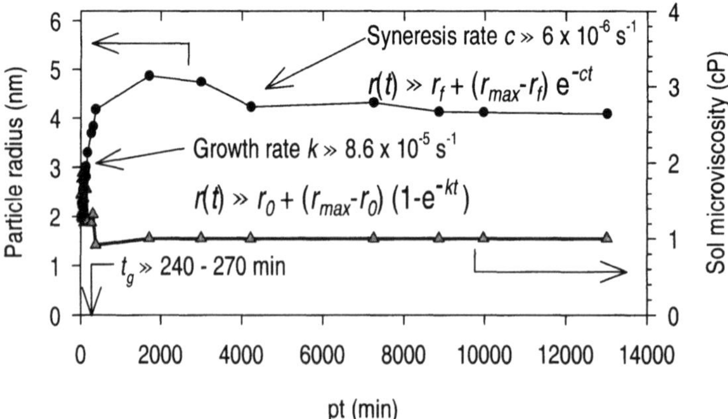

Fig. 3.3. Typical silica hydrogel particle growth and sol microviscosity at pH < 1 during a polymerisation time pt

produce secondary clusters of 4–5 nm maximum mean radius. Simple geometrical considerations predict that a secondary particle contains ~13 primary particles. These trends have been confirmed over a wide range of sols at pH < 1 [8, 9]. The formation of the primary particles is too fast to be resolved under these conditions.

3.6
Alkaline Hydrogels

Moving now to the alkaline region of stability at higher pH. Once again the extreme pH can cause dye instability, in this case deprotonation, but using rhodamine 700 as a probe Fig. 3.4 shows the particle growth kinetics at pH 10 assuming a microviscosity of 1 cp. This pH region represents destabilisation of the initial silicate (pH $\geq$ 13) to produce gelation under alkaline conditions and consequently, the particles would be expected to be highly negatively charged and mutually repulsive. Hence, unlike at low pH < 1, growth at high pH would be expected to be more by monomer addition and Ostwald ripening rather than particle aggregation [1, 25]. Fig. 3.4 obtained at pH 10 shows a dramatic difference from Fig. 3.3 at pH 0.9 in respect of the former showing a clear bi-modal growth. We have modelled the particle growth in Fig. 3.4 successfully using a function of the form:

$$r = r_0 + \left(r_p - r_0\right)\left(1 - \exp\left[-k_p t\right]\right) + \left(r_s - r_p\right)\left(1 - \exp\left[-k_s t\right]\right) \tag{3.20}$$

This function implies the growth of two different entities, one a precursor to the other, which reach different limiting radii r_p and r_s. Conventional thinking would ascribe k_p as the rate of growth of a primary particle species due to monomer addition and k_s as the growth of secondary particles, again by monomer addition, all slowed up sufficiently to be measurable by the mutual negative repulsive charge causing reaction limiting conditions. However, a single growth mechanism begs the question as to why two entities with distinct rates of formation are observed at

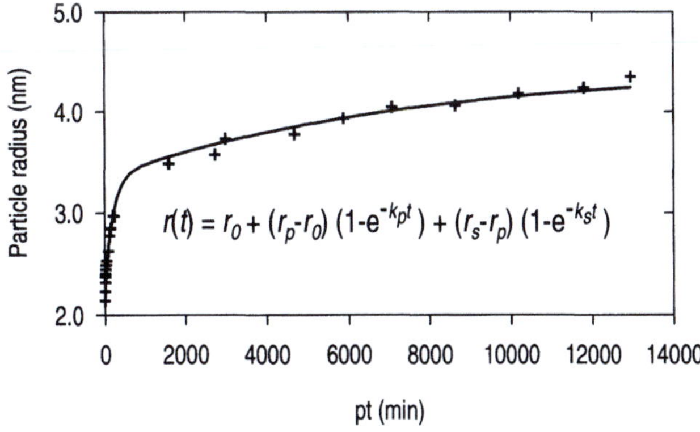

Fig. 3.4. Silica hydrogel particle growth at pH 10 for a 2% SiO_2 sol of $t_g > 250$ hr during a polymerisation time pt

all and from the fluorescence evidence the role of cluster-cluster aggregation cannot be ruled out, particularly at early times, since the growth rate (k_p) is close to that of 8.6×10^{-5} s^{-1} observed for secondary particles under acidic conditions at pH < 1 (see Fig. 3.3).

Fitting to Eq. 3.20 gives $r_0 = 2.3$ nm, $r_p = 3.3$ nm, $r_s = 4.6$ nm, $k_p = 9.2 \times 10^{-5}$ s^{-1} and $k_s = 1.9 \times 10^{-6}$ s^{-1}. According to this model r_0 simply reflects the mean primary particle size in our first measurement. k_s is just less than the syneresis rate measured under acidic conditions (from Fig. 3.3 $\sim 6 \times 10^{-6}$ s^{-1}) and might reflect a net growth limited by intra-particle syneresis.

3.7
Alkoxide Alcogels

As was already mentioned the alkoxide "alcogel" route (Eqs. 3.5–3.8) is the one that has been mostly used for research into sol-gel processes because it has better defined reactants and is typically simpler to prepare than the hydrogel route. The polymerisation of tetramethyl-orthosilicate (TMOS) has been studied previously using rhodamine 6G [14] and phase fluorometry and the anisotropy decay interpreted solely in terms of viscosity changes. We have recently re-investigated this system using a slowly gelling 21.9 % (w/w) SiO_2 sol at pH 2.3, i.e., close to the iso-electric point (c.f. Fig. 3.2 and [29]). At $20°$ C t_g is $\sim 6 \times 10^4$ min and this gives plenty of time to maximise the anisotropy statistical precision during measurements in which negligible sol changes occur. The measurement precision was further enhanced by using a fluorometer capable of two-photon excitation, increasing the initial anisotropy (c.f. Eq. 3.13 and [30]). Rhodamine 6G seems to be as close as possible to an ideal dye for this type of work. The merits of rhodamine 6G are that it is stable over a wide pH range ($\sim$1–12), is highly fluorescent, has a high two-photon absorption cross-section [31, 32] and $R_{02} \sim 0.5$ [30], is a well characterized isotropic rotor, is readily taken up by silica and has a fluorescence lifetime ≈ 4 ns, which is compatible with nanometre particle rotational times.

One of the advantages of the measurement techniques we have developed is that Eq. 3.10 also gives the microviscosity of the sol as well as the particle mean radius. In contrast to our study of silica hydrogels using JA120 we found the fraction of fluorescence f due to dye bound to the particles for the TMOS sol remained constant at ≈ 30 % during the polymerisation. Rhodamine 6G is taken up very efficiently by silica, so on this evidence alone we cannot rule out the possibility that in fact all the dye is taken up and that the "free" dye rotation ($1 - f$ in Eq. 3.10) actually refers to dye tethered and wobbling on the particle. However, τ_{r1} is $\sim$300 ps, i.e., compatible with that for the free dye in the water/alcohol mixture in the sol and for the purpose of calculating the hydrodynamic mean radius the model given by Eq. 3.10 is independent of these alternative interpretations of f. Moreover, since $r \sim \eta^{-3}$ the viscosity value used is less critical. Here we have used the viscosity determined from τ_{r1} assuming the theoretical hydrodynamic radius for rhodamine 6G

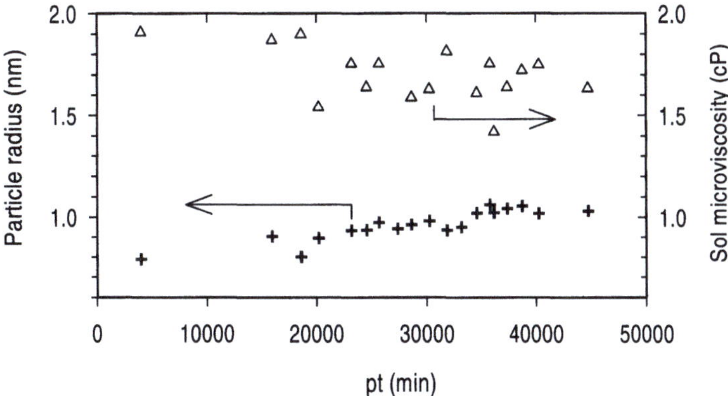

Fig. 3.5. Silica particle radius, +, and sol microviscosity, Δ, as a function of polymerisation time, pt, for a 21.9 % (w/w) SiO_2 TMOS sol at pH 2.3. The errors in particle radius were typically $\pm\,0.1$ nm

of 0.56 nm [33, 14]. Our own study of rhodamine 6G in a range of known viscosity solvents gave a consistent hydrodynamic radius of 0.53 ± 0.03 nm.

Fig. 3.5 shows how the calculated viscosity and particle mean radius change with polymerisation time, pt. The viscosity values are similar in magnitude and show a slight downward trend with time, similar to that shown in Fig. 3.3 for a silica hydrogel [8, 9] at low pH, reflecting here the polymerisation of the sol and the expulsion of methanol. The most notable feature of Fig. 3.5 is the smaller radius than anything we have so far detected, growing from ~ 0.8 to 1.1 nm and illustrating the high-resolution fluorescence anisotropy can offer. Clearly the small size reflects the different growth mechanism associated with monomer-monomer and monomer-particle addition in TMOS under these conditions rather than secondary particle aggregation. We fitted the growth of silica hydrogels at pH < 1 to a radius $r \sim 1 - e^{-kt}$ and attributed this to particle-particle aggregation. For TMOS under these conditions Fig. 3.5 can clearly be approximated by the limit of slow exponential growth at short times, i.e., $r \sim kt$.

3.8
Wet Pore Metrology

The measurement of nm pores in wet silica gels presents the final piece in the fluorescence jigsaw of unravelling the "kinetic life history". Unlike traditional techniques for this purpose, such as mercury porosimetry or nitrogen adsorption, fluorescence is compatible with *in-situ* studies of wet gels and this gives it a major advantage. As an example we have used the new donor–acceptor distribution analysis presented in Sect. 3.4 in FRET studies of the drying of a 12.6% SiO_2 sol at pH 0.8 with a t_g of 240–270 mins. For this work we chose 10^{-5} M rhodamine

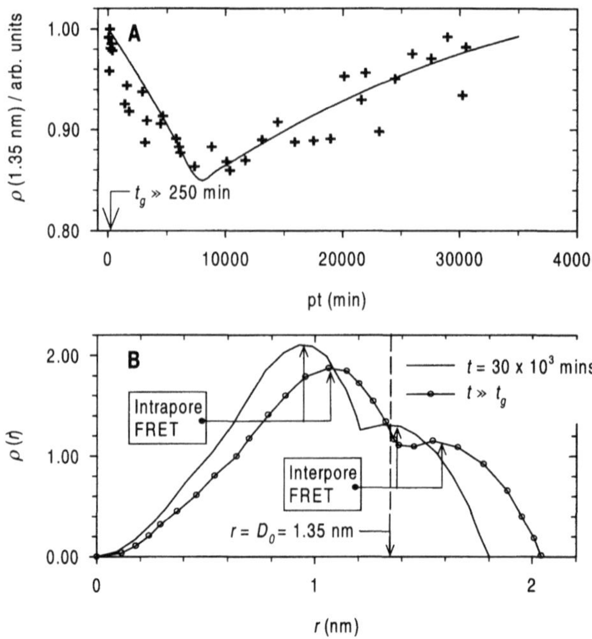

Fig. 3.6. A Measured donor-acceptor distribution function $\rho(r)$ at 1.35 nm for rhodamine 800-Cu^{2+} for an acidic gel drying at room temperature. The t_g was 240–270 mins. **B** Theoretical $\rho(r)$ for a two pore model, assuming the donor is located on the wall of one pore and the acceptors are randomly distributed throughout the same pore and an adjacent pore. Both intra- and inter-pore FRET are depicted in both curves as the gel dries and shrinks. The separation $r = 1.35$ nm corresponding to Fig. 3.6A is also shown

800 as the donor and 0.01M Cu^{2+} as the acceptor. This donor-acceptor pair has a D_0 value of 1.35 nm which is compatible with the pore size expected from the 4 nm particles in this sol (see Fig. 3.3), is minimally intrusive and FRET has been previously demonstrated in metal ion sensors [34, 35].

Fig. 3.6 shows how the donor-acceptor distance distribution function $\rho(1.35$ nm), determined using the procedures outlined in Sect. 3.4, changes during polymerisation, most of the time being occupied by the gel drying well after t_g. Essentially, the measurement sweeps out distribution information on a sphere of radius 1.35 nm and this is shown in Fig. 3.6A. A simulation of the theoretical $\rho(r)$ for a two pore model assuming the donor is located on the wall of one pore and the acceptors are randomly distributed throughout the same pore and an adjacent pore is shown in Fig. 3.6B as the gel shrinks. The dip in $\rho(1.35$ nm) shown in Fig. 3.6A can be interpreted as a reduction in the intra-pore acceptor site density detected as the measurement window at 1.35 nm moves outside the donor pore and into the wall between the pores and the increase at $t > 10,000$ mins reflects more acceptors in the adjacent pore moving into the measurement window. Fig. 3.7 depicts the two pore model we have used and the sizes obtained on analysing Fig. 3.6 A, assuming an exponential decrease as the gel dries and shrinks. The mean

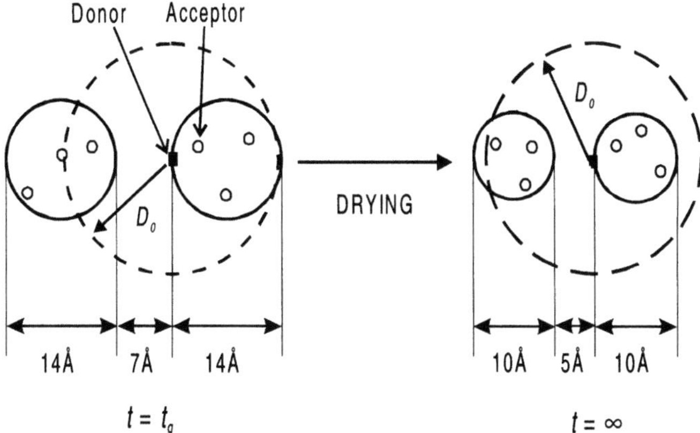

Fig. 3.7. Pore and wall dimensions found from the analysis of Fig. 3.6A as the gel dries

pore diameter is seen to reduce from 14 Å to 10 Å and the intervening silica wall thickness from 7 Å to 5 Å. The shrinkage rate constant is found to be ~ 4.8×10^{-7} s^{-1} and the volume change agrees well with that found from nitrogen adsorption measurements.

3.9
Conclusions

In this chapter we have illustrated the scope fluorescence offers for studying sol-gel processes under a wide range of experimental conditions and for different types of sols. Yet we have only scratched the surface of the rich science to be found in sol-gel transitions and much remains to be done. For example, everything we have reported here concerns a distribution of sizes of particles and pores and for each size a different rate equation pertains. We have so far been only able to address the mean and of course the distribution will influence the overall rates. Moreover, many processes can be simultaneous depending on conditions for example, hydrolysis and condensation, monomer addition and dissolution, particle aggregation and syneresis, etc.

Fluorescence nanoparticle metrology in sol-gels is only just beginning, but the signs are already promising that the full "kinetic life-history" of silica gel can be resolved over its range of reaction pathways. This should in turn lead to closer chemical control and if not new gel compositions then certainly better defined ones will be possible.

Acknowledgement. The authors wish to acknowledge the support of EPSRC and Ineos Silica Ltd.

References

1. Brinkler CJ, Scherer GW (1990) Sol-gel science: The physics and chemistry of sol-gel processing, Academic Press, New York.
2. Boonstra AH, Meeuwsen TPM, Baken JME, Aben GVA (1989) A two-step silica sol-gel process investigated with static and dynamic light-scattering measurements. J Non-Cryst Solids 109:153–163
3. Himmel B, Gerber Th, Burger H (1987) X-ray diffraction investigations of silica-gel structures. J Non-Cryst Solids 91:22–136
4. Orcel G, Hench LL, Artaki I, Jonas J, Zerda TW (1988) Effect of formamide additive on the chemistry of silica sol gels 2 – Gel structure. J Non-Cryst Solids 105:223–231
5. Winter R, Hua D W, Thiyagarajan P, Jonas J (1989) A SANS study of the effect of catalyst on the growth-process of silica-gels. J. Non-Cryst. Solids 108:137–142
6. Thompson NL (1991) Fluorescence correlation spectroscopy. In: Lakowicz JR (ed) Topics in fluorescence spectroscopy, vol 1: Techniques, Plenum, New York, pp 337–410
7. Birmingham JJ, Hughes NP, Treloar R (1995) Diffusion and binding measurements within oral biofilms using fluorescence photobleaching recovery methods. Phil Trans Royal Soc Lon B: Bio Sci 350:325–343
8. Birch DJS, Geddes CD (2000) Sol-gel particle growth studied using fluorescence anisotropy: An alternative to scattering techniques. Phys Rev E 62:2977–2980
9. Geddes CD, Birch DJS (2000) Nanometre resolution of silica hydrogel formation using time-resolved fluorescence anisotropy. J Non-Cryst Solids 270:191–204
10. Steiner RF (1991) Fluorescence anisotropy: theory and applications. In: Lakowicz JR (ed) Topics in fluorescence spectroscopy, vol 2: Principles, Plenum, New York, pp 1–51
11. Dunn B, Zink JI (1997) Probes of pore environment and molecule – matrix interactions in sol-gel materials. Chem Mater 9:2280–2291
12. Winter R, Hua DW, Song X, Mantulin W, Jonas J (1990) Structural and dynamical properties of the sol-gel transition. J Phys Chem 94:2706–2713
13. Dunn B and Zink JI (1991) Optical properties of sol-gel glasses doped with organic molecules. J Mater Chem 1:903–913
14. Narang U, Wang R, Prasad PN, Bright FV (1994) Effect of aging on the dynamics of rhodamine 6G in tetramethylorthosilicate-derived sols. J Phys Chem 98:17–22
15. Qian G, Wang, M (1999) Study on the microstructural evolution of silica gel during sol-gel-gel-glass conversions using fluorescence polarization of rhodamine B. J Phys D: Appl Phys 32:2462–2466
16. Blumen A, Klafter J, Zumofen G (1985) Influence of restricted geometries on the direct energy transfer. J Chem Phys 84:1397–1401
17. Pines-Rojanski D, Huppert D, Avnir D (1987) Pore-size effects on the fractal distribution of adsorbed acceptor molecules as revealed by electronic energy transfer on silica surfaces. Chem Phys Lett 139:109–115
18. Levitz P, Drake J M (1987) Direct energy transfer in restricted geometries as a probe of the pore morphology of silica. Phys Rev Lett 58:686–689
19. Levitz P, Drake JM, Klafter J (1988) Critical evaluation of the application of direct energy transfer in probing the morphology of porous solids. J Chem Phys 89:5224–5236

20. Rolinski OJ, Birch DJS (2000) Determination of acceptor distribution from fluorescence resonance energy transfer: theory and simulation. J Chem Phys 112:8923–8933
21. Rolinski OJ, Birch DJS, McCartney LJ, Pickup JC (2001) Sensing metabolites using donor-acceptor nanodistributions in fluorescence resonance energy transfer. Appl Phys Lett 78:2796–2798
22. Rolinski OJ, Birch DJS, McCartney LJ, Pickup JC (2001) Molecular distribution sensing in a fluorescence resonance energy transfer based affinity assay for glucose. Spectrochim Acta 57:2245–2254
23. Rolinski OJ, Birch DJS, McCartney LJ, Pickup JC (2001) Fluorescence nanotomography using resonance energy transfer: demonstration with a protein-sugar complex. Phys Med Biol 46:221–226
24. O'Hagan WJ, McKenna M, Sherrington DC, Rolinski OJ, Birch DJS (2001) MHz LED source for nanosecond fluorescence sensing. Meas Sci Technol (in press)
25. Iler RK (1979) The Chemistry of Silica, John Wiley and Sons Inc, New York
26. Birch DJS (2001) Multiphoton excited fluorescence spectroscopy of biomolecular systems. Spectrochim Acta A 57:2313–2336
27. Drexhage KH, Marx NJ, Arden-Jacob J (1997) New fluorescent probes for the red spectral region. J Fluorescence 7:91S–93S
28. Birch DJS, Imhof RE (1991) Time domain fluorescence spectroscopy using time-correlated single-photon counting. In: Lakowicz JR (ed) Topics in fluorescence spectroscopy, vol 1: Techniques, Plenum, New York, pp 1–95
29. Karolin J, Geddes CD, Wynne K, Birch DJS (2001) Nanoparticle metrology in sol-gels using multiphoton excited fluorescence. Meas Sci Technol (in press)
30. Volkmer A, Hatrick DA, Birch DJS (1997) Time-resolved nonlinear fluorescence spectroscopy using femtosecond multiphoton excitation and single-photon timing detection. Meas Sci Technol 8 (11):1339–1349
31. Fischer A, Cremer C, Stelzer EHK (1995) Fluorescence of coumarins and xanthenes after two-photon absorption with a pulsed titanium-sapphire laser. Appl Opt 34:1989–2003
32. Albota MA, Xu C, Webb WW (1998) Two-photon fluorescence excitation cross sections of biomolecular probes from 690 to 960 nm. Appl. Opt. 37:7352–7356
33. Porter G, Sadkowski PJ, Tredwell CJ (1977) Picosecond rotational diffusion in kinetic and steady state fluorescence spectroscopy. Chem Phys Lett 49:416–420
34. Birch DJS, Rolinski OJ, Hatrick D (1996) Fluorescence lifetime sensor of copper ions in water. Rev Sci Instrum 67:2732–2737
35. Birch DJS, Rolinski OJ (2001) Fluorescence resonance energy transfer sensors. Res Chem Int 27:425–446

CHAPTER 4

Integrated Supramolecular Systems: From Sensors to Switches

J.-P. MALVAL, I. GOSSE, J.-P. MORAND, R. LAPOUYADE

Integrated supramolecular systems with a receptor built in a photo- or electroactive unit have been reviewed with the focus on their particular electronic properties and different photochemical and electrochemical processes which make them suitable for cation sensing or switching. The fluoroionophores with an electron donating ionophore have been the most investigated and their initial weakness related to cation decoordination in the excited state. The small blue-shift of the fluorescence spectrum and the slight change of the emission quantum yield upon cation complexation, have now been overcome by a careful combination of several donor and acceptor units, which provide new low-lying excited states decoupled from the complexed ionophore and by using TICT probes where the electronic coupling between the D and A parts is too small to induce decoordination of the cation during the excited state lifetime. On the contrary the switching action requires that the binding ability of the ionophore be lowered or increased on a larger time scale. This has been done by electrochemical oxidation and by insertion of the ionophore into a photochromic system. Differences in binding ability of three to four orders of magnitude have been obtained and it is our belief that integrated supramolecular systems combining an ionophore and a photochromic moiety (photoionochromics) will be for cation switching as successfull as integrated fluoroionophores have been for sensing cations.

4.1
Introduction

Supramolecular photochemistry and photophysics are at the crossroad of two major fields of research: supramolecular chemistry and the use of light as an external stimulus for sensing and switching and they benefit from their spectacular and rapid growth [1–4]. But far from simply adding up the progress of each discipline, they perform new functions, some of which are emerging, as molecular machines and motors [5] and others, as chemical sensors, are already emerging from traditional areas of chemistry, including inorganic, organic, physical, analytic, polymer, and biological ones, to practical applications in monitoring our health, and environment on a continuous basis [6].

Chemical sensing is accomplished from responsive supramolecular systems obtained by assembling a recognition subunit (receptor) with an active subunit through a link, which control the degree of communication between them. Selectivity is derived from receptors and the transduction of the chemical event of a molecular (ion) recognition into an easily measurable signal is ensured, most of the time, by an electro- or photoactive subunit. As a matter of fact, these supramolecular systems behave as chemical sensors when there is a modification of their electrochemical or photophysical properties upon molecular (ion) recognition. Several recent reviews have appeared on this general topic [6, 7], which encompass the last strategies for the recognition step and the diverse transduction methods.

Electrochemical sensing requires an efficient communication between a redox-active subunit and the binding site so that the recognition event can change the redox potential. It should be noted that electrodes represent one of the best ways to interface molecular-level systems with the macroscopic world and, while the switching action of the fluorescent probes is limited to the excited state lifetime, permanent electrochemical switching is obtained with systems that display binding strength modulated by their oxidation state [8].

Fluorescent systems are suitable sensors, particularly for biomolecules, being sensitive down to the single molecule [9], and lasers provide the opportunity of working in a very short time domain and very small space via microscopic imaging. An excellent and exhaustive review of de Silva et al., reporting the different mechanisms of the transduction mechanisms of discrete and stoichiometric recognition events into fluorescence signals, appeared in 1997 [10]. If we oversimplify the picture that emerges, one can distinguish the supramolecular systems, where a direct interaction of the guest with the fluorophore accounts for the photophysical effects (integrated systems, also called intrinsic systems), from those where it is the interaction between two components of the supramolecular system (fluorophore – spacer – receptor systems) that changes the luminescence properties upon guest recognition [11] (see Fig. 4.1).

Within the latter system the photoinduced electron transfer (PET) process has been extensively used because of the modular composition and the foreseeable properties of the relevant probes. Particularly the large change in fluorescence in-

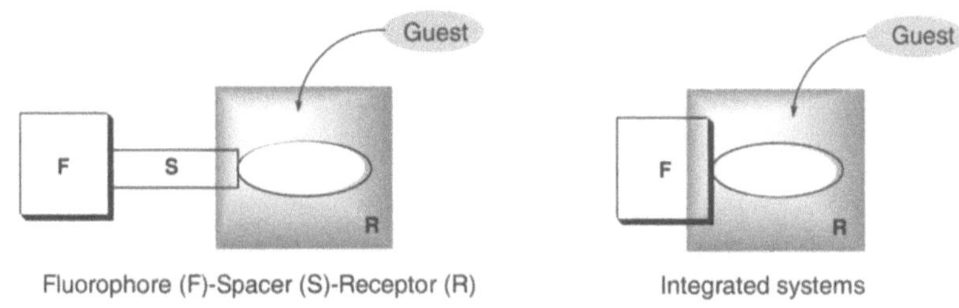

Fig. 4.1. Fluorescent supramolecular systems

tensity upon ion recognition led to call them "off–on" or "on–off" fluorescent sensors [12]. But because of the absence of interaction in the ground state, between the components of the system and between the fluorophore and the guest, and because the long-range interaction of the fluorophore in the excited state (energy transfer or electron transfer) is a pure competitive process with the fluorescence rate constant, guest recognition does not lead to any shift of the fluorescence or excitation spectra, which precludes the possibility of intensity-ratio measurements at two wavelengths to evaluate the association level. There are at least two major exceptions in these multicomponent systems where we can use the above-mentioned ratiometric method. First, when there are two fluorescent groups, the Förster resonance energy transfer (FRET) provides a sensitive tool to yield distance information in the range 10–100 Å. This method has found a wide range of applications in the biological field [13]. One of the fluorescent groups can be a luminescent lanthanide chelate leading to the development of resonance energy transfer using lanthanides as donors (LRET), which presents a number of technical advantages [14]. Second, when the fluorescent system contains two groups that can form excimer (like pyrene, anthracene, etc.), if their relative distance is brought to (or deviated from) the excimer distance (3–4 Å) by guest recognition, the concentration of the guest can be monitored by the monomer/excimer fluorescence intensity ratio [15].

In integrated supramolecular systems the guest directly interacts with the chromophore lowering or increasing the energy of the highest occupied molecular orbital (HOMO) and/or lowest unoccupied molecular orbital (LUMO) and as a result leads to a shift of the absorption spectra [16]. The absorption of a photon changes the electronic distribution of the chromophore and, accordingly, the strength of the interaction with the guest that should shift the luminescence spectra, but differently from the absorption spectra [17, 18]. On the other hand, the excited and ground states of the probe should exhibit different affinity for a given guest providing a light-driven switch of guest binding. But the lifetime of the excited states is essentially short and the switching action will be quickly reversed as the system looses its excitation energy and returns to the original ground state. However, it is possible to design more permanent switching from two stable forms of the supramolecular system (bistable systems) displaying different binding strength for a given guest. For this purpose photochromic and redox-switchable systems were obvious targets.

While the integrated cation sensors have been recently reviewed [17–20], this topical review will focus on their particular electronic properties and different photochemical and electrochemical processes that make them suitable for cation sensing or switching and our emphasis will be on new results.

4.2
Optical Detection of Ion–Ground State Probe Interaction

The first chromoionophores were obtained by combining a crown ether with a phenolic chromophore [21, 22]. With these crown-dyes (A, B, see Fig. 4.2) the selective cation complexation is accompanied by the phenol deprotonation and therefore is made visible by a color change. These compounds were first prepared as selective alkali-metal extractants. More recently chiral crown ethers (B) have allowed the differentiation between enantiomers of chiral amines by UV-VIS spectroscopy [23] and the optical effect for chiral recognition has been intensified by introduction of the crown dyes into a cholesteric liquid-crystal matrix because the wavelength of reflection for incident light depends on the pitch of the helical structure in the liquid crystalline phase. This liquid-crystal system can be considered the first visual optical sensor for chirality [24]. Of particular importance for analytical purpose (sensitivity) is the coupling of a specific, molecular recognition event into a collective property of the host. It is now a trend in the field of chemical sensors.

From the spectroscopic point of view the deprotonation of the phenol group leads to a better donor group and therefore to a red-shifted intramolecular charge transfer (ICT) transition.

Recently, naked-eye detection of anions in dichloromethane have been reported [25] from a new class of anion sensors combining calix[4]pyrrole as the receptor directly linked to anthraquinone as the withdrawing group. The anion recognition, through H-bonds leads to an increased electron donor ability of calix[4]pyrrole and the appearance of a CT band in the visible region.

Vögtle has developed another interesting family of chromoionophores from π--conjugated systems, particularly stilbene or azobenzene (C), substituted at the two ends by nitro and amino groups so that the first electronic transition involves an ICT process. By replacing the amino group by an azacrown the electronic properties are not affected but the complexation of a cation leads to a stabilization of the nitrogen electron pair that is decoupled from the conjugated system leading to the disappearance of the CT transition [16].

These series of chromoionophores illustrate the optical signaling of ion recognition from ICT probes when ion interaction with the electron-donating group suppresses or increases the ICT transition. They allow transportation of ions to less hydrophilic solvents. But these compounds with ionic or low-lying $n\pi*$ states do not fluoresce because of efficient nonradiative channels.

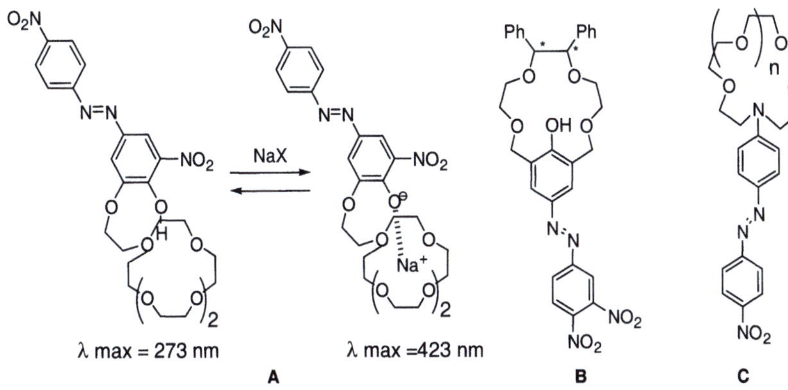

Fig. 4.2. Chemical structures of chromoionophores

4.3
Cation Sensing from Fluorescent Photoinduced Intramolecular Charge Transfer (PICT) Sensors

Many of the structural and electronic features which influence fluorescence efficiency have been delineated: double bond twisting, low energy $n\pi*$ states, photoinduced electron (charge) transfer (PE(C)T), electronic energy transfer, etc. Their perturbation by a cation guest can be monitored by fluorometric analysis [26].

The integrated fluoroionophores usually derive from a π-conjugated D-A arrangement where D is an electron-rich donating and A an electron-accepting substituent. Upon light absorption these compounds undergo a photoinduced intramolecular charge transfer (PICT) process and lead to broad, structureless, red-shifted emission bands in polar solvents, which make them suitable as polarity probes [27]. In most of the integrated fluoroionophores it is the electron donor group that is the ionophore and upon cation binding the absorption spectra are strongly blue-shifted while the fluorescence spectra are much less affected [28, 29]. When the ionophore is on the acceptor side, purely electrostatic interactions should lead to a stabilization of the polar excited state and a red shift of the emission. This has been observed from cation interaction with the carbonyl group of an electron-accepting ester group [30] and by reversing the electronic character of a phenyl azacrown by cation recognition [31] (see Fig. 4.3).

From the probes with an electron-donor ionophore, the relatively small blue-shift of the main fluorescence band, upon cation complexation, has been attributed to a decoordination reaction in the excited complex as a consequence of the PICT process, which involves a shift of negative charge from the ionophore and as a result leads to electrostatic repulsion between the cation and the resulting positively charged ionophore [32, 33]. As a matter of fact, from the calcium complex of 4--(N-monoaza-15-crown-5)-4'-cyano stilbene (DCS-crown) a subpicosecond pump-probe experiment provided the kinetics of three transient absorption spectra of the

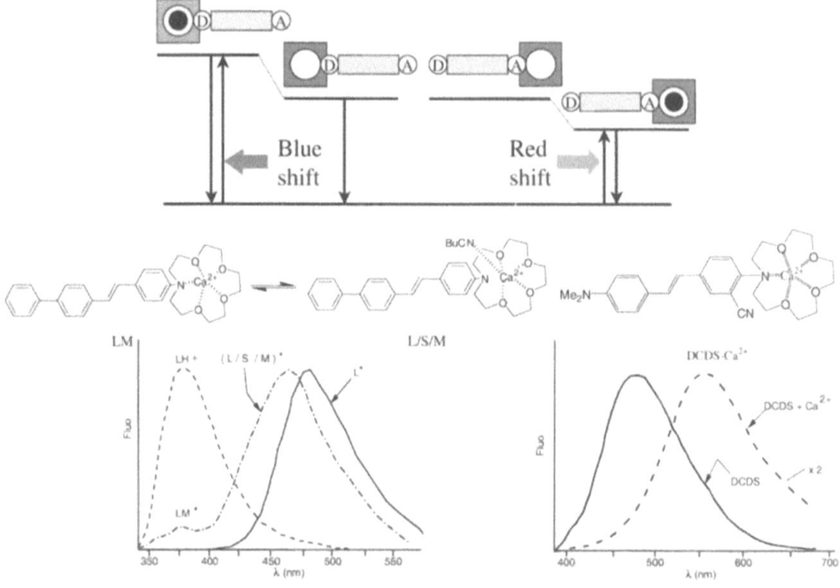

Fig. 4.3. Cation spectral shifts with an electron-donor or electron-acceptor ionophore

successive intermediates, on the decoordination pathway, that we assigned to the Franck-Condon state $(LM)^*$, a cation-probe pair (L^*M) and a solvent separated pair $(L^*/S/M)$ [34]. This assignment followed from the time resolved, red-shifted, successive emissions and the concomitant blue shift of the transient absorptions and by analogy with the well-characterized contact ion-pairs and solvent separated ion pairs in the ground state of benzylic anions [35], but also in the excited state of β-naphtholate [36].

The photodisruption of the complex in integrated systems with the ionophore as the electron donating group is now well accepted and accounts for the small shift in emission and also for the relatively small change in the quantum yield of fluorescence upon ion binding. We have discussed the photophysical results of the complexation of cations from fluoroionophores with an azacrown as the ionophore, but the chelators of the BAPTA series, their electronically equivalent which are used for the recognition of cytosolic cations and in particular calcium ions, behave the same way, and for analytical purposes dual excitation wavelengths monitoring and ratioing of the emission intensity must be used [37]. Fura-2 is one of the most popular derivatives of the BAPTA series from which a detailed study of the fluorescence decay with global compartmental analysis has been realized. It follows that the stability constant of the calcium complex in the excited state is more than three orders of magnitude smaller than in the ground state but that there is negligible interference of the excited state reaction with the determination of the stability constant and the concentration of calcium from fluorimetric titration curves [38]. Regarding the exception of Indo-1, which gives a well-resolved new

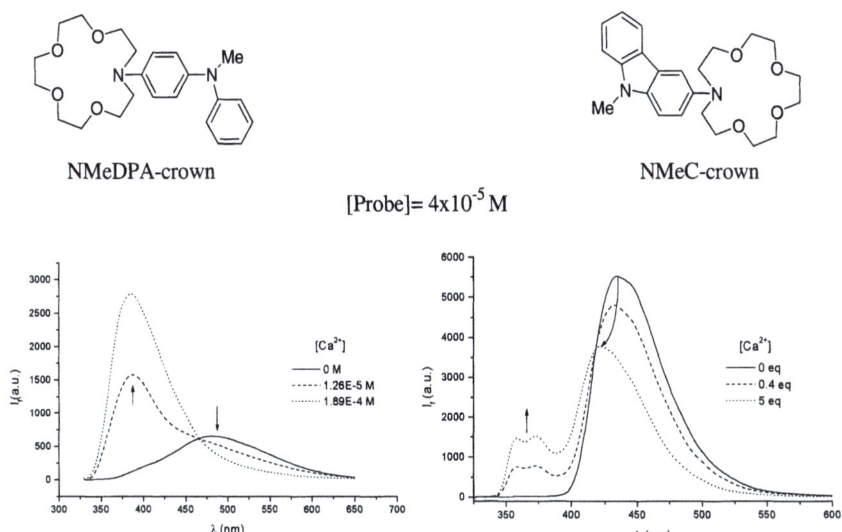

Fig. 4.4. Calcium effect upon the fluorescence of TICT and ICT probes in acetonitrile

emission band upon calcium association, the donor–acceptor character of the excited state is not strong enough to cause the probe–Ca bond breaking.

Indeed the experimental stability constants of several series of crown fluorochromoionophores with Na^+ and Ca^{2+} correlate with the calculated charge on the whole dimethylamino group of the dimethylamino analogs in the ground state [39]. It is more difficult to foresee the image of the fluorescence spectra when the probe partially decoordinates in the excited state: well-resolved two-bands behavior giving evidence of the emission from the excited complex or very slight blue shift of the emission from a disrupted complex. As for all processes of the excited state, the quantum efficiency of fluorescence is the result of a competition; the shift from the "one band" to the "two-band" systems is determined, at the excited Franck-Condon complex, by the ratio of the rate constants of emission and of decoordination. But from ICT probes, the decoordination is directly related to the ICT formation so that the kinetics and therefore the mechanism of the ICT formation is decisive. We have recently obtained fluoroionophores from derivatives of the aniline dimer, one of which, NMeDPA-crown, leads to an emissive twisted intramolecular charge transfer (TICT) state (μ_{TICT} = 20.6 D) while another one, NMeC-crown, has a planar polar excited state (μ_{TICT} = 9 D). The calcium complex of the former, in acetonitrile, has its own emission band [40] while the analogous complex of the latter leads to several emissions because of the stepwise decoordination of the cation during the lifetime of the excited state (Fig. 4.4) [41].

This means that the decisive parameter for differentiating PICT probes, which will give a specific emission for the complexes from probes, that will decoordinate from the cation, is not the value of the PICT or the dipole moment of the excited state (if it was the case TICT probes would be among the most cation releasing), but the kinetics of the charge transfer formation. Actually, DMABN, the archetype of TICT compounds, has been derivatized into the azacrown (DMABN-crown)

[42] and tetraazacrown (DMABN-cyclam) [43] derivatives which signal alkali, al-kaline-earth metal cations and transition metal ions, respectively, by the decrease of the TICT emission and the concomitant increase of the short wavelength emis-sion from the precursor LE state. More generally, all destabilizing interactions of the TICT excited states favour the strongly emissive, less polar, precursor LE state. Introduction of a cation into the electron-donor ionophore is one way of de-stabilizing the TICT state, but lowering the solvent polarity is another way and in-deed several of the TICT compounds, often with a non-emissive TICT state, have been used as sensors for environmental polarity [44].

To return to NMeDPA-crown and NMeC-crown, the two probes have a high association constant with Ca^{2+} (log K_a = 5.3 and 6.7, respectively) because the ionophore is linked to an electron-rich chromophore. Upon light absorption NMeDPA leads to a TICT state but with a cation in the ionophore the rotational relaxation to the TICT state is reduced well below the radiative rate constant of the first excited state, the LE state, and the ratio of the two emissions provide a sensi-tive signaling of the cation concentration. From NMeC-crown, light absorption re-organizes the electronic distribution of the carbazole group and the decrease of the electronic density on the nitrogen atom of the azacrown leads to the breaking of its interaction with the cation, faster than the radiative emission. The emission of at least two new bands illustrates the time-evolution of cation–probe interaction [41].

Complexation of cations by D–A or D–D probes with the ionophore in the ac-ceptor part of the molecule has been less studied. However, in that case, there is a bathochromic shift in absorption but also in emission [30, 31, 45] because the complexes are more stable in the excited state than in the ground state, no decoor-dination occurs upon light absorption. Nevertheless, the interaction of a cation with the electron-accepting group leads sometimes to a decrease of the emission intensity [31], but this drawback has been overcome by the design of D-A-D probes where the complexation of the cation changes the relative energetic posi-tion of the different excited species and, on the contrary, extraordinary high fluo-rescent enhancement has been achieved [46]. Most of the examples cited above involve integrated polar probes with a D or A ionophore that participates in the optical transition and the spectroscopic effect of cation recognition can be ana-lyzed by a change of the energetic positions of the lowest excited states from purely electrostatic interactions [47]. The guest compounds directly interact with the chromophore of integrated probes, so that the understanding of the photophys-ics of the fluorophore and the knowledge of the nature of the interaction is a pre-requisite for the interpretation of the cation-induced effects.

We recently obtained tripodal ligand incorporating the TICT molecules, di-methylaminobenzamide (T-DMABA) and dimethylaminobenzene sulphonamide (T-DMABS) that provided new ratioable fluorescent probes for Cu^{2+} (see Fig. 4.5) [48].

The spectroscopic effect of copper on the fluorescence of the sulfonamide de-rivative parallels the effect of the deprotonation, but 4 pH units lower, and there-fore can be ascribed to a decrease of the electron-accepting ability of the sulfona-mide anion relative to the neutral sulfonamide group. The amide derivative coordi-nates to the copper cation from the oxygen atom of the carbonyl function [48] and

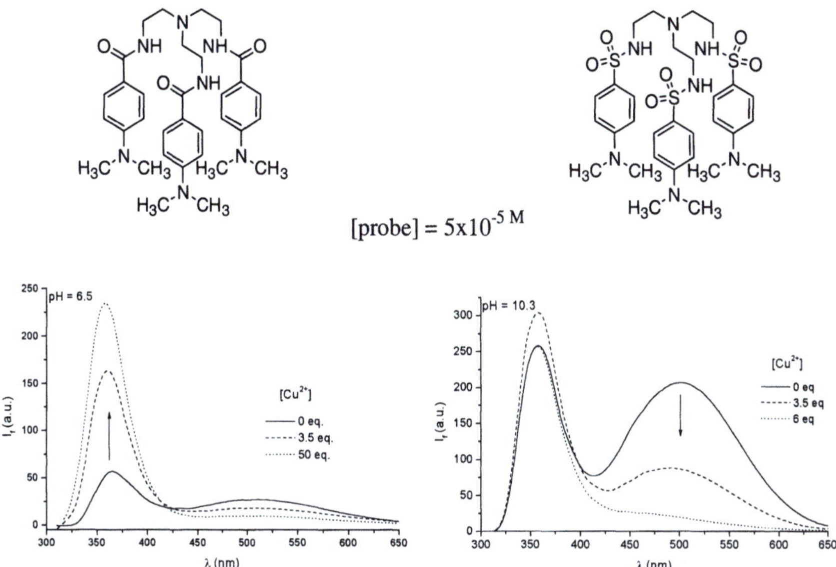

Fig. 4.5. Effect of Cu^{2+} on the fluorescence spectra of TICT amide and sulfonamide in acetonitrile/water (1v/1v)

the mesomeric form with a carbon–nitrogen double bond is favored and the amide function is less electron attracting, as shown by the blue shift of the absorption spectrum.

These two new examples of the interaction of a cation with the electron-accepting ionophore show the opposite effects that were expected from purely electrostatic interactions, that is to say, a stabilization of the more polar excited state and an increased red shift of the corresponding emission. This is understandable from the nature of the cation–receptor interaction described above.

There are only few PICT sensors for neutral molecules, with the notable exception of aromatic boronic acid derivatives for saccharides [49]. After complexation with a sugar molecule, boronic acid changes from the neutral form to the anionic one and consequently changes its electronic character from acceptor to donor. The combination of an electron-donating group with the boronic acid group leads to a CT emission which is quenched by saccharide addition while probes with an electron-accepting group and a boronic acid give a CT emission upon saccharide addition. These probes provide a new optical and ratiometric approach for the analysis of sugar [50].

In short, a desirable feature of these probes for an efficient sensor property is to keep, or increase, the association constant in the excited state in order that a new, well-separated band emission signals the recognition event. This has been realized from D–A probes with the ionophore on the acceptor side, but also from fluoro-ionophores with an electron–donor ionophore (D) when the D–A parts are weakly coupled as in TICT state or in exciplex [51] and more generally, by an ingenious tuning of the donor–acceptor substitution pattern [46]. Cation-probes emitting in the near-infrared [52] and bifunctional fluoroionophores, signalling metal ions, either independently or cooperatively [53] have recently been described and the integration of these probes into conjugated polymer should enhance their sensitivity [54, 55].

The decoordination in the excited state occurs mainly from probes with planar, strongly polar, excited state associated, on the electron donor side, with s-block alkali and alkaline-earth metal ions by essentially electrostatic interactions. This behavior led some authors to propose these probes for photorelease of cations to produce fast and spatially controllable concentration changes, but the mean diffusion length of the cation is limited by the excited state lifetime. The mean diffusion lengths for Li^+ and Ca^{2+} have been calculated to be 20.5 Å and 17.3 Å, respectively, in water, at 25°C during the 2 ns of the excited state [56]. This behaviour is reminiscent of the pH jump method for photogenerating proton or hydroxide pulses [57, 58].

4.4
Optical and Electrochemical Release of Cations

Brief, localized fluctuations of intracellular free Ca^{2+} are believed to control neurosynaptic transmission, hormone secretion, muscle contraction and other physiological functions. The ability to generate fast but permanent localized rises in free $[Ca^{2+}]$ would be a powerful tool to study these biochemical processes. This was carried out by different photochemical reactions: photorearrangement of o--nitrobenzhydrols to nitrosobenzophenones in 1,2-bis(o-aminophenoxy)ethane (BAPTA) derivatives [59], photocleavage of the chelator [60, 61] but these reactions are irreversible and, involving several steps, they are also relatively slow. Fast, permanent, photoreversible cation recognition were realized from different photochemical reactions, such as photocycloaddition and photoisomerization in functional supramolecular systems. The more common photoactive units are photochromic bisanthracenes [62], azobenzenes [63], diarylethenes [64, 65], styryl dyes [66], spirobenzopyrans and derivatives [67].

We have prepared a photochromic crown ether (E) composed of dithienyl perfluorocyclopentene substituted at positions 5 and 5' by phenylaza-15-crown-5 as the ionophore and a formyl group as a strong electron-withdrawing group, respectively. While the ionophore and the electroactive group are not conjugated in the open form of the photochromic unit, they are in the relation of an integrated photoionochromic in the closed form and the electronic effect of the formyl group decreases the cation complexing ability by a factor close to 1×10^4 (Fig. 4.6).

Another method to decrease the electronic density on the ionophore and therefore its cation binding ability was to bring it to the radical cation stage. Since the first example of electrochemical recognition of cations by Saji in 1986, redox changes have been used to enhance (or diminish) the binding affinity of ligands for ions or guest molecules [68–70]. An electrochemically switchable system is usually described by a simple square system in which the redox equilibrium is

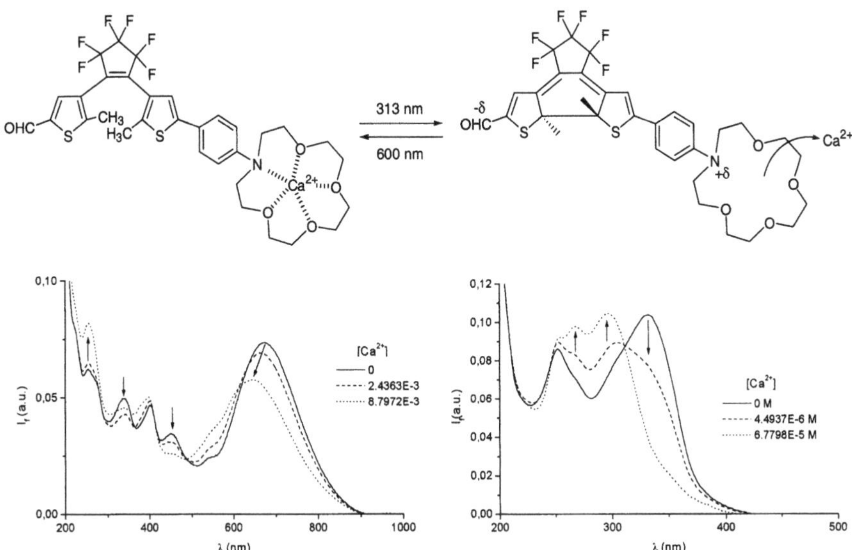

Fig. 4.6. Cation binding property of the two forms of an integrated photoionochromic in acetonitrile

coupled with the reversible binding reaction. Experimentally when two separate voltametric waves corresponding to the redox processes of the free and complexed species are observed, the binding enhancement from one oxidation state to the other oxidation state can be obtained from the shift between the half-wave reduction potentials [70].

We reasoned that from the integrated, electron-rich fluoroionophore NMeC-crown, which decoordinate in the excited state, the electrochemical oxidation should decrease the electronic density on the ionophore and accordingly lower the cation-binding ability. Indeed, cyclic voltametry of this probe shows a new oxidation peak upon addition of calcium perchlorate that is assigned to the oxidation of the complex. The magnitude of the shift in redox potential (0.183 V) is one of the highest reported for organic supramolecular systems and it corresponds to a binding inhibition on oxidation 8×10^{-4} at the radical cation state. This result, with the long lifetime of the radical cation, argues for a complete release of the cation [71].

4.5
Conclusions

Integrated supramolecular systems with a receptor built in a photo- or electro-active unit provide the necessary structural feature for a direct interaction with the guest and accordingly for an effective spectroscopic modification (sensor function) while photochemical or electrochemical inputs alter the binding strength of the receptor (switching function).

The sensor function, towards cations, of integrated systems derived from a π-conjugated D–A arrangement has been widely and successfully exploited.

The fluoroionophores with an electron donating ionophore have been the most investigated and their initial weakness related to cation decoordination in the excited state, small blue-shift of the fluorescence spectrum and slight change of the emission quantum yield, has been overcome by a careful combination of several donor and acceptor units which provide new low-lying excited states decoupled from the complexed ionophore and by using TICT probes, where the electronic coupling between the D and A parts is too small to induce the decoordination of the cation during the excited state lifetime.

On the contrary, the switching action requires that the binding ability of the ionophore be lowered or increased on a larger time scale. This has been done by electrochemical oxidation and by insertion of the ionophore into a photochromic system, the two forms of which are exhibiting very different binding ability. Differences in binding ability of three to four orders of magnitude have been obtained and it is our belief that integrated supramolecular systems combining an ionophore and a photochromic moiety (photoionochromics) will be as successful for cation switching as integrated fluoroionophores have been for sensing cations.

References

1. Lehn JM (1988) Angew Chem Int Ed 27:89
2. Sauvage JP, Hosseini MW (1996) (eds) Comprehensive supramolecular chemistry. Pergamon Press, Oxford
3. Balzani V (1987) (ed) Supramolecular photochemistry. Reidel, Dordrecht
4. Weiss S (1999) Science 283:1676
5. Stoddart JF (2001) Molecular machines, Acc Chem Res 34:410
6. Ellis AB, Walt DR (2000) Chemical sensors. Chem Rev 100:2477
7. Desvergne JP, Czarnik AW (1997) (eds) Chemosensors of ion and molecule recognition. NATO ASI Series. Kluwer Academic Publishers, Dordrecht
8. Kaifer AE, Mendoza S (1996) In: Sauvage JP, Hosseini MW (eds) Comprehensive supramolecular chemistry. Pergamon Press, Oxford, p 701
9. Moerner WE, Orrit M (1999) Science 283:1670
10. De Silva AP, Gunaratne HQN, Gunnlaugsson T, Huxley JM, McCoy CP, Rademacher JT, Rice TE (1997) Chem Rev 97:1515
11. Bissel RA, de Silva MP, Gunaratne HQN, Lynch PLM, Maguire GEM, McCoy CP, Sandanayake KRAS (1992) Chem Soc Rev 21:187
12. Bissel RA, de Silva MP, Gunaratne HQN, Lynch PLM, Maguire GEM, Sandanayake KRAS (1993) Topics in Current Chemistry 168:224
13. Selvin P (1995) In: Sauer K (ed) Methods in enzymology, Academic Press, Orlando, pp 300–344
14. Sammes PG, Yahioglu G (1996) Modern bioassays using metal chelates as luminescent probes 13:1
15. Bouas-Laurent H, Castellan A, Daney M, Desvergne JP, Guinand G, Marsau P, Riffaud M-H (1986) J Am Chem Soc 108:315

16. Löhr HG, Vögtle F (1985) Acc Chem Res 18:65
17. Valeur B (1994) In: Lakowicz JR (ed) Topics in fluorescence spectroscopy. Plenum Press, New York, p 21
18. Rettig W, Lapouyade R (1994) In: Lakowicz JR (ed) Topics in Fluorescence Spectroscopy. Plenum Press, New York, p 109
19. Valeur B, Leray I (2000) Coord Chem Rev 205:3
20. De Silva AP, Fox DB, Allen JM, Huxley AJM, Moody SM (2000) Coord Chem Rev 205:41
21. Takagi M, Nakamura H, Ueno K (1977) Anal Lett 10:1115
22. Vögtle F (1980) Pure Appl Chem 52:2405
23. Kaneda T, Hirose K, Misumi S (1989) J Am Chem Soc 111:742
24. Nishi T, Ikeda A, Matsuda T, Shinkai S (1991) J Chem Soc Chem Commun 339
25. Miyaji H, Sato W, Sessler JL (2000) Angew Chem Int Ed 39:1777
26. Czarnik AW (1995) Chem Biol 2:423
27. Valeur B (1993) In: Molecular luminescence spectroscopy, part 3. Wiley, New York, p 25
28. Bourson J, Valeur B (1989) J Phys Chem 93:3871
29. Létard JF, Lapouyade R, Rettig W (1993) Pure Appl Chem 65:1705
30. Bourson J, Pouget J, Valeur B (1993) J Phys Chem 97:4552
31. Delmond S, Létard JF, Lapouyade R, Mathevet R, Jonusauskas G, Rulliere C (1996) New J Chem 20:861
32. Martin MM, Plaza P, Meyer YH, Badaoui F, Bourson J, Lefevre JP, Valeur B (1994) J Phys Chem 100:6879
33. Dumon P, Jonusauskas G, Dupuy F, Pée P, Rullière C, Létard JF, Lapouyade R (1994) J Phys Chem 98:10391
34. Mathevet R, Jonusauskas G, Rullière C, Létard JF, Lapouyade R (1995) J Phys Chem 99:15709
35. Smid J (1972) In: Szwarc M (ed) Ions and ion pairs in organic reactions. Wiley Interscience, New York, chap 3
36. Soumillion JP, Vandereecken P, Van Der Auweraer M, De Schryver FC, Schanck A (1989) J Am Chem Soc 111:2217
37. Grynkiewicz G, Poenie M, Tsien RY (1985) J Biol Chem 260:3440
38. Van den Bergh V, Boens N, De Schryver FC, Ameloot M, Steels P, Gallay J, Vincent M, Kowalczyk A (1995) Biophys J 68:1110
39. Rurack K, Sczepan M, Spieles M, Resch-Genger U, Rettig W (2000) Chem Phys Lett 320:87
40. Crochet P, Malval JP, Lapouyade R (2000) Chem Commun 289
41. Malval JP, Chaimbault C, Fischer B, Morand JP, Lapouyade R (2001) Res Chem Intermed 27:21
42. Létard JF, Delmond S, Lapouyade R, Braun D, Rettig W (1995) Rec Trav Chim Pays-Bas 114:517
43. Collins GE, Choi LS, Callahan JH (1998) J Am Chem Soc 120:1474
44. Kosower EM, (1982) Acc Chem Res 15:266
45. Roshal AD, Grigorovich AV, Doroshenko AO, Pivovarenko VG, Demchenko AP (1998) J Phys Chem A 102:5907
46. Rurack K, Rettig W, Resch-Genger U, (2000) Chem Commun 407
47. Rettig W, Rurack K, Sczepan M (2001) In: Valeur B, Brochon JC (ed) New trends in fluorescence spectroscopy. Springer, Berlin, p 125

48. Malval JP, Lapouyade R (2001) Helv Chim Acta, 84:2439
49. James TD, Sandanayake KRAS, Shinkai S (1996) Angew Chem Int Ed 35:1911
50. DiCesare N, Lakowicz JR (2001) J Phys Chem A 105:6834
51. Arimori S, Bosch LI, Ward CJ, James TD (2001) Tetrahedron Lett 42:4553
52. Rurack K, Bricks JL, Reck G, Radeglia R, Resch-Genger U (2000) J Phys Chem 104: 3087
53. Rurack K, Koval'chuck A, Bricks JL, Slominskii JL (2001) J Am Chem Soc 123:6205
54. Swager TM (1998) Acc Chem Res 31:201
55. McQuade DT, Pullen AE, Swager TM (2000) Chem Rev 100:2537
56. Martin MM, Plaza P, Meyer YH, Badaoui F, Bourson J, Lefèvre JP, Valeur B (1996) J Phys Chem 100:6879
57. Clark JH, Shapiro SL, Winn KRC (1979) J Am Chem Soc 101:746
58. Gutman M, Huppert D, Pines E (1981) J Am Chem Soc 103:3709
59. Adams SR, Kao JPY, Grynkiewicz G, Minta A, Tsien RY (1988) J Am Chem Soc 110: 3212
60. Ellis-Davis GCR, Kaplan JH (1988) J Org Chem 53:1966
61. Warmuth R, Grell E, Lehn JM, Bats JW, Quinckert G (1991) Helv Chim Acta 74:671
62. Bouas Laurent H, Castellan A, Desvergne JP, Lapouyade R (2000) Chem Soc Rev 29: 43
63. Dürr H, Bouas-Laurent H (1990) Photochromism, molecules and systems. Elsevier, Amsterdam
64. Irie M, Uchida K (1998) Bull Chem Soc Jpn 71:985
65. Irie M (2000) Chem Rev 100:1685
66. Alfimov MV, Gromov SP, Fedorov YV, Fedorova OA, Vedernikov AI, Churakov AV, Kuz'mina LG, Howard JAK, Bossmann S, Braun A, Woerner M, Sears Jr DF, Saltiel J (1999) J Am Chem Soc 121:4992
67. Crano JC, Guglielmetti RJ (1999) Organic photochromic and thermochromic compounds. Plenum, New York
68. Saji T (1986) Chem Lett 275
69. Boulas PL, Gomez-Kaifer M, Echegoyen L (1998) Angew Chem Int Ed 37:216
70. Beer PD, Gale PA, Chen GZ (1999) J Chem Soc Dalton Trans 1897
71. Miller SR, Gustowski DA, Chen ZH, Gokel GW, Echegoyen L, Kaifer AE (1988) Anal Chem 60:2021

CHAPTER 5

Ratiometric Probes: Design and Applications

A. P. DEMCHENKO, A. S. KLYMCHENKO, V. G. PIVOVARENKO, AND S. ERCELEN

With the aim of substantial improvement of solvatochromic and electrochromic fluorescence probes by coupling their response with an excited-state reaction, a series of 3-hydroxychromone derivatives have been synthesized. They demonstrate two well-separated emission bands that originate from normal and phototautomer forms due to excited-state intramolecular proton transfer (ESIPT) reaction. In the studies of solvent polarity effect, electrochromism (internal Stark effect) and red edge effect a strong amplification of probe response is achieved by recording of relative intensity ratios of these fluorescence bands instead of their spectral shifts. This result is an illustration of a new principle in design of fluorescence probes, which may allow the achievement of almost perfect on–off switching behavior.

5.1
Introduction

Most of the information that the human brain obtains from the surrounding world is acquired by visual perception. Colored vision and characterization of visual objects in reality is an important contribution to this process. Therefore, it is not surprising that in spectroscopy and in its combination with microscopy the attempts to obtain information in real colors in the visible range of spectrum are in the minds of many researchers. This situation was not changed with the application of electronic photometric devices. The spectroscopic information based on the changes in energy of emitted quanta is often more unambiguously interpreted than the information obtained in the studies of emission intensity, anisotropy, and lifetimes. Therefore, the fluorescent probes that change the color of emission are in great demand.

Usually the probes that exhibit the changes in emission spectra are called "ratiometric" because they allow a convenient quantitative characterization of changes in fluorescence emission by evaluation of intensity ratio at two selected wavelengths. These are usually the points on wavelength scale at which the most significant changes of fluorescence intensity are observed. According to the effects that are produced in original spectra, the ratiometric probes can be classified into two groups. One group combines the probes in which the ratiometric effect is produced by the shifts of original spectra. Usually, in the background of these shifts is the transfer of electronic charge in the excited-state fluorophore of the probe molecule. It is induced by the changes in interaction energies of the fluorophore with surrounding molecules and groups of atoms (solvatochromic shifts), interaction with external electric fields (electrochromic shifts) and dielectric relaxations in the fluorophore environment (dielectric relaxational shifts) [1]. The response of polarity-sensitive probes is based on solvatochromic shifts, while the biomembrane potential-sensitive dyes respond by electrochromic shifts [2]. In all these cases the ratiometric response extends to a certain limit, which is determined by the properties of fluorophore (mainly by changes of dipole moment upon excitation, the width of the emission spectrum) and the extent of produced perturbation. This limit determines the limit in ratiometric response.

The other group combines the probes, the fluorescence spectra of which consist of two or more bands, and the changes of relative intensities between these bands are generated. This can occur as a result of excited-state reaction with the participation of the probe fluorophore. The reaction should proceed between discrete energy states, so that both initial and product states should emit fluorescence with the shifts in energy. A number of fluorescent probes exhibiting such reactions is currently applied in research [1, 3]. In the probes with intramolecular charge transfer (CT) a reaction may involve considerable transfer of electronic charge with the formation of a new band, which can be coupled with molecular isomerization (twisting). The reactions of excited-state intramolecular proton transfer (ESIPT) results in tautomerization leading to substantial redistribution of electronic density, and this may lead to dramatic changes in the energy of emitted quanta. In the excited states, new molecular complexes can be formed, which do not exist in the ground state, and they are accompanied with a dramatic change of spectra (excimers and exciplexes). If the electronic spectra of two molecules overlap, a distance and orientation dependent transfer of electronic energy may occur between them with the decrease of emission of one fluorophore (donor) and increase of emission of the other fluorophore (acceptor). In principle all these reactions may exhibit an on–off behavior, which means that under the effect of a specific perturbation during the probe operation one form can be switched into the other form in emission spectra.

The high level of efficiency in response of fluorescence probes and sensors is very difficult to achieve. The present communication aims at description and exploration of an idea that photophysical mechanisms that are in the background of ratiometric response based on spectral shifts, while coupled with the excited-state reaction, can produce a dramatic influence on this reaction. As a result, a much stronger effect of variation of relative intensities of two separated bands, belonging to reactant and product of this reaction, can be achieved. This may produce a strong amplification effect with a dramatic change in color of emitted light.

5.2
3-Hydroxyflavones and Other 3-Hydroxychromone Derivatives. New Compounds and Their Properties

3-Hydroxychromones and 3-hydroxyflavones are the compounds with remarkable fluorescence properties. They exhibit the ESIPT reaction with the possibility of observing the emission of initially excited N* form (which in some derivatives can achieve the strong CT properties) and a tautomer T* form, which is the product of proton transfer reaction [4–8]. These two excited isomers return to the ground state with the emission of photons with a substantial separation in energies, and this gives rise to two well resolvable and often highly emissive fluorescence bands. The positions and the ratio of intensities of these bands exhibit strong dependence on solvent polarity and hydrogen bonding with the solvent molecules. These particular features suggest a variety of applications of these compounds as fluorescent probes in different fields of research.

In view of potential applications we made an attempt to improve the spectroscopic properties of these molecules by proper modifications. The introduction of an electron donor group in the 4'-position in 3-hydroxyflavone is known to increase the charge-transfer nature of the N* form and makes the proton transfer reaction N*→T* more sensitive to medium polarity [5–8]. The synthesis of its azacrown derivative FCR (Fig. 5.1) allowed to increase the probe solubility in polar solvents, to make it more suitable for biomembrane studies [9] and to the investigations of complexation with bivalent cations [10, 11].

On the next step an attempt was made [12, 13] to substitute the benzene ring by benzofurane or naphthofurane heterocycles (see Fig. 5.1). These substituents are known as π-electron-abundant systems. They simultaneously increase the length

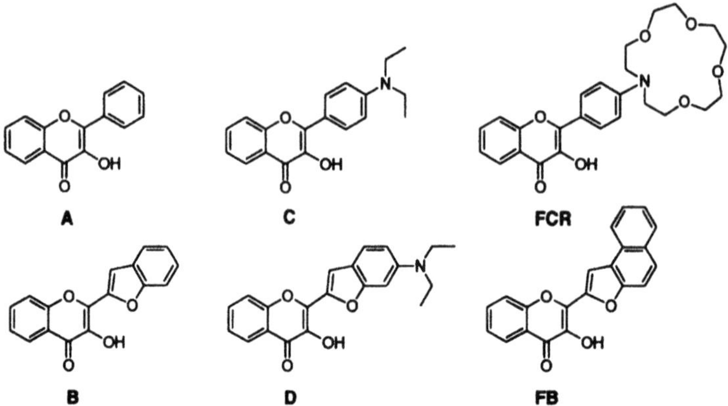

Fig. 5.1. The structures of 3-hydroxyflavone (**A**), its derivatives (**C, FCR**) and new 3-hydroxychromone derivatives (**B, D, FB**)

of chromophore unit. Therefore, it was expected that such a substituent in position 2 of the chromone ring increases the transfer of electron density to the carbonyl group in the excited state and the dipole moment of the molecule, which should lead to the enhancement of absorbance and the stabilization of T* form.

This effect can be additionally modulated by electron–donor substituents in benzofurane ring. Furthermore, having shifted the absorption and emission bands to longer wavelengths, such heterocyclic systems are expected to serve better in biological systems, where the excitation at longer wavelengths is always preferable.

Absorption and fluorescence properties of some of the new compounds B and D in comparison with well-known A and C are presented in Fig 5.2. For compound D we achieved a significant shift of absorption spectrum (approaching ~450 nm), the increase of molar absorption coefficient up to about 40 000 mL mol^{-1} cm^{-1} and fluorescence quantum yield between 22 and 36 % in the studied aprotic solvents. Furthermore, expansion of conjugation compared with C leads to increase of band separation between N* and T* forms (up to 105 nm or 3 370 cm^{-1} in toluene).

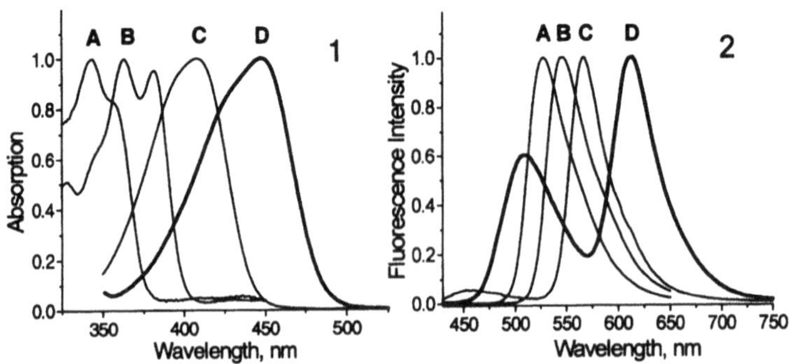

Fig. 5.2. Absorption (**1**) and fluorescence (**2**) spectra of compounds A–D

5.3
A New Level of Sensitivity to Solvent Polarity

The ratio of intensities at two wavelength maxima (I_{N*}/I_{T*}) may serve as a sensitive indicator of solvent polarity, hydrogen bonding ability or the local structure of molecules surrounding the probe. We tried this parameter as an indicator for solvent polarity and unexpectedly observed that none of our probes can cover the whole polarity range. Instead, they can operate in a rather narrow range of polarity values, but can resolve substantial differences among the solvents that are very close in properties. The results are illustrated in Fig. 5.3.

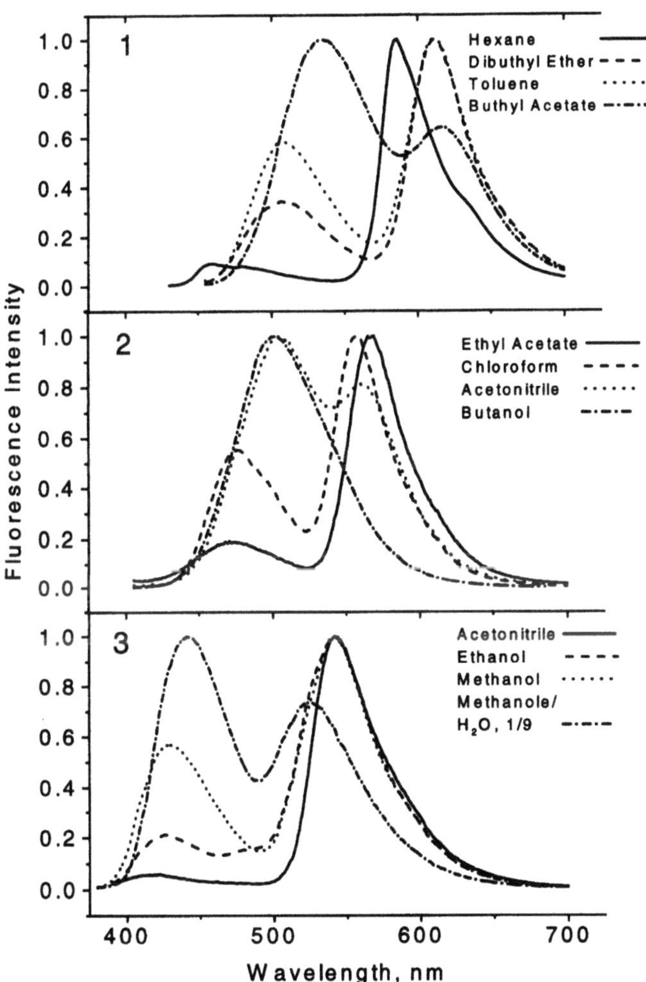

Fig. 5.3. Fluorescence spectra recorded in solvents of different polarities with probes D (**1**), F (**2**), and B (**3**)

It can be seen that the range of sensitivity of probe D is within the low-polar solvents that are more hydrophobic than ethyl acetate. The amplification is apparent. Thus, the difference between hexane and toluene, which is relatively small when tested with common polarity probes, in our case is expanded to almost half of the full scale. This suggests the application of this probe for precise analysis of quality of oil products. On the other extreme is probe B. It shows high sensitivity to the solvents that are more polar than ethanol and which allows to characterize successfully and with a high precision the ethanol–water and even methanol–methanol mixtures. In the case of aprotic solvents of medium and high polarity the most efficient is analog of probe C, 4'-dimethylamino-3-hydroxyflavone (probe F). Its range of maximal sensitivity (where the two bands are similar in magnitude) is observed in solvents from ethyl acetate to dimethylformamide, and also in phospholipid membranes. Therefore, it can be used in the studies of membrane structures and dynamics. In general, in the studies of micro-heterogeneous systems such as proteins and biomembranes the full coverage of the entire polarity scale is commonly not required, and the properties of the binding sites can be often predicted. But the high sensitivity to a narrow range of polarity may be a great advantage in characterizing the properties of interfaces and in monitoring the conformational changes in polymers and biopolymers in which the probe binding sites are involved.

In order to achieve a more complete characterization of solvent properties, in addition to I_{N*}/I_{T*} band intensity ratio the other more traditional probe parameters can also be used – the spectral positions of absorption and of two fluorescence bands, and the quantum yields. This new possibility of multiparametric analysis of solvent properties can be useful.

We observe also that our probes being dissolved in aprotic solvents are extremely sensitive (on the level of 0.01–1 M) to the presence of protic co-solvents. This suggests their application as molecular sensors of hydroxylic compounds as impurities in different solvent mixtures and also of hydration effects in phospholipid membranes [14, 15]. In a related study [16] the probe F was suggested for the measurement of water content in acetone.

5.4
Amplification by ESIPT of Electrochromic Effects

Electrochromism (or Stark effect) is the phenomenon of shifts of electronic spectra under the influence of electric fields. It found an important application in the design of probes sensitive to biomembrane potential [2]. For the design of molecular sensors, and in particular for the sensors of ions the exploration of internal Stark effect (ISE), the shifts of absorption bands under the influence of proximal charges within the same molecule, is most prospective. Usually, this effect is relatively small, and its amplification is very desirable.

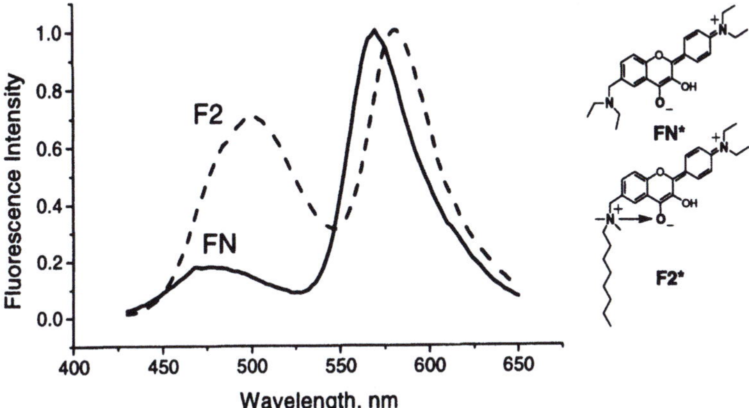

Fig. 5.4. Fluorescence spectra of flavones **FN** and **F2** in ethyl acetate. * The structures illustrate distribution of charges in excited state

In order to study the possible involvement of this effect in modulating the intramolecular proton transfer reactions in the excited state, we designed several derivatives of 3-hydroxyflavone, which contain neutral and positively charged substituents in position 6 of the chromone ring. These compounds were studied in solvents of different polarities [17]. In these experiments the shifts of absorption spectra and of both normal and tautomer fluorescence bands are clearly seen in a manner predicted by Stark effect theory (Fig. 5.4). A dramatic suppression by introduced charge of tautomer fluorescence is observed when the positive charge is located from the side of chromone benzene ring. These observations may represent a new phenomenon – the coupling between internal Stark effect and ESIPT reaction with a strong amplification of response.

Thus, a new principle of design of fluorescence sensors for charged analytes can be realized. It does not require an electronic coupling between the chromophore and the sensor group, but only their electrostatic interaction through space. The design and synthesis of new ion-sensitive probes based on this principle is in progress.

The results of these studies suggest also the possibility for design of a new generation of probes sensitive to membrane potential. In these probes the response of the sensor could be in the form of electrochromic effect, and this effect could be amplified by ESIPT reaction. Preliminary results on probe F2 in phospholipid vesicles (with diffusion potential formed by K^+) are encouraging.

5.5
Molecular Order and Dynamics in Phospholipid Membranes

The phase transition in dipalmitoyl phosphatidylcholine from gel to liquid-crystalline state is observed as the strong increase of intensity and decrease of the I_{N*}/I_{T*}

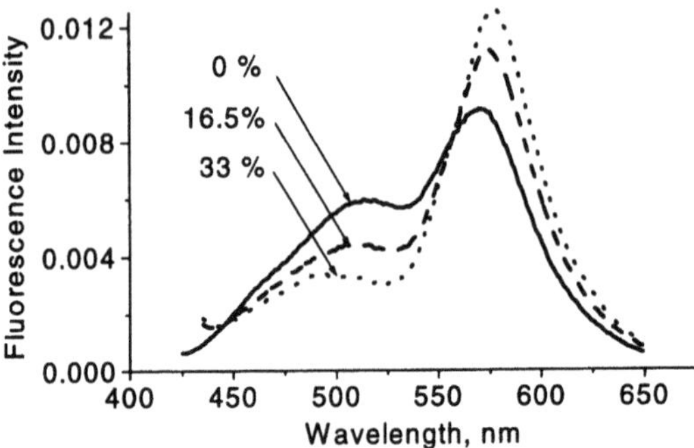

Fig. 5.5. The effect of cholesterol on fluorescence spectra of probe F in egg yolk lecithin. The spectra were normalized by integral intensity

ratio in fluorescence spectra of probe F [18]. Probably, during the phase transition the probe changes its location from that being close to the biomembrane surface to that embedded deeper into low-polar part of the bilayer. On increase of temperature within liquid crystalline state the I_{N*}/I_{T*} ratio increases, which may be the result of higher mobility and higher hydration of the probe binding site.

These results were confirmed and extended in the studies of vesicles made of egg yolk phosphatidylcholine. It was found that the addition of cholesterol produces the effect opposite to that of higher temperatures – the decrease of the I_{N*}/I_{T*} ratio. Thus, the probe is sensitive to phospholipid dynamics. Probably, this sensitivity is mediated by the changes in phospholipid hydration at the sites of binding of the probe.

In development of these studies it was found, that the profile of emission spectrum of probes F and F2 depend strongly on the nature of phospholipid, and in particular on the lipid charge [14, 15]. Thus, the I_{N*}/I_{T*} ratio is low for negatively charged phosphatidylserine and becomes substantially higher for neutral and cationic lipids.

5.6
Amplification by ESIPT of Site-selective Red Edge Effect

For aromatic fluorophores embedded into different rigid and highly viscous media the shifts in fluorescence spectra can be observed as a function of excitation wavelength. Due to the existence of some disorder in interactions between the molecules and their groups of atoms the electronic spectrum is always broadened, reflecting the distribution on these molecular interactions. The shifts of fluorescence spectra (the red edge effects), can be coupled with the lifetime of emission and contain the information about not only the static, but also the dynamic disorder. These observations found extensive application in different areas of research: colloid and polymer science, photophysics and molecular biophysics [19, 20]. However, the spectral shifts are often very small. Can they be amplified by coupling with proton-transfer reaction?

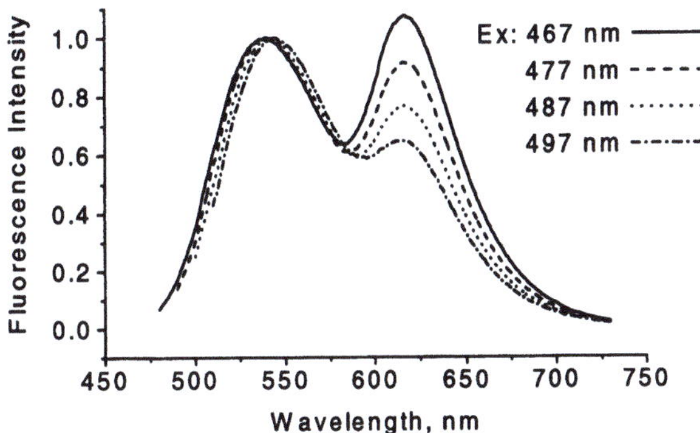

Fig. 5.6. Fluorescence spectra of chromone D in BSA at different excitation wavelengths with probe/BSA ratio 1/2. The spectra were normalized at the N* band maxima

With the new flavone and chromone derivatives this effect of amplification was clearly demonstrated in different systems: polymer films, phospholipid membranes, and protein molecules [20]. One of these examples is presented in Fig. 5.6. Chromone D exhibits a strong binding to the bovine serum albumin molecule. The binding site is rigid, which originates the red edge effects. Their common observation is the shift of the N* band. We may note, that this shift is small, and the much stronger effect appears as the variation of intensity of the T* band.

While both N* and T* forms are present at the main-band excitation, the transition in excitation to the red edge results in elimination of T* band from the spectrum. This means, that dielectrically stabilized species that are selectively excited in rigid environment can not exhibit this reaction.

This effect is also strong when observed for probes F and F2 in phospholipid membranes [15]. It demonstrates not only a high level of molecular disorder, but also the slow rate of dielectric relaxations in this system. In polymers it can be efficiently applied for probing the molecular order and dynamics of local motions [20]. Thus, the coupling the site-photoselection with excited-state reaction can provide a significant amplification of the effects of molecular disorder.

5.7
Conclusions

Thus, we suggest a new principle of design of ratiometric fluorescence probes for molecular and cellular research. It is based on the ability of 3-hydroxyflavone and 3-hydroxychromone derivatives of coupling the solvatochromic, electrochromic or site-selective wavelength-shift response with much stronger modulating effect on the excited-state intramolecular proton transfer reaction. The latter features the redistribution of intensity between two well separated (by up to 100 nm) peaks of fluorescence spectra. This coupling allows to provide a dramatic amplification of response and to achieve an almost perfect on–off switching behavior. The probes are conveniently excited in the wavelength range 400–460 nm with an easily detected change of color of fluorescence emission from blue-green to orange-red. Based on this approach the series of new solvatochromic and electrochromic probes are synthesized and tested for various applications. These applications include the studies of polarity, and conformational flexibility of binding sites in proteins and the sensing of structural perturbations, hydration and membrane potential in model and biological membranes.

References

1. De Silva AP, Gunaratne HQN, Gunnlaugsson T, Huxley AJM, McCoy CP, Rademacher JT, Rice TE (1997) Chem Rev 97:1515
2. Gross E, Bedlack RS, Loew LM (1994) Biophys J 67:208
3. Valeur B (1993) Fluorescent probes for evaluation of local physical and structural parameters. In: Schulman SG (ed) Molecular luminescence spectroscopy. Methods and applications, part 3. Wiley-Interscience, New York, p 25
4. Sengupta PK, Kasha M (1979) Chem Phys Lett 68:382
5. Swinney TC, Kelley DF (1993) J Phys Chem 99:211
6. Chou P-T, Martinez ML, Clements J-H (1993) J Phys Chem 97:2618
7. Ormson SM, Brown RG, Vollmer F, Rettig W (1994) J Photochem Photobiol A Chem 81:65
8. Pivovarenko VG, Tuganova AV, Klymchenko AS, Demchenko AP (1997) Cell Mol Biol Lett 2:355
9. Bondar OP, Pivovarenko VG, Rowe ES (1998) Biochim Biophys Acta 1369:119
10. Roshal AD, Grigorovich AV, Doroshenko AO, Pivovarenko VG, Demchenko AP (1998) J Phys Chem A 102:5907
11. Rochal AD, Grigorovich AV, Doroshenko AO, Pivovarenko VG, Demchenko AP (1999) J Photochem Photobiol A Chem 127:89
12. Klymchenko AS, Özturk T, Pivovarenko VG, Demchenko AP (2001) Can J Chem 79:358
13. Klymchenko AS, Özturk T, Pivovarenko VG, Demchenko AP (2001) Tetrahedron Lett 42/45:7967
14. Duportail G, Klymchenko AS, Mely Y, Demchenko AP (2001) FEBS Lett 508:196
15. Duportail G, Klymchenko AS, Mely Y, Demchenko AP (2002) J Fluorescence (in press)
16. Liu W, Wang Y, Jin W, Shen G, Yu R (1999) Anal Chim Acta 383:299
17. Klymchenko AS, Demchenko AP (2002) Proc. SPIE – Int Soc Opt Eng (in press)
18. Klymchenko AS, Özturk T, Pivovarenko VG, Demchenko AP (1999) In: Kotyk A (ed) Fluorescence spectroscopy and fluorescence probes. Espero Publishing, Prague, p 153
19. Demchenko AP (2001) Luminescence 17:19
20. Demchenko AP, Ercelen S, Klymchenko AS (2002) Proc. SPIE – Int Soc Opt Eng (in press)

CHAPTER 6

Binding of Ethidium to Yeast tRNA^{Phe}: A New Perspective on an Old Bromide

M. TRAMIER, O. HOLUB, J. C. CRONEY, T. ISHI, S. E. SEIFRIED, AND D. M. JAMESON

We have reinvestigated the binding of ethidium bromide (EB) to yeast tRNA^{Phe} using frequency domain fluorometry and Global Analysis. Previous fluorescence investigations of EB – tRNA interactions, carried out for more than 30 years, have indicated a "strong" binding site with a lifetime near 26 ns and one or more "weak, non-specific" binding sites with reduced lifetimes. In our study, under specific conditions in which only one EB is bound, a fluorescence lifetime of 27 ns was obtained. However, as the EB/ tRNA ratio increased, shorter lifetime components appeared. Global Analysis of the lifetime data was consistent with a model in which the second EB molecule bound has a lifetime of only 5.4 ns. Global Analysis also indicated that this second binding event leads to a reduction in the lifetime of the first EB bound, namely from 27 ns to 17.7 ns. The lifetime decrease associated with the "strong" binding site could be due to a quenching process arising either from energy transfer between EB molecules or from alterations in the conformation of the tRNA, or both. These results are considered in light of recent NMR observations on an EB/tRNA system. We also investigated the effect of ionic strength on the lifetime and relative affinities of these two binding components and found that NaCl levels up to 900 mM did not significantly affect the results.

6.1
Overview

Spectroscopic investigations of ethidium bromide (EB) (CAS Number/Name: 1239-45-8/Phenanthridinium-3,8-diamino-5-ethyl-6-phenyl-bromide) interactions with transferRNA (tRNA) have been carried out for more than 30 years [1–21]. The spectroscopic techniques utilized include NMR, X-ray and optical methods such as circular dichroism, absorption and fluorescence. The consensus of these studies is that one or more EB intercalates into the acceptor stem of tRNA [1, 4, 6, 7, 19]. X-ray [8] and NMR [4, 6] studies have furthermore identified a specific region of the acceptor stem of tRNA with the "strong" binding site.

We reinvestigated the interaction of EB with yeast tRNA^{Phe} using time-resolved fluorescence, specifically utilizing multifrequency phase and modulation fluorometry coupled with Global Analysis [22]. This approach allowed us to resolve several classes of fluorescence lifetimes, that vary in intensity as the EB/tRNA ratio is changed. Global Analysis of the intensity decay data supports a model in which the lifetime of the EB bound to the "strong" binding site is significantly reduced upon binding of subsequent EB molecules to the EB/tRNA complex.

6.2
Experimental

6.2.1
Sample Preparation

Yeast tRNA^{Phe} was obtained from Sigma (St. Louis, MO) and used without further purification. EB was obtained from Molecular Probes (Eugene, OR). Unless otherwise indicated, the buffer utilized was pH 7, 20 mM HEPES, 2 mM MgCl$_2$, 0.1 mM EDTA, 100 mM KCl.

6.2.2
Multifrequency Phase and Modulation Fluorometry

Intensity decay data were obtained using an ISS K2 multifrequency phase and modulation fluorometer utilizing a Spectra-Physics model 2045 Argon-Ion laser as the excitation source. Samples were excited at 514 nm and emission >550 nm was viewed through a Schott OG87 cut-on filter. Phase and modulation data were collected over a frequency range of 0.8 MHz to 80 MHz and analyzed using Global software. Fig. 6.1 shows phase and modulation curves for four of the seven EB/tRNA ratios utilized (three of the data sets used in the analysis are omitted for clarity). The raw data clearly show changes in the lifetime properties as the EB/tRNA ratio increases.

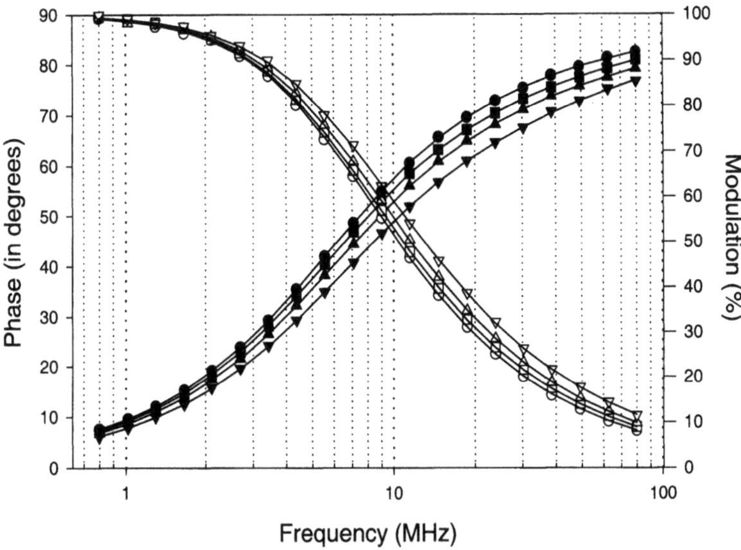

Fig. 6.1 Multifrequency phase (filled symbols) and modulation (open symbols) data for EB/tRNA ratios of 0.27 (circles), 1.34 (squares), 2.41 (triangles) and 4.05 (inverted triangles)

6.2.3
Three Component Analysis

Our initial approach to analyzing this lifetime data was to use three components: one corresponding to free EB, plus two other discrete exponentials. Figs. 6.2 and 6.3 show two such lifetime analyses of this data set – lifetime components and fractional intensities are displayed as a fucntion of the EB/tRNA ratio. In the first lifetime analysis (Fig. 6.2) a short (1.8 ns) component corresponding to free EB is fixed but the other two components are free to vary. One notes that lifetime 2 fluctuates between ~ 5.7–7.9 ns whereas lifetime 3 decreases monotonically from about 27 ns to about 24 ns. In the second analysis (Fig. 6.3), lifetime 2 is linked (but not fixed) between the data sets and is about 7.2 ns; in this case lifetime 3 still decreases monotonically from ~27 ns to ~24 ns.

In both cases, the data fits appear good as judged by the reduced chi-square values which were both ~0.3 (assuming standard errors of 0.2° in phase angle and 0.004 in the modulation ratio).

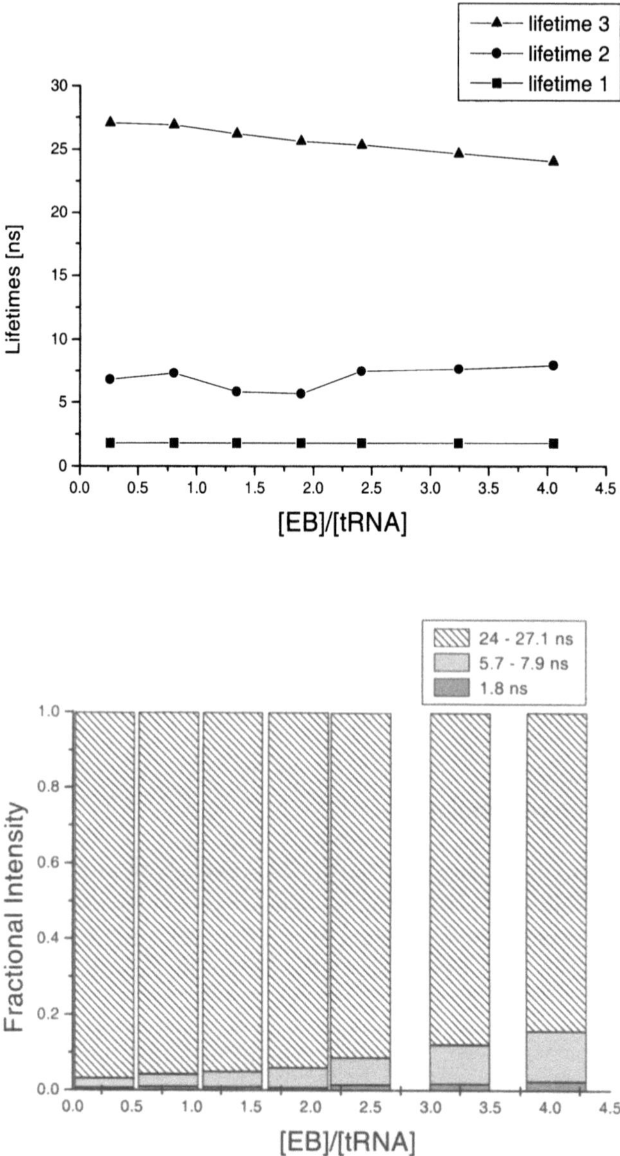

Fig. 6.2. Three component analysis of lifetime titration data. No components are linked but the 1.8 ns component is fixed

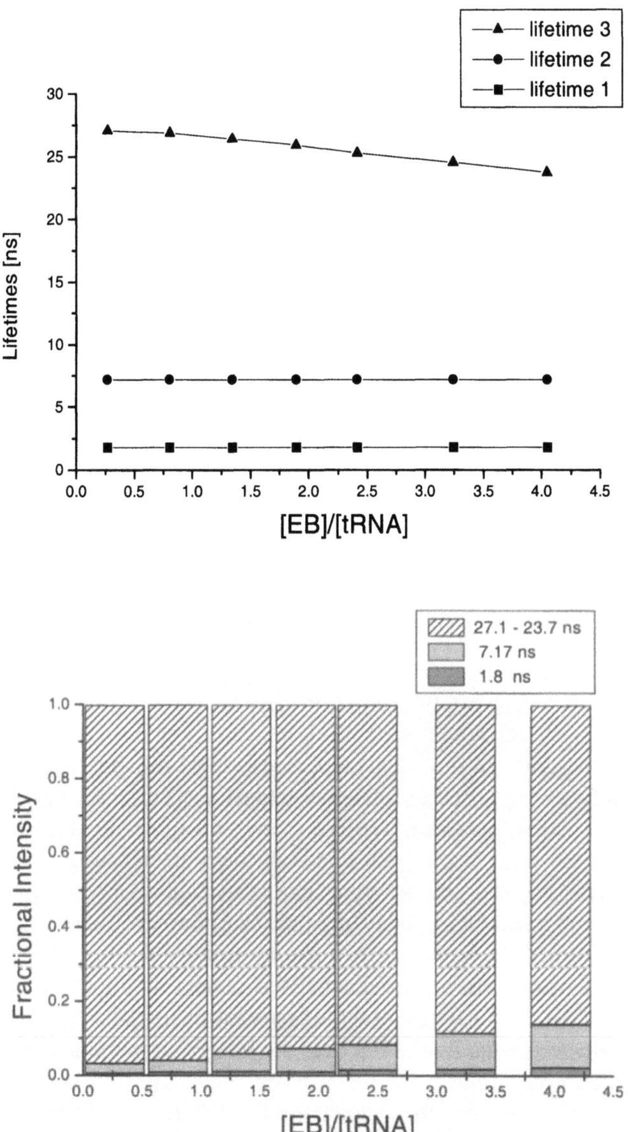

Fig. 6.3. Three component analysis of lifetime titration data. The intermediate lifetime component is linked and the 1.8 ns component is fixed

6.3
New Analysis

As mentioned above, the quality of the data fit in both of the preceding three-component analyses was good. However, except for fixing one component at 1.8 ns which corresponds to the known lifetime of free EB, these analyses did not take into account the nature of the multiple equilibria one expects in such a system. Specifically, as the titration progresses one expects some of the tRNA molecules to have only one EB bound and the lifetime of this species should be invariant. We thus carried out a three component analysis, with the long lifetime component linked throughout all data sets, and with a 1.8 ns component fixed, but with the second, shorter lifetime component free to vary and with the fractional intensities associated with each component free to vary. In this scenario, lifetime 2 increased from ~3.2 to 11.4 ns and the quality of the fit was reduced (chi-square ~0.8).

We then carried out a four-component analysis (Fig. 6.4) in which lifetime 1 was fixed to 1.8 ns (corresponding to free EB), while lifetimes 2, 3 and 4 were free to vary but were linked across the data sets; the fractional intensities associated with each component were free to vary. The quality of the fit to this model was very good (chi-square ~0.25). In this analysis, the resolved lifetime components are 5.4 ns, 17.7 ns and 27 ns, with standard errors of approximately 10–15% in each case. The fractional intensity corresponding to the long (27 ns) component decreases as the EB/tRNA ratio increases while the fractional intensities corresponding to the other three components all increase (Fig. 6.4).

We note that one can also analyze time-resolved data in terms of the pre-exponential factors as opposed to fractional intensities. To relate the pre-exponential factors to the actual number of different emitting molecules, however, requires knowledge of the relative quantum yields of the molecular species, at the excitation wavelength utilized and the emission wavelengths observed. In the present case, we have no information regarding the absorption and emission spectra of the 5.4 ns component. Nor do we know if the spectra for the longer lifetime component remain the same after the second EB is bound. We note that the excitation wavelength utilized (514 nm) and the emission wavelengths observed (>550 nm) were chosen to weigh the bound EB over the free EB to facilitate resolution of the bound EB lifetimes.

We determined, however, upon excitation at 501 nm (the approximate isosbestic wavelength for free and bound EB) and observing all emission >530 nm, that the enhancement in the yield of the EB bound was approximately 9-fold. This enhancement was measured using a large excess of tRNA compared to the EB concentration so that no tRNA should have more than one EB bound. We note this value to be less than one might expect from the ratio of the free and bound lifetimes (e.g., 27 ns/1.8 ns = 15). Given the fact that the absorption and emission spectra of free and bound EB differ substantially, however, there is no *a priori* reason to expect the ratio between lifetimes and quantum yields to be identical.

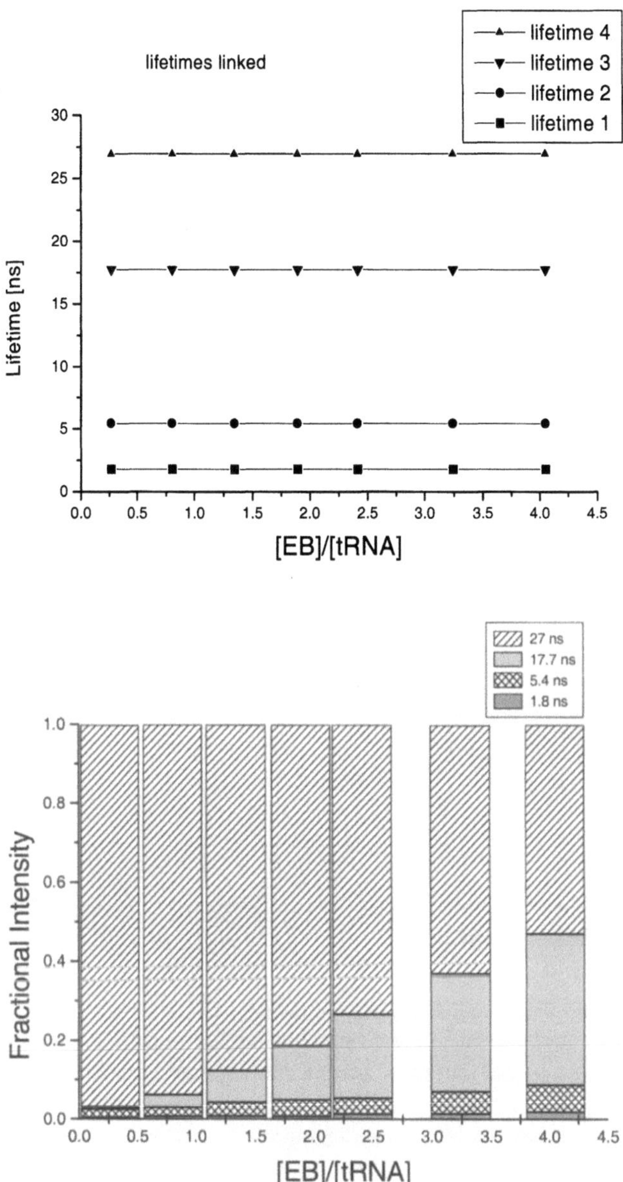

Fig. 6.4. Four component analysis of lifetime titration data. All lifetime components are linked (the 1.8 ns component is fixed) but the fractional intensities are free to vary

6.4
The Model

Our interpretation of this analysis is that EB in the "strong" binding site (which has a dissociation constant of ~1–2 µM based on steady-state titration data (not shown) has a lifetime of 27 ns if the "weak" binding site is not occupied. The lifetime of EB bound to the "weak" binding site (which has a dissociation constant in the range of 20–40 µM; data not shown) is 5.4 ns. However, when the "weak" binding site is occupied, the lifetime of EB in the "strong" binding site is reduced to 17.7 ns. This model is illustrated in Fig. 6.5.

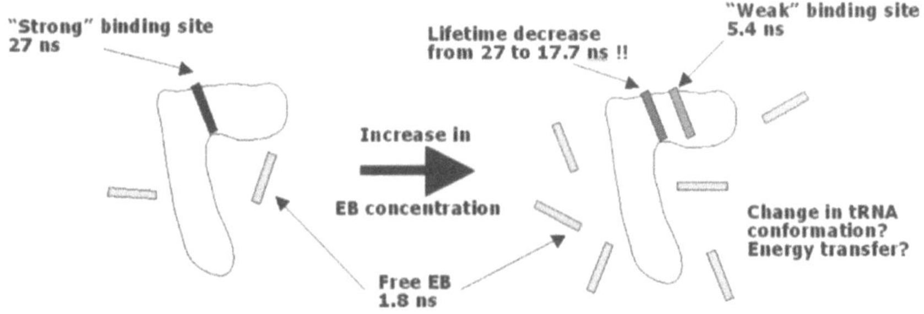

Fig. 6.5. Schematic model depicting the various lifetimes of EB bound to tRNA

6.5
Effect of Ionic Strength

It has previously been reported that increasing ionic strength, specifically NaCl, abolishes non-intercalative interactions between EB and tRNA [19]. If the 5.4 ns component is due to EB bound in a non-intercalative manner, one would expect that increasing ionic strength would reduce the relative amount of this component. Therefore, we studied the effect of varying levels of NaCl on the lifetime data at an EB/tRNA ratio of 4. As shown in Fig. 6.6, the effect of NaCl up to 900 mM was negligible as regards the actual lifetimes recovered and the relative proportions of the lifetime components.

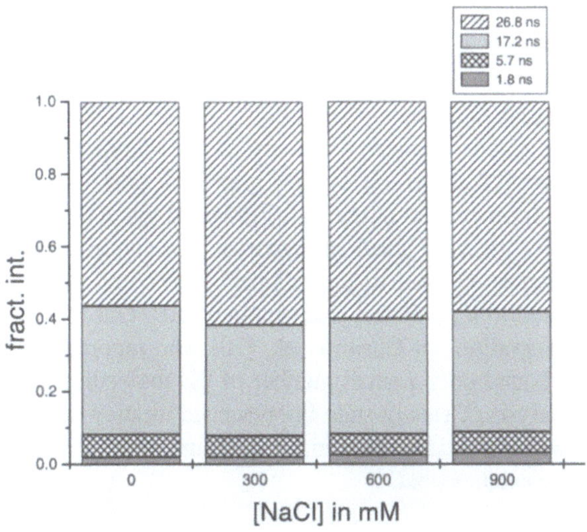

Fig. 6.6. Results of four component linked lifetime analysis of 100µM EB and 400µM tRNA in standard buffer in the presence of increasing NaCl concentrations

6.6
Conclusions

Previous measurements of the fluorescence lifetime of EB associated with tRNA (at low EB/tRNA) ratios, generally found values in the range of 26–28 ns [5, 7, 18, 21, 23]. Our present results are consistent with these earlier observations, i.e., at low EB/tRNA ratios we find a single exponential decay of 27 ns. As the EB/tRNA ratio increases we note a distinct decrease in the "average" lifetime of the system. Global Analysis of multiple data sets, obtained at increasing EB/tRNA ratios, supports a model in which binding of additional EB leads to a decrease of the lifetime of the EB in the "strong" binding site. Specifically, the lifetime of the initially bound EB decreases from 27 ns to 17.7 ns.

Proton NMR experiments of Jones and Kearns [6] suggested that the "strong" EB binding site was located near base pairs AU6 and AU7 in the amino acceptor stem. More recently, Chu et al., [4] used 19F NMR (by incorporation of 5-fluor-ouracil into specific locations in the tRNA) and proton NMR to study EB binding to *Escherichia coli* tRNA^{Val}. They also report the "strong" EB binding to be be-tween base pairs A6:U67 and U7:A66, in agreement with the observations of Jones and Kearns. Chu et al. [4] also found evidence for a second, weaker EB binding site in the amino acceptor stem, near U4:A69 and G5:C68. This result suggests the two EB binding sites may be quite close to one another. Hence, the decrease we observed in the lifetime (27 ns to 17.7 ns) of EB bound to the "strong" site could be due either to a change in the tRNA conformation near the "strong" binding site or energy transfer from EB in the "strong" site to EB in the "weak" site. A change in the conformation of the tRNA near the "strong" binding site could, for example, increase the accessibility of water to the excited EB, which could decrease the quantum yield [9].

Clearly, these results suggest the analysis and interpretation of steady-state fluorescence titration data on this system must take into account the decrease in the fluorescence enhancement of the first EB bound when the second site is occupied. Additionally, the very small influence of salt concentration on the "weak" binding site is interesting since it suggests the 5.4 ns component is due to an intercalated, as opposed to a non-intercalative, bound EB. Our results are also consistent with the observations of Ghribi et al., [20] who reported that at higher ionic strengths tRNA bound only a small number of EB molecules. Finally, we wish to point out that analysis of steady-state fluorescence titrations on EB/tRNA systems must take into account the decrease in the quantum yield of the first bound EB as subsequent EB molecules bind.

Acknowledgements. D.M.J. acknowledges support from NSF grant MCB9808427. The Laboratory for Fluorescence Dynamics is an NIH Research Resource (RR03155). T.I. was supported by a Predoctoral Fellowship from the American Heart Association. M.T. acknowledges travel support from the Institut Curie and also a European Union Fellowship (BIO4 CT97 2177).

References

1. Bittman R (1969) Studies of the binding of ethidium bromide to transfer ribonucleic acid: absorption, fluorescence, ultracentrifugation and kinetic investigations. J Mol Biol 46:251–268
2. Burns VW (1969) Fluorescence decay time characteristics of the complex between ethidium bromide and nucleic acids. Arch Biochem Biophys 133:420–424
3. LePecq JB, Paoletti C (1967) A fluorescent complex between ethidium bromide and nucleic acids. Physical-chemical characterization. J Mol Biol 27:87–106
4. Chu WC, Liu JC, Horowitz J (1997) Localization of the major ethidium bromide binding site on tRNA. Nucleic Acids Res 25:3944–3949
5. Hazlett TL, Johnson AE, Jameson DM (1989) Time-resolved fluorescence studies on the ternary complex formed between bacterial elongation factor Tu, guanosine 5'-triphosphate, and phenylalanyl-tRNAPhe. Biochemistry 28:4109–4117
6. Jones CR, Kearns DR (1975) Identification of a unique ethidium bromide binding site on yeast tRNAPhe by high resolution (300 MHz) nuclear magnetic resonance. Biochemistry 14:2660–2665
7. Jones CR, Bolton PH, Kearns DR (1978) Ethidium bromide binding to transfer RNA: transfer RNA as a model system for studying drug-RNA interactions. Biochemistry 17: 601–607
8. Liebman M, Rubin J, Sundaralingam M (1977) Nonintercalative binding of ethidium bromide to nucleic acids: crystal structure of an ethidium-tRNA molecular complex. Proc Natl Acad Sci USA 74:4821–4825

9. Olmsted JD, Kearns DR (1977) Mechanism of ethidium bromide fluorescence enhancement on binding to nucleic acids. Biochemistry 16:3647–3654

10. Sela I (1969) Fluorescence of nucleic acids with ethidium bromide: an indication of the configurative state of nucleic acids. Biochim Biophys Acta 190:216–219

11. Sturgill TW (1978) Thermodynamic characterization of ethidium bromide binding to a unique site on yeast tRNAphe. Biopolymers 17:1793–1810

12. Surovaya AN, Borissova OF (1976) Conformational peculiarities of tRNAMetf from E. coli as revealed by fluorescent methods. Mol Biol Rep 2:487–495

13. Torgerson PM, Drickamer HG, Weber G (1980) Effect of hydrostatic pressure upon ethidium bromide association with transfer ribonucleic acid. Biochemistry 19:3957–3960

14. Tritton TR, Mohr SC (1971) Relaxation kinetics of the binding of ethidium bromide to unfractionated yeast tRNA at low dye-phosphate ratio. Biochem Biophys Res Commun 45:1240–1249

15. Tritton TR, Mohr SC (1973) Kinetics of ethidium bromide binding as a probe of transfer ribonucleic acid structure. Biochemistry 12:905–914

16. Urbanke C, Romer R, Maass G (1973) The binding of ethidium bromide to different conformations of tRNA. Unfolding of tertiary structure. Eur J Biochem 33:511–516

17. Van Nuland Y, Snauwaert J, Heremans KA (1974) Proceedings: Influence of pressure on the relaxation kinetics of the binding of ethidium bromide to yeast tRNA. Arch Int Physiol Biochim 82:780

18. Tao T, Nelson JH, Cantor CR (1970) Conformational studies on transfer ribonucleic acid. Fluorescence lifetime and nanosecond depolarization measurements on bound ethidium bromide. Biochemistry 9:3514–3524

19. Wells BD, Cantor CR (1977) A strong ethidium binding site in the acceptor stem of most or all transfer RNAs. Nucleic Acids Res 4:1667–1680

20. Ghribi S, Maurel MC, Rougee M, Favre A (1988) Evidence for tertiary structure in natural single stranded RNAs in solution. Nucleic Acids Res 16:1095–1112

21. Thomas JC, Schurr JM, Hare DR (1984) Rotational dynamics of transfer ribonucleic acid: effect of ionic strength and concentration. Biochemistry 23:5407–5413

22. Beechem JM, Gratton E, Ameloot M, Knutson JR, Brand L (1991) The global analysis of fluorescence intensity and anisotropy decay data: Second generation theory and programs. In: Lakowicz JR (ed) Topics in fluorescence spectroscopy II, Volume 5. Plenum, New York pp. 241–305

23. Ferguson BQ, Yang DC (1986) Localization of noncovalently bound ethidium in free and methionyl-tRNA synthetase bound tRNAfMet by singlet-singlet energy transfer. Biochemistry 25:5298–304.

CHAPTER 7

Experimental Aspects of DNA Computing by Blocking: Use of Fluorescence Techniques for Detection

K. A. SCHMIDT, C. V. HENKEL, G. ROZENBERG, AND H. P. SPAINK

The suitability of fluorescence techniques for the experimental implementation of a DNA based algorithm was studied. Two different assays based on the use of PicoGreen and FRET, respectively, have been used for the detection of hybridisation. The limitations of both assays are discussed.

7.1
Introduction

Biomolecular computing studies the use of deoxyribonucleic acid (DNA) (or other biomolecules) for solving various sorts of computational problems [1]. DNA is especially suitable for dealing with optimisation problems for which the computing time increases exponentially with the problem size: since billions of DNA molecules can act as billions of parallel processors, the potential to reduce computing time is enormous. Fluorescence techniques seem very promising for the experimental implementation of DNA computing because they are non-destructive, offer a high sensitivity and may be used for implementing fast screening assays.

One of the problems that has been often considered in the framework of biomolecular computing is the so-called satisfiability (SAT) problem (see e.g., [2, 3]). Briefly, the problem asks an assignment (a set of values for variables) that satisfies a logical formula. Recently, a new type of DNA-based algorithm has been described for solving the SAT problem [4]. First, all library molecules are synthesized, i.e., a mixture of DNA molecules encoding all candidate solutions for a given problem. Then a set of so-called blocker molecules is created that encode the falsifying assignments for each clause. These blockers are used to inactivate (block) those library molecules that do not contribute to (finding) a solution. Addition of the blocker molecules to a mixture of all library molecules results in blocking all "wrong" assignments, i.e., only those library molecules, which do represent a solution are left unblocked, and hence, are available for further processing. For the experimental implementation of this algorithm the authors proposed a procedure based on polymerase chain reaction (PCR) in combination with the use of peptide nucleic acid (PNA) oligonucleotides.

In this study, we have investigated an alternative procedure for implementing a blocking algorithm, which relies on the detection of hybridisation between library molecules and blockers (both encoded as single stranded DNA molecules). For this purpose the suitability of two different fluorescence techniques was tested.

7.2
Experimental

Oligonucleotides were purchased from Isogen Bioscience (Netherlands) and Eurogentec (Belgium). Details of the sequence design and encoding will be reported elsewhere (C. Henkel, in preparation). All hybridisation experiments were performed in 1x SSC buffer (150 mM NaCl, 15 mM sodium citrate, pH 7.0). For wellplate experiments 3.2 mM NaOH was added to the hybridisation buffer. Fluorescence measurements were performed with a LS50B Luminescence Spectrometer (Perkin Elmer). For quantification of hybridisation in wellplates PicoGreen® dsDNA quantification reagent (Molecular Probes) was used (stock solution 16000 x diluted). The library molecules (0.5 μM in SSC) were distributed in 16 different wells of the wellplate (Packard View Plate) and a mixture containing the blockers (each 0.5 μM in SSC) was added. After addition of PicoGreen the fluorescence emitted at 520 nm was measured using 485 nm excitation. Images of the wellplate were taken with a FluorS™ MultiImager (BioRad) with UV excitation and LP 520 detection. For FRET measurements two different combinations of 5' fluorescein-labelled library molecules (1.4 μM in SSC) and a 5' TAMRA-labelled blocker molecule (1.6 μM in SSC) were used. Fluorescence emission spectra were recorded at different temperatures using 460 nm excitation.

7.3
Results and Discussion

PicoGreen is a base-intercalating dye, which is commonly used for the quantification of double stranded (ds) DNA molecules. The dye is essentially non-fluorescent when free in solution. Consequently, formation of dsDNA, i.e., hybridisation, in a mixture of blocker and library molecules should result in an increase of the fluorescence emitted from the solution. In order to test the applicability of this approach we used a small instance of the SAT problem, $F = b \vee c \vee {\sim}d$. 16 different oligonucleotides encoding the library molecules were distributed in 16 different wells of a wellplate.

Fig. 7.1A shows an image of the wellplate after addition of a mixture containing the two blocker molecules and PicoGreen. Due to a rather high background fluorescence signal of the wellplate (probably caused by reflection and scattering especially at the borders of the wells), it is difficult to identify the solutions from the image of the wellplate alone. Therefore, the fluorescence intensities emitted at 520 nm (485 nm excitation) were used as a measure for the amount of hybridisation.

In the fluorescence intensity diagram (Fig. 7.1B) the two wells in which blocking (hybridisation) occurs (labelled 0001 and 1001) exhibit a higher fluorescence than the other wells. However, for the non-blocking combinations the measured fluorescence intensities show a rather high variation. Partially, this might be due to background fluorescence of the wellplate. Alternatively, the higher fluorescence emission from a number of the non-blocking combinations may indicate that some hybridisation occurred under the experimental conditions used.

In order to develop a more specific method for hybridisation detection, a second assay based on fluorescence resonance energy transfer (FRET) was applied. The library molecules were labelled at their 5' end with fluorescein (donor) and the blocker molecules at their 5' end with TAMRA (acceptor). Using 460 nm excitation, i.e., in the fluorescein absorption band, hybridisation between library and blocker molecules should result in a decrease of fluorescein emission (quenching) as well as increase of the emission from TAMRA. In a first experiment, 3' labelling was used for the blocker molecules. However, this approach resulted in a strong quenching of both dyes, most likely due to exciton interaction [5].

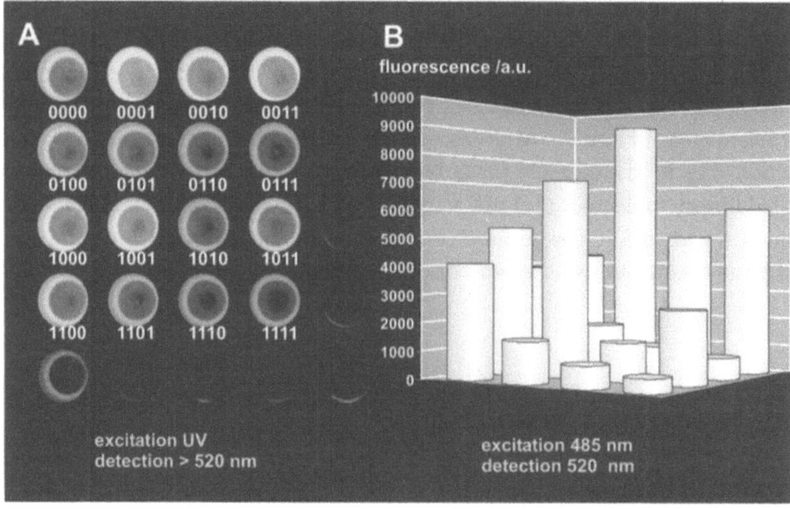

Fig. 7.1. Quantification of blocking with PicoGreen. A small instance of the SAT problem was used: $F = b \vee c \vee \sim d$, falsified by $[a\ b\ c\ d] = 0001 \vee 1001$. **A** Image of the wellplate taken with UV excitation and detection at > 520 nm. The blocking combinations are located in the wells labelled 0001 and 1001. **B** Fluorescence intensity diagram using 485 nm excitation and 520 nm detection

For testing the FRET assay two different library molecules and one blocker molecule were synthesized: Library 4 5'-TCT TCA TCT TCT TC-3', library 7 5"-TCT TCA TCA TCA TC-3' and blocker B 5'-GAA GAA GAT GAA GA-3'.

One of the constraints used for the sequence design was to limit the number of guanine residues in order to avoid quenching of TAMRA (unpublished data, see also [6]).

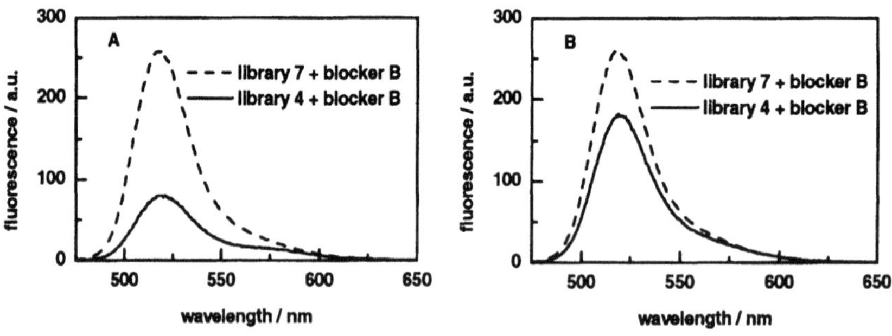

Fig. 7.2. Fluorescence emission spectra of two different combinations of blocker and library molecules at 37°C (**A**) and 52°C (**B**) using 460 nm excitation. Dashed lines: library 7 plus blocker B, non-blocking combination. Solid lines: library 4 plus blocker B, blocking combination

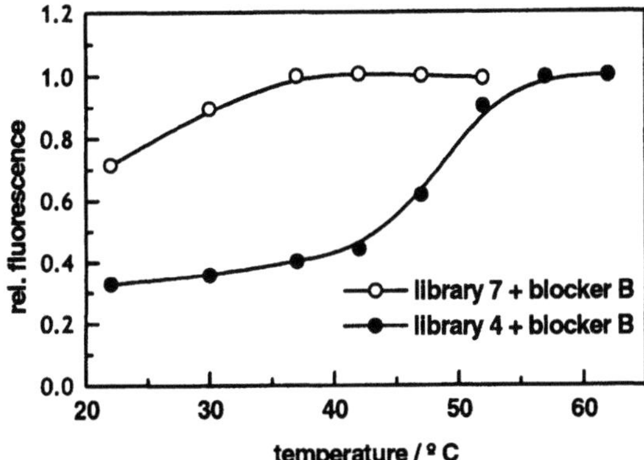

Fig. 7.3. Relative fluorescein emission at different temperatures measured at 520 nm (460 nm excitation). Open circles: library 7 plus blocker B, non-blocking combination. Solid circles: library 4 plus blocker B, blocking combination

Fig. 7.2 depicts fluorescence emission spectra for the two library/blocker combinations recorded at 37 and 52°C, respectively. Under permissive conditions (37°C, Fig. 7.2A) quenching of fluorescein emission was only observed for the blocking combination (library 4 and blocker B) but not for the non-blocking combination (library 7 and blocker B). In contrast to what is expected for ideal energy transfer behaviour, only a very small increase of the emission from TAMRA occurred. This lack of an increase in acceptor fluorescence may be due to quenching of the excited state of the fluorophore [7]. Under melting conditions (52 °C, Fig. 7.2B) the emission from fluorescein was only slightly quenched in the blocking combination. The corresponding melting curves are shown in Fig. 7.3.

7.4
Conclusion

The experiments described in this paper show that both the PicoGreen and the FRET assay are in principle suited for hybridisation detection and, therefore, can be used for the experimental implementation of a blocking algorithm. However, both approaches are susceptible to a number of experimental errors and proper controls are needed. The successful experimental implementation of a blocking algorithm requires accurate detection of blocking in a mixture containing all blockers at the same time. For problems exceeding the proof-of-principle scale this simultaneous detection may be difficult to achieve.

In DNA computing hybridisation is generally seen as a simple "yes" or "no" decision. However, thermodynamic studies have shown that hybridisation of oligonucleotides does not depend only on experimental conditions like temperature or salt concentration but also on the DNA sequence itself [8]. For small problems, these difficulties may be overcome by applying corresponding constraints for sequence design (C. Henkel, unpublished results), but attacking problems of realistic scale will most likely require the development of new algorithms, which are not based on hybridisation only.

References

1. Paun G, Rozenberg G, Salomaa, A (1998) DNA Computing - New Computing Paradigms. Springer, Berlin, Heidelberg, New York
2. Liu Q, Wang L, Frutos AG, Condon AE, Corn RM, Smith LM (2000) DNA computing on surfaces. Nature 403: 175–179
3. Faulhammer D, Cukras AR, Lipton RJ, Landweber LF (2000) Molecular computation: RNA solutions to chess problems. Proc Natl Acad Sci USA 97: 1385–1389
4. Rozenberg G, Spaink HP (2001) DNA computing by blocking. Theoretical Computer Science, in press
5. Bernacchi S, Mély Y (2001) Exciton interaction in molecular beacons: a sensitive sensor for short range modifications of the nucleic acid structure. Nucl Acids Res 29: e62
6. Knemeyer J-P, Marmé N, Sauer M (2000) Probes for detection of specific DNA sequences at the single-molecule level. Anal Chem 72: 3717–3724
7. Gelfand CA, Plum GE, Mielewczyk S, Remeta DP, Breslauer KJ (1999) A quantitative method for evaluating the stabilities of nucleic acids. Proc Natl Acad Sci USA 96: 6113–118
8. SantaLucia J (1998) A unified view of polymer, dumbbell, and oligonucleotide DNA nearest-neighbor thermodynamics. Proc Natl Acad Sci USA 95: 1460–465

Part 2
Fluorescence Spectroscopy of Single Molecules and Molecular Assemblies

Multiparametric Detection of Fluorescence Emitted from Individual Multichromophoric Systems

M. COTLET, J. HOFKENS, M. MAUS, AND F. C. DE SCHRYVER

While few years ago the main goal in room temperature single molecule fluorescence spectroscopy (SMS) was to visualize individual molecules, nowadays experiments are designed such that multiparametric observation of different fluorescence characteristics of the investigated molecular systems is allowed. The simultaneous observation of fluorescence characteristics such as spectral peak position, fluorescence decay times, polarization properties or wavelength integrated intensity of fluorescence during the survival time of the investigated molecules leads to a more detailed picture of the molecular states and environment changes experienced by the probed molecules. In this contribution a diffraction-limited scanning stage confocal microscope set-up allowing real time multiparametric observation of fluorescence detected from single molecules immobilized in thin polymer matrix is described. By using pulsed excitation in combination with burst integrated fluorescence lifetime (BIFL) type detection, the simultaneous acquisition of fluorescence spectra, wavelength integrated fluorescence intensity time traces and time-resolved decay curves is demonstrated for synthetic as well as biological systems. By using continuous wave excitation and BIFL type detection, two dimensional fluorescence intensity time traces containing photons resolved in time with an accuracy of 50 ns and carrying polarization information can be recorded from individual immobilized molecules. In combination with specific analysis procedures, the SMS set-up is proved to be a suitable tool for the identification of different emitting species as well as for monitoring dynamical processes at the single molecule level.

8.1
Introduction

In recent years the possibility of detecting laser-induced fluorescence from individual molecules has opened new horizons in the field of ultrasensitive analysis in life science [1, 2]. Detection, identification and counting of single molecules has become possible nowadays [3–7]. By analyzing the time or spectrally resolved fluorescence detected from individual molecules, fluorescence characteristics such as the spectral peak position, the spectral shape, the fluorescence decay times, the triplet lifetime or intersystem crossing quantum yield can be determined and their fluctuations can be monitored. By probing individual members of a population via single molecule fluorescence spectroscopy (SMS), the statistical averaging inherent in ensemble experiments is eliminated and eventual heterogeneities in the molecular and kinetic properties of the investigated population can be revealed.

A major drawback of the SMS experiments is the relatively low ratio between the detected fluorescence and the background signal, the last one mainly determined by Raman and Rayleigh scattering of the excitation laser light, limiting the applicability of the method only to specific fluorophores. However, different approaches were proposed in order to increase the signal-to-noise ratio (S/N) of the detected signal by collecting and detecting as much as possible from the fluorescence emitted by the molecule and by suppressing the background signal [1, 8–13]. Nowadays, the most common SMS experiment implies a epi-illuminated microscope equipped with a high numerical aperture objective lens for efficient collection of fluorescence, spectral filters for the suppression of both Raman and Rayleigh scatter, confocal detection for the minimization of the excitation volume leading to background suppression and high efficient detectors. Investigation of both synthetic and biological systems by means of SMS can be performed both in solution as well as in fixed matrices. While in the first case the time interval in which fluorescence can be detected from an individual molecule is limited by the diffusion time of the molecule through the laser focus, in the case of immobilized molecules the detection time of fluorescence is limited by photobleaching of the single molecule. Because of these limitations, resulting in a low amount of detected photons, multiparametric approaches leading to the simultaneous observation of different fluorescence characteristics have been developed in the field of SMS in order to gain as much as possible information [5, 14–17]. For solution experiments a real time two-dimensional spectroscopic technique based on correlation spectroscopy but using pulsed excitation, the burst integrated fluorescence lifetime (BIFL) technique, was recently introduced and demonstrated as a valuable tool for monitoring conformational dynamics of individual biomolecules [14, 18]. For immobilized molecules, the simultaneous acquisition of wavelength-resolved fluorescence emission (fluorescence spectra) and time-resolved fluorescence decays was used to discriminate between different emitting species in individual multichromophoric synthetic as well as biological systems [15, 20].

We report here on the simultaneous detection of two-dimensional (two-color or polarization) wavelength integrated fluorescence intensity time traces (fluorescence transients), time-resolved decay curves and fluorescence spectra from individual molecules immobilized and isolated in polymer matrices by using a diffraction limited confocal microscope in combination with pulsed excitation and BIFL type detection [14]. By using low power continuous wave excitation (CW) in combination with the BIFL type detection, time-resolved detection with 50 ns accuracy of the fluorescence emitted from individual molecules embedded in matrices is demonstrated. Point process global analysis of the fluorescence intensity traces reconstructed using the time-resolved detected photocounts leads to the recovery of rate constants associated with on–off processes in individual molecules. The multiparametric approach introduced here is demonstrated on two multichromophoric systems that were recently investigated at the single molecule level from the point of view of their photophysical properties [15, 19], a dendritic molecule consisting of hexaphenylbenzene building blocks and bearing eight perylenecarboximide units as chromophores (g2) and the B-phycoerythrin protein.

8.2
Materials and Methods

8.2.1
Sample Preparation

The g2 dendrimer was synthesized as previously reported [20] while the B-phycoerythrin (B-Pe) from the red algae *Porphyridium cruentum* was commercially available as lyophilized powder from Sigma. For g2, the samples for SMS experiments were prepared by spin coating a solution of dendrimer (5×10^{-10} M) in toluene containing 3 mg/mL Zeonex on a cover glass at 2500 rpm while for B-Pe, a 10 mM phosphate buffered saline solution (PBS pH 7.4) of the protein (10^{-10} M) was mixed with a PBS solution containing 1 wt-% polyvinylalcohol (PVA, Agfa Gevaert, M_w-25 000) and then spin coated on a cover glass at 2500 rpm.

8.2.2
Experimental Set-up

A schematic diagram of the SMS set-up used in our experiments is depicted in Fig. 8.1. Pulsed excitation at 488 nm wavelength using 8.13 MHz repetition rate was achieved by using a laser system that was previously described [21]. In brief, by combining a picosecond regenerative mode-locked Ti:Sapphire (Spectra Physics Tsunami 3950-D, 82 MHz, FWHM of 1.2 ps) laser pumped by a CW Ar-ion laser (Spectra Physics Beam Lock 2080) with an optical parametric oscillator (OPO Spectra Physics ps-KDP) and a frequency doubling crystal (FD-BBO), laser lines over the UV-NIR spectral range can be delivered. An acousto-optic modulator (AOM) present in the beam path after the OPO allowed the reduction of the repetition rate of the excitation light down to 8.13 MHz. CW excitation at 543 nm wavelength was achieved by using a He-Ne laser (Melles Griot).

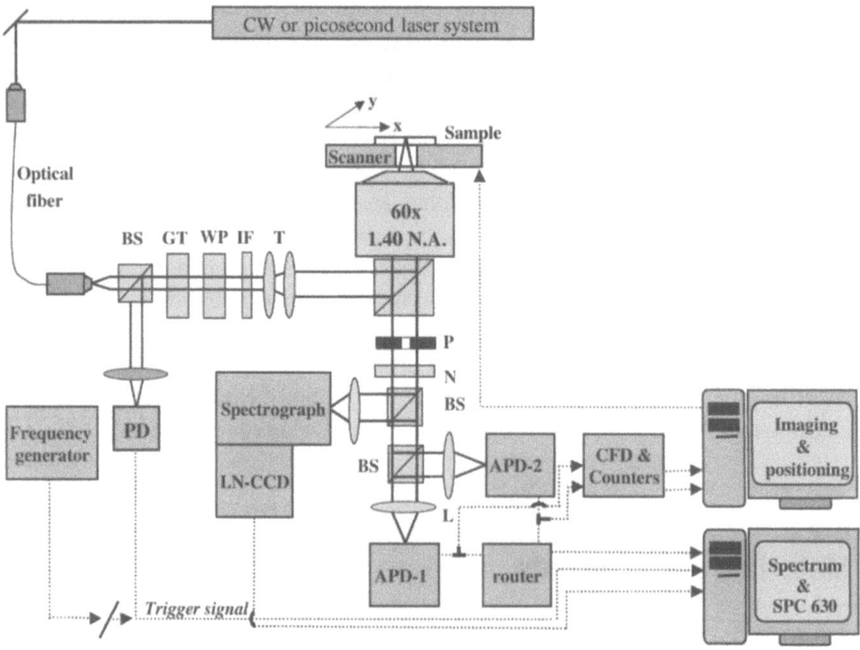

Fig. 8.1. Schematic diagram of the confocal epi-illuminated microscope setup: continuous wave laser (CW), beam splitter (BS), Glan Thomson polarizer (GT), wave plate (WP), interference band pass filter (IF), telescope (T), lens (L), ns photodiode (PD), pinhole (P), notch (N), liquid nitrogen cooled camera (LN-CCD), avalanche photodiode (APD), constant fraction discriminator (CFD)

For both pulsed and CW excitation regimes, before entering the microscope, the laser light was collimated by passing it through a single mode optical fiber. After the fiber, vertically linear polarization was achieved by means of a Glan--Thomson polarizer and a half-wave plate. By using a narrow interference band--pass filter (CVI), excitation lines with full width at the half-maximum (FWHM) of less than 4 nm could be achieved. After reflecting on a dichromatic long-pass beam splitter (DM-Omega Filters), the laser beam was focused on the sample by means of an oil immersion objective lens (Olympus, 1.4 NA, 60×) mounted into an inverted epi-illuminated microscope (Olympus IX 70). In order to fulfill the back aperture of the objective lens, the laser beam was expanded in diameter, before the microscope, by means of a telescopic lens system. A feedback-controlled scanning stage (Physical Instruments, Waldbrown Germany, 1 nm resolution) mounted on the top of the microscope and carrying the sample allowed raster scanning areas up to 100 μm and positioning by means of two piezo drivers controlled with a PC card (Computerboards DAQ). A low vacuum applied on the sample holder prevented eventual sliding of the cover glass with respect to the objective lens. Alignment and control of focusing onto the sample was possible by using a charge-coupled device (CCD) camera (Bischke 4012P Germany) coupled through an objective lens system to the photocamera port of the microscope. Using adequate neutral density filters, the excitation laser power could be adjusted at the desired value, typically from 0.5 to 1 kW/cm^2 at the sample.

Laser-induced fluorescence was collected by the same objective lens and, after passing the dichromatic beam splitter, reflected by a total reflection prism into the detection path of the SMS set-up via the side port of the microscope. Rejection of the excitation light from reaching the detection path was done by the dichromatic beam splitter and by suitable holographic notch filters (Kaiser Optics-Notch Plus). Diffraction-limited confocality was achieved by the insertion, in the image plane of the objective lens situated at the side port of the microscope, of a turret containing pinholes (Olympus OSP-TUX) with diameters varying from 10 up to 300 μm. An additional eyepiece allowed, by means of a total reflective prism inserted into the beam path, the alignment of the pinhole into the fluorescence beam.

After the pinhole, fluorescence was split by a 50:50 nonpolarizing beam splitter cube (Newport) in two beam paths. The right angle reflected fluorescence was focused into a 150 L/mm polychromator (Spectra Pro 150 Acton Research Corporation) coupled to a back-illuminated liquid nitrogen-cooled CCD camera (LN/CCD-1340SB, Princeton Instruments) in order to record fluorescence spectra with resolution down to 1 nm. The transmitted fluorescence was split, by means of a 50:50 per cent non-polarizing beam splitter cube, in two components which were focused afterwards into the active areas of two single photon counting avalanche photodiodes (APD-AWQ 151, EG&G Canada). By placing suitable spectral filters in front of the APDs, two-color detection could be achieved. Replacing the non-polarizing beam splitter cube with a polarizing one, polarization-sensitive detection of the fluorescence could be performed.

The detected single photon counting signals from both APDs were collected into a two-channel router (HRT82 Becker & Hickl GmbH) coupled to a time-correlated single photon counting PC card (SPC 630 Becker & Hickl GmbH) working in the reversed start–stop mode. The SPC 630 PC card was used in the "first in – first out" (FIFO) mode, allowing the registering, for each detected photocount, of the position with respect to the detector (routing signal), the arrival time with respect to the previous detected photocount (time lag) with an accuracy of 50 ns as well as the arrival time with respect to the laser pulse with an accuracy down to 12 ps. In parallel, the signals from the APDs were broadened by means of a constant fraction discriminator (CFD, QUAD-EG&G Canada) and input into a data acquisition PC counter board (Computerboards DAQ) used for the construction of the fluorescence image of the scanned area. Depending on the experimental arrangement, either two-color or polarization resolved fluorescence images of a scanned areas of the sample can be recorded. In order to localize the fluorescing individual molecules, typical areas of 10×10 μm^2 were raster scanned and the corresponding 128×128 pixel fluorescence images were recorded via the APD/counter board with a typical integration time per pixel of 5 ms.

For SMS experiments performed using pulsed excitation, the starting signal for the SPC 630 PC card was delivered by a ns photodiode (Newport) illuminated by a part of the excitation laser light (see Fig. 8.1). The time-resolved photocounts were detected in 256 channels. The instrumental response function (IRF) was measured by using a droplet of aqueous solution of erythrosine (90 ps lifetime). Since the fluorophore has a lifetime sizeably shorter than the time-resolution of the APD (>300 ps), this measurement gives a good representation of the IRF. Measuring the IRF with a blank polymer or only with a cover glass would distort the IRF since for the detection of Rayleigh photons at least the notch filter should be removed from the optical path.

A similar experimental arrangement was used for CW excitation, except for the replacement of the ns photodiode with a frequency generator (Hewlett-Packard) delivering a TTL signal with a frequency of 8 MHz used as start signal for time to amplitude converter (TAC) of the SPC 630 PC card driven in the FIFO mode. Since the triggering of the single photon timing experiment is not delivered by a signal correlated with the excitation laser light, the position in ps timescale of the detected photons is not anymore resolved. However, during the timescale of the experiment, the detected photons are located in time with an accuracy of 50 ns determined by the synchronization of the trigger signal and the duration of the TAC window T.

8.3
Data Processing and Analysis

Independent on the excitation regime, 50% of the emitted fluorescence from any probed single molecule is spectrally resolved in the polychromator and detected as a fluorescence emission spectrum. Depending on the signal-to-noise ratio of a particular experiment, the integration time applied for each spectrum can be as short as 500 ms. In this way, for each probed molecule a spectral run consisting of several fluorescence spectra is obtained. The measured spectra are corrected by removing the spikes occurring due to cosmic radiation, by removing the background spectrum obtained in conditions of nonirradiation and by the wavelength dependency introduced by the presence of optical elements in the detection path of the setup. By fitting the corrected spectra with Gaussian functions, spectral parameters such as peak position, intensity and FWHM can be extracted and their evolution can be monitored during the survival time of the probed molecule. The remaining 50% of the emitted fluorescence from the probed single molecule is detected using the BIFL-type technique [14].

8.3.1
Pulsed Excitation

When pulsed excitation is impinged on an individual molecule, the BIFL measurement yields a two-dimensional temporal information with respect to the detected photocounts [14]: the macroscopic position of a particular event in the time scale of the experiment by means of the measured time lag between consecutive detected photocounts with an accuracy of 50 ns and the arrival time of that event with respect to the excitation pulse measured by single photon timing technique with an accuracy up to 12 ps. Hence, the BIFL type detection opens up the possibility to monitor the evolution of the fluorescence intensity emitted from single molecules in macro-timescale via the reconstructed wavelength integrated fluorescence intensity time trace (fluorescence transient) as well as in micro-timescale via the reconstructed ps-fluorescence decay histogram. Moreover, if two-color BIFL type detection is used, beside the two-dimensional temporal information, each detected photon count will be also spectrally resolved since BIFL can store routing signal information. After reconstructing the fluorescence transient using a specific bin time, the ps-fluorescence decay histograms are build up from photocounts belonging to constant levels of intensity within the transient. In this way, during the survival time of the probed molecule, several decay histograms related to different intensity levels within the fluorescent transient can be reconstructed. Analysis of these decays will yield parameters that will characterize the investigated molecule only for a specific macro-time interval, permitting the monitoring of their evolution during the survival time of the molecule. Because the duration of the constant levels of intensity within a fluorescence transient can be as short as few hundreds of milliseconds, the number of the photocounts building the corresponding ps-decay histograms are usually very low. The maximum likelihood estimation method was proven to deliver reliable results when decay histograms with less than 200 total photocounts spread over 200 channels were analyzed [22, 23]. In our approach, each reconstructed decay is rebinned, if necessary, at most four times to 64 channels in order to get at least 80 photocounts in the peak. Analysis is performed in a global fashion by using the same experimental IRF and the same correlated background for all the decay histograms reconstructed from the probed molecule. The experimental IRF is convoluted with an exponential model fit function m (see Eq. 8.1) containing in addition a time shift, a constant background c accounting for noncorrelated photocounts and a scaling factor γ accounting for the correlated photocounts, i.e., scattered Rayleigh and/or Raman photons [24]:

$$m_i = \sum_i (p_{ji} \cdot \exp(-jT/k\tau_i)) + c + \gamma \cdot bg_j, \quad j = 1 - k \text{ and } i = 1,2 \quad (8.1)$$

Here T is the preset time window of the TAC, k represents the number of channels taken into account in analysis and τ and p are the fluorescence decay time and amplitude, respectively. The photocounts detected after the bleaching of the molecule are used to construct the decay histogram accounting for the correlated background. Depending on the duration of the macro-time region corresponding to each reconstructed decay histogram, different scaling factors are used. The goodness of the fits is judged by the values of the quality factor $2I^*$ [24] as well as of the inspection of the residual graphs for each analyzed decay histogram. Finally, for each single molecule the values of the recovered decay times and their corresponding amplitudes are obtained. In this way the evolution in the macro-timescale of the decay times associated with a particular investigated molecule can be monitored.

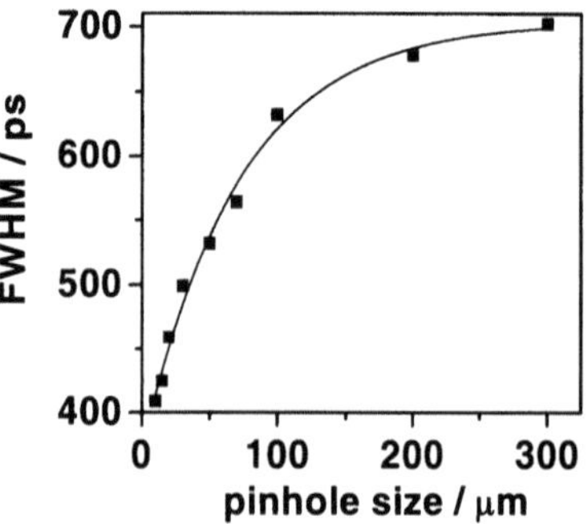

Fig. 8.2. The dependence of the FWHM of the experimental IRF on the size (diameter) of the detection pinhole for the experimental set-up depicted in Fig. 8.1

When convolution is used in order to recover fluorescence decay times, the accuracy of the obtained values strongly depends on the FWHM of the experimental IRF. Moreover, convolution requires that the IRF is measured in the same experimental conditions as the decay from the sample [25]. For a time response of about 350 ps for the APD and a FWHM of 2 ps for the laser pulse, the value of FWHM for IRF is mainly sensitive to the optical alignment:

$$\text{FWHM}_{\text{IRF}} = [(\Delta t_{\text{detector}})^2 + (\text{FWHM}_{\text{laser}})^2 + [(\Delta t_{\text{optics}})^2]^{1/2} \tag{8.2}$$

Assuming a good alignment of both optical excitation and detection paths of the confocal microscope, i.e., both paths are focused into the same diffraction-limited spot, the most crucial influence on the value of the FWHM of IRF comes from the magnitude of the pinhole used in the detection path for achieving confocality. Fig. 8.2 displays the dependence of the FWHM of the experimental IRF on the size of the detection pinhole in the case of our SMS setup. In our experimental conditions, i.e., numerical aperture of the objective lens NA = 1.4, magnification $M = 60\times$ and excitation wavelength around 500 nm, the best pinhole for achieving confocality with high signal-to-noise ratio (*S/N*) would have a value of the diameter according to [26]:

$$D_{\text{pinhole}} = [2M\,(0.61)\lambda] / \text{NA} \cong 25\mu\text{m} \tag{8.3}$$

corresponding to a FWHM of the IRF of about 470 ps (see Fig. 8.2). However, the presence of variable size pinholes in the detection path of a SMS set-up offers some advantages. Depending on the application, a compromise between *S/N* and FWHM of IRF can be achieved. For example, when fluorescence is detected with high *S/N* and fast processes are investigated at single molecule level, the use of a small pinhole would permit to resolve eventual fast (subnanosecond) decay time components present in the detected time-resolved fluorescence.

8.3.2
CW Excitation

In the CW excitation regime the BIFL type detection delivers, for each detected photocount, only the macrotime information by means of the measured time lag between consecutive detected photocounts. The absence of pulsed excitation leads to the loss of the micro-time information. If the experiment is performed in a two--channel detection configuration, the following information is stored for each photocount: either the spectral position for two-color detection or the polarization information for polarization sensitive detection. However, because the time lag between consecutive photocounts is measured with an accuracy of 50 ns, the detected fluorescence will be resolved in time with an accuracy of 50 ns. When polarization sensitivity BIFL type detection is used, two macro-traces are obtained, i.e., the parallel (I_{par}) and perpendicular (I_{perp}) polarized fluorescence components. The I_{par} and I_{perp} components can be afterwards used for the reconstruction of the total fluorescence transient (I_{tot}):

$$I_{tot} = I_{par} + 2 \cdot G \cdot I_{perp} \tag{8.4}$$

as well as of the polarization degree time trace (p):

$$p = (I_{par} - G \cdot I_{perp}) / (I_{par} + G \cdot I_{perp}) \tag{8.5}$$

Here G is a factor accounting for the difference in sensitivity of the two APDs when polarized detection is used [27].

Fluorescence transients detected from immobilized molecules contain real-time information about the on–off processes taking place at single molecule level [28, 29]. The discrete jumps in intensity from a high ("on") level to the background ("off") level and back to the high level during the survival time of a single molecule are usually caused by quantum jumps to a nonemissive, i.e., dark state [28]. By analyzing the fluorescence transients, information about the rate constants and quantum yields associated with particular on–off processes can be obtained. Up to now of recovery of these parameters was done by using the duration histogram analysis method which is based on a three-electronic-state model accounting for the on–off behavior of the molecule (see Fig. 8.3a) for which the average duration of the on (τ_{on}) and off (τ_{off}) levels can be expressed as [28]:

$$\frac{1}{\tau_{on}} = k_{exc} \cdot \frac{k_{on-off}}{(k_{on-off} + k_{fluoresc} + k_{IC})} \quad ; \qquad \frac{1}{\tau_{off}} = k_{dark} \tag{8.6}$$

where k_{exc} relates to the excitation rate. By measuring the duration of the on (on--time Δt_{off}) and off levels (off-time Δt_{off}) within the fluorescence transient detected from an individual molecule, by building up the corresponding $H(\Delta t_{off})$ and $H(\Delta t_{on})$ histograms and fitting them with single exponential decaying model functions, the τ_{off} and τ_{on} values can be recovered. Moreover, by measuring the value of the mean intensity of the on level (I_{on}), the quantum yield of the on–off process can be roughly estimated as [28]:

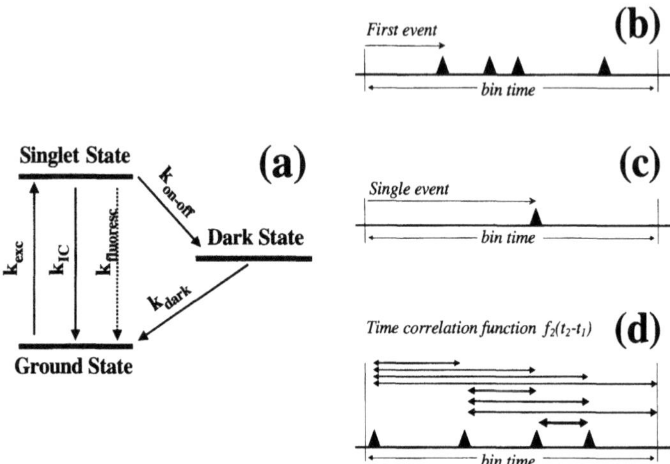

Fig. 8.3. a Schematic representation of a three-level model accounting for the on–off process taking place at single molecule level. Schematic representation for the determination of **b** first event, **c** single event, and **d** time correlation for the detected photocounts within the user-defined bin time

$$\Phi_{\text{on-off}} = \frac{\Phi_{\text{fluoresc}} \cdot k_{\text{det}}}{\tau_{\text{on}} \cdot I_{\text{on}}} \quad \text{with} \quad I_{\text{on}} = k_{\text{exc}} \frac{k_{\text{fluoresc}}}{k_{\text{on-off}} + k_{\text{fluoresc}} + k_{\text{IC}}} k_{\text{det}} \quad (8.7)$$

if one knows the detection efficiency k_{det} of the set-up and the fluorescence quantum yield Φ_{fluoresc} of the molecule. However, when the signal-to-noise ratio between the on and off levels is relatively low, the results delivered by this method strongly depends on the applied threshold for the discrimination between the two levels. Moreover, on–off processes faster than the experimental bin times restrict the applicability of the method.

Recently, an alternative method was proposed for the determination of the molecular parameters from fluorescence transients measured upon the CW excitation of individual immobilized molecules [30, 31]. The method makes use of the existing link between the statistical characteristics of the excitation and detected photons via the statistical characteristics of the single emitting molecule, as well as of the possibility of measuring the arrival time of each detected photon, i.e., the case of the BIFL type detection. Considering a well-stabilized laser working in CW regime as excitation source, the excitation photons are distributed in time according to a Poisson distribution [32]. Because of the excited state processes taking place in the single molecule, the detected photocounts will follow a temporal distribution distorted as compared to the one of the excitation photons, carrying information about the single molecule. Since the excitation photons appear as well defined "points" on the time axis, the probed molecule can be regarded as a system of point process transformation and the entire SMS experiment can be described in terms of the theory of random point processes [33, 34]. The mathematical formalism of generating functionals (GFL) is an appropriate way for describing point processes [34]. In brief, the GFL formalism leads to a set of functional equations in which the excitation photons (regarded as input) are linked to the detected photocounts (regarded as output) via the characteristics of the single molecule. Building up a GFL including information about the excitation photons, the detection efficiency of the setup and the characteristics of the probed molecule, several statistical characteristics of the output process, i.e., detected photocounts, can be obtained by functional differentiation of the GFL (see Fig. 8.3b–d) [31]. The probability density distributions of the arrival times of the first photocount (first event $a_1\,(t;\Omega)$) and of the single photocount (single event $\pi_1(t;\ \Omega)$), the photon counting distribution per bin time ($P_n(\Omega)$) and the time correlation function between the arrival times of the photocounts in the bin time (time correlation function $f_2\,(t_1 - t_2)$) can be obtained. Since by means of BIFL type detection the arrival times of the detected photocounts can be recorded with an accuracy of 50 ns, user defined bin times with specific time length and containing time-resolved photocounts can be constructed and, moreover, all the above mentioned statistical characteristics can be build up for specific user-defined bin times. By choosing an appropriate model function for the description of the excited state dynamics involving the single molecule, usually a three-level model (see Fig. 8.3a) described by a function of the following type [30, 31]:

$$q(\Delta) = [(1 - p)/\tau_{\text{singlet}}]\exp(-\Delta/\tau_{\text{singlet}}) + (p/\tau_{\text{dark}})\exp(-\Delta/\tau_{\text{dark}}) \qquad (8.8)$$

with Δ the time that the molecule remains in the excited state, p the probability to find the molecule in the dark state (related to the quantum yield of the singlet to dark state transition), τ_{singlet} and τ_{dark} the decay times of the two excited states, specific model fit functions can be constructed for the proposed probability density distributions describing the detected photocounts. Global fitting of the experimental histograms of all four statistical characteristics for specific user defined bin times will deliver the appropriate molecular parameters, i.e., off-time and quantum yield of the on–off process. Finally, the contribution of the background radiation is either taken into account in the GFL as a superposition of two different processes or can be estimated from the region after the bleaching of the single molecule [30].

8.4
Results and Discussion

The g2 dendrimer, chosen here as a probe to demonstrate the capabilities of the proposed multiparametric SMS approach, was previously investigated with respect to its photophysical properties by means of ensemble as well as single molecule spectroscopy [15, 35]. On the basis of ensemble stationary fluorescence, single photon timing and fluorescence upconversion measurements performed in solution a decay scheme for the excited state dynamics taking place in g2 was proposed [35]. In this model, optical excitation in the first single excited state of g2 can lead to emission from isolated, i.e., noninteracting chromophores, decaying with a time constant of 4 ns or to the formation of an excited state complex, an excimer-like species having the emission red-shifted and decaying radiatively with a time constant of 8 ns. Energy transfer within 300 ps was also observed from isolated chromophores to the preformed dimer. In addition to the ensemble studies, a detailed SMS study was recently reported on immobilized g2 [15]. Fluorescence spectra, transients and decay histograms recorded from individual g2 molecules revealed, apart from the ensemble measurements, the presence of on–off processes involving all the chromophores, stepwise bleaching, transitions from less-structured spectra to blue-shifted and vibrationally structured spectra and changes in decay times. Because of the spectral overlap existent between the emission of the monomeric and the excimer-like species, simultaneously detection of fluorescence spectra and decays from individual molecules was performed. In this way structured/unstructured spectra were observed together with short/long decays times and associated, on the basis of the ensemble data, with emission from noninteracting monomer/excimer-like interacting chromophores within an individual g2 dendrimer.

8.4.1
Pulsed Excitation

We applied 488 nm pulsed excitation in combination with simultaneous detection of the fluorescence spectra and two-color BIFL type detection in order to distinguish between the two previously observed emitting species in individual g2 dendrimers [15]. Two-color detection was achieved by means of a of 550/70 nm band-pass filter ("blue" spectral region) and a 600 nm long-pass filter ("red" spectral region). The multiparametric data set including the results obtained from the analysis of the different fluorescence characteristics, simultaneously recorded from an individual g2 immobilized in Zeonex, is displayed in Fig. 8.4. Using the macro-time and spectral information from the BIFL dataset, the two fluorescence transients corresponding to the blue (Fig. 8.4a, *black colored trace*) and red (Fig. 8.4a, *gray colored trace*) spectral regions were build up using a bin time of 30 ms. By disregarding the spectral information from the BIFL dataset, the wavelength integrated fluorescence transient can be built up (Fig. 8.4b). Both the wavelength integrated and the two color fluorescence transients display typical features observed for multichromophoric systems: different emitting levels, on–off jumps between high intensity levels and background level accounting for the existence of dark states and collective effects involving the chromophores within the dendrimer [15, 19, 36, 37]. However, as compared with the wavelength integrated detection, in the case of two-color detection, an additional characteristic of the detected fluorescence is revealed, i.e., the anticorrelation in the time evolutions of the blue and red transients, especially in the beginning of the emission (see Fig. 8.4a from 0 up to 3 seconds). This accounts for fluorescence emitted from different spectral regions during the survival time of the probed molecule either because of eventual spectral jumping or of the presence of different emitting species. Using the micro--time information from the BIFL data set, fluorescence decay histograms were re-

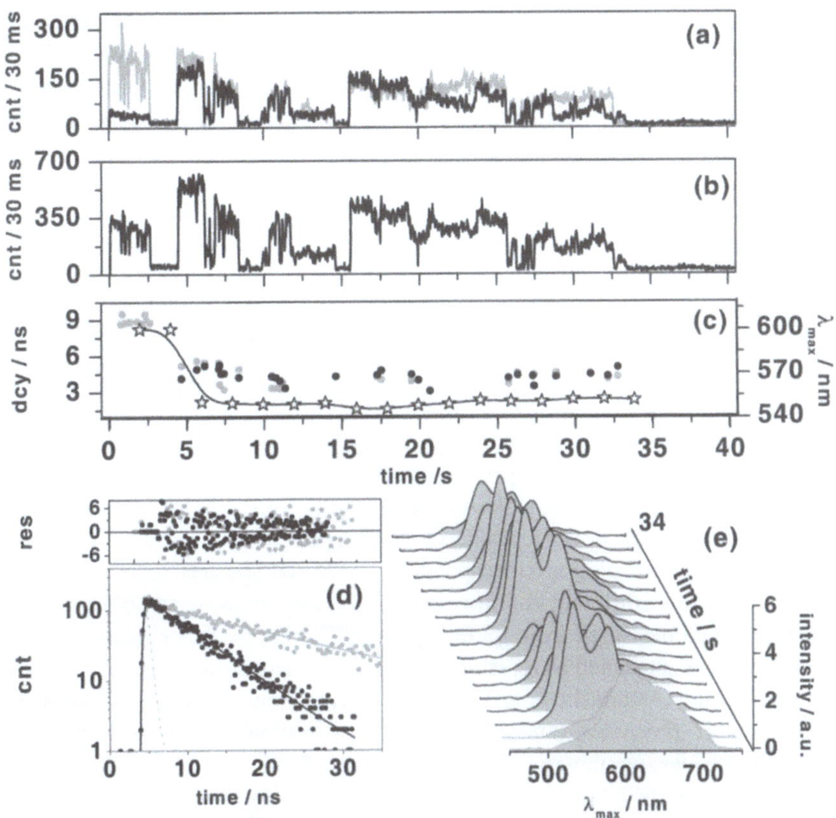

Fig. 8.4. Fluorescence characteristics detected simultaneously from an individual g2 dendrimer immobilized in Zeonex upon 488 nm pulsed excitation: **a** two-color fluorescence transients reconstructed with a bin time of 30 ms from the macro-time and spectral information contained in the measured BIFL data set. The *gray* and *black* colored traces correspond to the red and blue spectral regions, respectively; **b** wavelength integrated fluorescence transient reconstructed with a bin time of 30 ms from the BIFL data set without taking into account the spectral information (see text); **c** time evolutions of the recovered nanosecond decay times for the blue (*black dots*) and red (*gray dots*) spectral regions as well as of the spectral peak position determined from the fluorescence spectra (*line + open stars*); **d** Decay curves reconstructed from the microtime and spectral information contained in the measured BIFL data set by using photocounts detected from 0 to 0.75 s in the red channel (*gray dots*) and from 4.5 to 6.2 s in the blue channel (*black dots*) together with the corresponding fit and residual graphs; **e** time evolution of the fluorescence spectra recorded simultaneously with the BIFL data set using an integration time of 2 seconds

constructed using photocounts belonging to constant regions of intensity within the blue and red transients (see Fig. 8.4a). The time evolution of the recovered nanosecond decay times from the global analysis of the reconstructed decay histograms from both blue and red spectral regions are displayed in Figure 8.4c. For the blue spectral region the decays could be fitted monoexponentially with time constants varying between 3.5 and 5.5 ns, except for the beginning of the emission (0 to 3 s) where a fast subnanosecond component was detected (not shown). For the red spectral region, apart from fluorescence decaying mono exponentially with similar time constants having values between 3.5 to 5.5 ns, fluorescence decaying biexponentially with a long component of about 9 ns and accompanied by a short one of 0.4 ns was observed, mainly in the beginning of the emission (0 to 3 s).

Moreover, the long decay time observed in the beginning of the emission in the red channel was accompanied in the blue channel by a subnanosecond component. Two typical decay histograms together with the corresponding fit and residual graphs and accounting for the short and long nanosecond decay times detected in both channels are depicted in the Fig. 8.4d. For the red spectral region (*gray dots*) the decay histogram was reconstructed by using photocounts detected from 0 to 0.75 s while for the blue one (*black dots*) a time region from 4.5 to 6.2 s was used. While the fluctuations of the recovered decay times between 3.5 and 5.5 ns observed for both red and blue spectral region can account for conformational dynamics of the emitting chromophores and dynamics of the local environment, the jump from about 9 ns down to around 5 ns observed in the beginning of emission in the decay time evolution corresponding to the red spectral region can be explained in terms of different emitting species. Relying on the ensemble single photon timing data measured from g2 in solution [35], the long/short nanosecond decay time can be attributed to red-shifted excimer-like/monomer emission while the subnanosecond component observed in the blue channel accounts for excitation delocalization processes (energy transfer from isolated chromophores to the dimer-like species as suggested in [35]) occurring in the dendrimer. However, the overlap between the emission of the monomer and excimer-like species makes the distinction between the fluorescence characteristics of the two observed species difficult. For this reason, the monomeric fluorescence, i.e., the subnanosecond and short nanosecond components, is observed also in the red channel.

The time evolutions of the fluorescence spectra (Fig. 8.4e; integration time 2 seconds) and of the corresponding spectral peak positions (Fig. 8.4c; *line and open stars*) reveal changes from unstructured spectra peaking at around 590 nm to blue shifted and vibrationally structured spectra peaking around 550 nm, the last ones not detectable in ensemble solution experiments [35]. Simultaneous detection of fluorescence spectra and decays reveals that the red shifted unstructured/structured spectra are observed together with long/short nanosecond decay times. By correlating these observations with those found by means of ensemble experiments, the fluorescence emitted from the dimer-like species was associated with the unstructured/long decay times while the fluorescence emitted from monomer was associated with structured/short ns decay times. Moreover, the anticorrelation between the red and blue transients observed in the beginning of the emission was associated with the presence of excimer-like species in the dendrimer and the transition from the excimer-like to the monomer emission was detected through a long-lived (about 1.5 s) dark state. By observing simultaneously different characteristics for the fluorescence emitted by an individual g2 dendrimer and relying on additional findings from the ensemble measurements, discrimination between two different species was possible even in the condition of a strong overlap between their emission spectra.

8.4.2
CW Excitation

Similar as in the case of pulsed excitation, a multiparametric data set can be obtained when CW excitation is used in combination with two-channel BIFL-type detection. Fig. 8.5 displays the fluorescence characteristics obtained from an immobilized g2 molecule upon 543 nm excitation when polarization sensitive detection is used. By using the macro-time information from the BIFL dataset, the parallel (Fig. 8.5a, *gray colored trace*) and perpendicular (Fig. 8.5a, *black colored trace*) polarized fluorescence components were reconstructed using a bin time of 30 ms and furthermore the total fluorescence transient (Fig. 8.5b) and the polarization degree (p) time trace (Fig. 8.5d) were built up by means of Eqs. (8.4) and (8.5). The G-factor was determined from an additional experiment by using a highly diluted aqueous solution of rhodamine 123. The dynamical processes observed in the wavelength integrated fluorescence transient detected upon pulsed excitation were found back also for the CW regime: different emitting levels, collective on-off behavior leading to long-lived dark states and stepwise bleaching.

Moreover, the polarization sensitive detection reveals the presence of an anticorrelation between the detected I_{par} and I_{perp} components, both in the beginning and at later times during emission. The polarization degree trace reveals much clearer the observed anticorrelation by means of the discrete jumps in the values of p from 0.4 down to –0.4 (see Fig. 8.5d). Simultaneous detection of fluorescence spectra (Fig. 8.5c) and polarized transients (Figs. 8.5a and 8.5d) from this individual molecule reveals the existence of a correlation between the shifts observed in the spectral peak position between 590 and 560 nm and the anticorrelation/discrete jumps observed in the polarized transients/p time traces for the time interval from 0 to 12 s. Since p yields information about the orientation in the in-plane of the transition dipole moment of the investigated molecule, jumps in its values can originate from molecular reorientation or different emitting chromophores/species. Relying on the observations from the pulsed single molecule experiments, the jump in the p-value from 0.4 down to 0.25 observed together with a red shift of the fluorescence peak from about 560 up to about 590 nm in the first 2 seconds of emission can be attributed to the formation of an excimer-like species from a monomer emitting one. Similar, the jump in the p-value from 0.25 down to about 0.1 observed together with a blue shift of the fluorescence peak from about 590 down to about 560 nm after 12 seconds can be attributed to the disappearance of the excimer-like species. The jumps observed in the p-values after 12 s from the beginning of emission but not accompanied by spectral shifts would account for emission either from different chromophores or for eventual molecular reorientations induced by the local environment. Previously, the detailed SMS study on g2 revealed the presence of collective on–off processes leading to dark states lasting

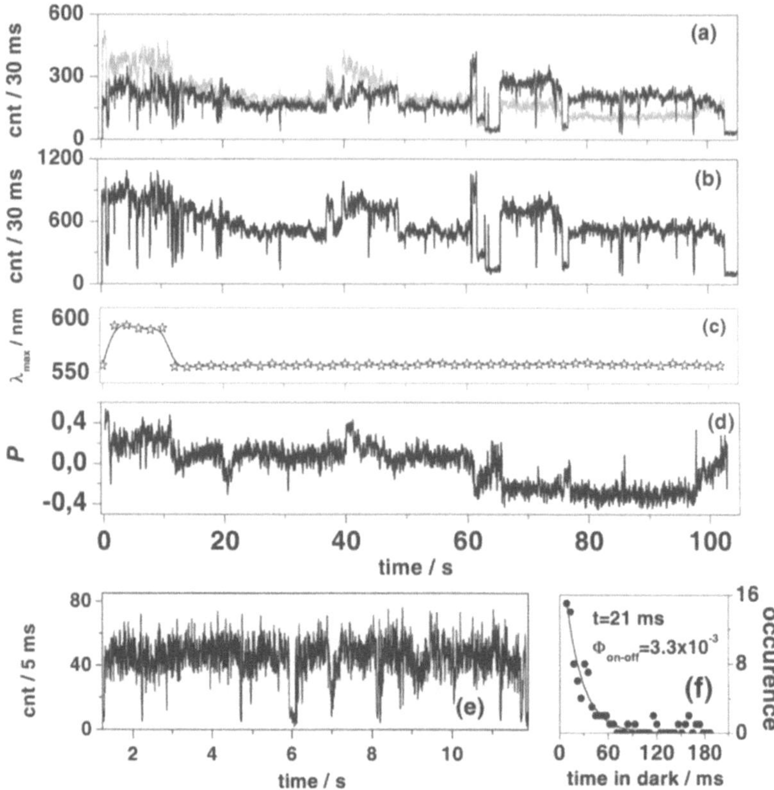

Fig. 8.5. Fluorescence characteristics detected simultaneously from an individual g2 dendrimer immobilized in Zeonex upon 543 nm CW excitation: **a** parallel (*gray*) and perpendicular (*black*) polarized fluorescence components reconstructed with a bin time of 30 ms from the macro-time and polarization information contained in the measured BIFL data set; **b** wavelength integrated fluorescence transient reconstructed from the parallel and perpendicular polarized components displayed in the upper panel according to Eq. (8.4); **c** time evolution of the spectral peak positions determined from the fluorescence spectra recorded simultaneously with the BIFL data set using an integration time of 2 seconds; **d** polarization degree time trace reconstructed from the parallel and perpendicular polarized components displayed in the upper panel according to Eq. (8.5); **e** zoom in the fluorescence transient displayed in Fig. 8.5b for a bin time of 5 ms; **f** results of the duration histogram analysis applied for the region displayed in Fig. 8.5e

from milliseconds to seconds [15]. Similar processes can also be observed in the total transients depicted in Figs. 8.4b and 8.5b. Because for the investigated molecule the S/N within the total transient is relatively high, the duration histogram analysis was applied for the time region where the dimer-like emission was observed (see Fig. 8.5e). An average off time of about 21 ms and a quantum yield for the on–off process of about 3×10^{-3} (see Fig. 8.5f) were recovered for this specific time region of the transient where the excimer-like emission is detected. Since this excited state species is the lowest energetically lying one from the probed dendrimer, the assumption stated in [15] that collective effects involving all the chromophores within one dendritic unit are determined by triplet states acting as nonradiative energy trapping sites is incontestably proved by these experiments. Once the excimer-like species disappears, the number of the detected on–off events becomes relatively small (see Fig. 8.5b) due to the low value of the quantum yield of the on–off processes involving only the monomeric form in g2 [38].

To test the applicability of the point process global analysis method, another multichromophoric system was chosen, the B-PE protein. A previous single molecule study evidenced the presence of collective on–off processes involving all the chromophores [19]. Off times in the timescale of milliseconds as well as high values for the quantum yield of the on–off processes were detected for single immobilized proteins in PVA matrix. We investigated here individual B-PE by using 543 nm CW excitation and polarization sensitive BIFL-type detection. The multiparametric data set obtained from an individual protein is depicted in Figure 8.6. As for the g2 dendrimer, the parallel (Fig. 8.6a, *gray colored trace*) and perpendicular (Fig. 8.6a, *black colored trace*) polarized fluorescence components, the total fluorescence transient and the polarization degree time trace (Fig. 8.6d) were reconstructed from the macrotime information contained in the BIFL dataset using a bin time of 30 ms. The high time resolution of the experiment is exemplified in Fig. 8.6d where a zoom into the total fluorescence transient, reconstructed using a bin time of 4 μs, is depicted. The multichromophoric behavior of B-phycoerithrin is revealed by the different emitting levels, the jumps between high intensity levels and the background level, the anticorrelation between the I_{par} and I_{perp} components concretized in discrete jumps in the values of P, usually accompanied by steps in fluorescence intensity.

In addition, long-lived dark states which are frequently observed in proteins at single molecule level [29, 39] were found for B-PE. Apart from these long dark states, a fast on–off process involving all the chromophores was detected in B-PE (see Fig. 8.6c). By applying the point process global analysis method, the histograms of the first event (Fig. 8.6e), the single event (Fig. 8.6f), the time correlation function (Fig. 8.6g) and the photon counting histogram (Fig. 8.6h) of the detected photocounts were built up for the time region displayed in Fig. 8.6c by using different experimental bins between 0.1 and 2 ms (see Fig. 8.6). Since the longest fluorescence decay time of B-PE detected in ensemble single photon timing experiments ($\approx$ 2.4 ns) [40] is much smaller than the accuracy of the experiment (50 ns), the experimental histogram were globally fitted with a monoexponential model function described by Eq. (8.2) which contained only the time constant related to the on–off process. A global value of 2.18 ms was recovered for the time constant describing the on–off process. In parallel, the same time trace was analyzed by the duration time histogram, method which yielded values between 1.4 to 2.7 ms for a bin time of 0.7 ms, depending on the applied threshold between the on and the off levels.

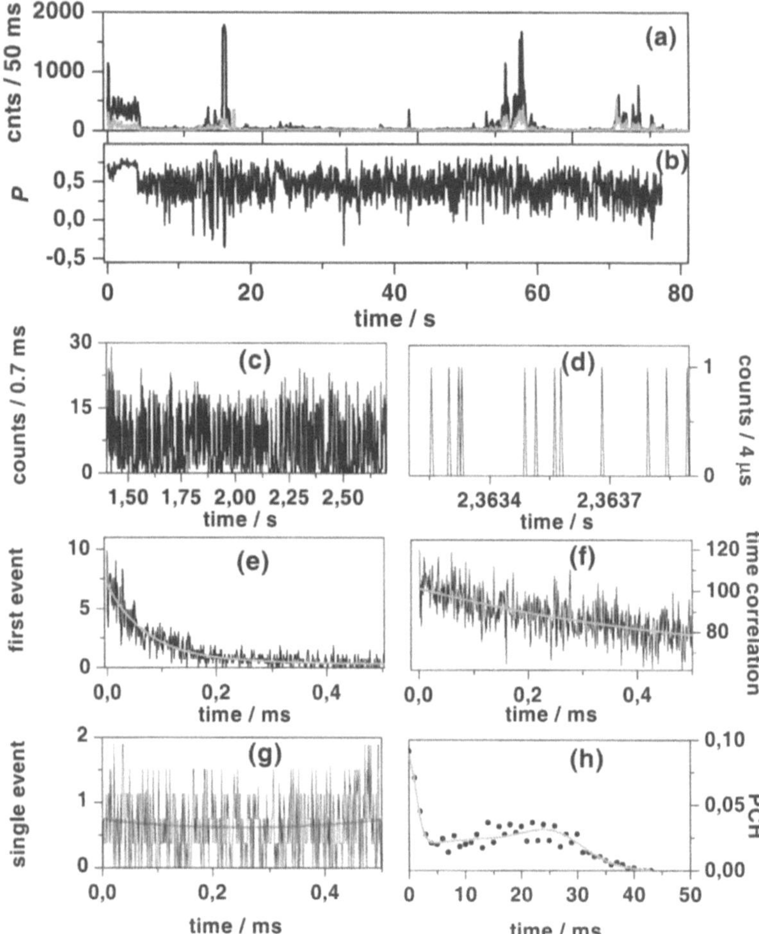

Fig. 8.6. Fluorescence characteristics detected simultaneously from an individual B-PE immobilized in PVA upon 543 nm CW excitation. **a** parallel (*gray*) and perpendicular (*black*) polarized fluorescence components, and **b** *P* trace reconstructed from the macro-time and polarization information contained in the measured BIFL data set (bin time 30 ms); **c** zoom in the wavelength integrated fluorescence transient reconstructed from the parallel and perpendicular polarized components displayed in Fig. 8.6a for a bin time of 0.7 ms; **d** zoom in the wavelength integrated fluorescence transient for a bin time of 4 μs. Fit results of the **e** first event, **f** time correlation, **g** single event and **h** photon counting histograms built up for the region displayed in Fig. 8.6c

The global point process analysis in combination with BIFL-type detection upon CW excitation appears as an alternative for the duration histogram method applied usually to fluorescence transients recorded with conventional counter boards in which the trace consists of a set of equal time intervals (bin times), each of them containing a number of detected photocounts. Since the duration histogram method gives reliable outputs for traces with high S/N and for on–off processes having time constants larger than the bin time of the trace, requirements such as high detection efficiency, molecular systems with high fluorescence quantum yield and undergoing on–off processes in a timescale larger than the accuracy of the experiment are needed. For low fluorescing systems, an alternative would be the increase of the excitation power, leading to fast photobleaching, or the re-binning of the recorded trace leading to loss of information. The method proposed here makes use of the arrival time information of the detected photocounts, offering the possibility to reconstruct the classical fluorescence transient without losing the temporal information inside the bin. Hence, both analysis methods can be applied, depending on the characteristics of the obtained trace. However, knowing the arrival time for each photocount, different statistical characteristics can be build up on the basis of the same model (see Eq. (8.8)) and globally fitted afterwards to yield molecular parameters related to the on-off process. Since the intensity within the bin time of the trace is not used in the analysis, except for the photon counting distribution, the method can be applied for the cases of low excitation power, i.e., long survival time of the investigated molecule, or for low detection efficiency/low emitting systems. Moreover, since the accuracy of the experiment goes up to 50 ns, on–off processes with time constants down to hundreds of ns can in principle be analyzed.

8.5
Conclusions

We have demonstrated that a conventional confocal epi-illuminated microscope equipped with spectrally resolved and BIFL-type detection can offer a multi-parametric observation of different characteristics of the fluorescence emitted from single immobilized molecules, both for the cases of pulsed and CW excitation. By using a tunable picosecond laser system over the VIS-NIR spectral range, simultaneous detection of two color fluorescence transients, fluorescence spectra and ps decay curves can be achieved at the single molecule level for a large variety of synthetic and biological systems. The use of two-color BIFL-type detection in combination with spectrally resolved detection for immobilized molecules allows the construction/analysis of the decay curves in such a way that different molecular states/emitting species can be monitored/distinguished and characterized during the survival time of an individual emitting molecule. In this way selective time-resolved fluorescence spectroscopy can be achieved. Moreover, because two-color detection is implied, FRET studies at the single molecule level can be performed [41]. The SMS set-up can be used without any additional modification for selective time-resolved single molecule studies in magic angle or polarization configuration by using the BIFL approaches proposed in [6, 14]. By using CW excitation, a similar multiparametric picture as for pulsed excitation can be obtained, except for the decay times. The simultaneous acquisition of temporal, spectral and polarization resolved information allows the monitoring of different emitting species, on–off processes, molecular reorientations. The use of BIFL-type detection in combination with CW excitation opens up the possibility to detect two-dimensional (two-color or polarized) fluorescence transients whose photocounts are time-resolved with an accuracy of 50 ns. Global point process analysis of temporal statistical characteristics built up using the arrival times of the detected photocounts delivers reliable molecular parameters for the investigated single molecules. Finally, since the BIFL data set delivers for each detected photocount the macro-time position with an accuracy of 50 ns, fluorescence correlation spectroscopy can also be applied [11].

Acknowledgments. The authors gratefully acknowledge the F.W.O, the Flemish Ministry of Education for the support through GOA/1/96 and GOA 2001/1, the support of DWTC (Belgium) through IUAP-IV-11 and the support of the K.U.Leuven through an IDO project. A very fruitful collaboration within the framework of an EC TMR SISITOMAS with the group of Prof. K. Mullen providing the dendritic molecules used here to describe the methodology.

References

1. Ambrose WP, Goodwin PM, Jett JH, Van Olden A, Werner JH, Keller RA (1999) Chem Rev 99(10):2929
2. Weiss S (1999) Science 283:1676
3. Shera EB, Seitzinger NK, Davis LM, Keller RA, Soper SA (1990) Chem Phys Lett 174: 553
4. Ambrose WP, Goodwin PM, Martin JC, Keller RA (1994) Phys Rev Lett 72:160
5. Prummer M, Hubner CG, Sick B, Hecht B, Renn A, Wild UP (2000) Anal Chem 72: 443
6. Schaffer J, Volkmer A, Eggeling C, Subramanian V, Striker G, Seidel CAM (1999) J Phys Chem A 103:332
7. Sauer M, Drexhage KH, Zander C, Wolfrum J (1996) Chem Phys Lett 254:223
8. Goodwin PM, Ambrose WP, Keller RA (1996) Acc Chem Res 29:607
9. Betzig E, Chichester RJ (1993) Science 262:1422
10. Soper SA, Davis LM, Shera EB (1992) JOSA B 9(10):1761
11. Eigen M, Rigler R (1994) Proc Natl Acad Sci USA 91:5740
12. Macklin JJ, Trautman JK, Harris TD, Brus LE (1996) Science 272:255
13. Dickson RM, Norris DJ, Tzeng YL, Moerner WE (1996) Science 274:966
14. Eggeling C, Fries JR, Brand L, Gunther R, Seidel CAM (1998) Proc Natl Acad Sci USA 95:1556
15. Hofkens J, Maus M, Gensch T, Vosch T, Cotlet M, Köhn F, Herrmann A, Mullen K, De Schryver FC (2000) J Am Chem Soc USA 122:9278
16. Tinnenfeld P, Herten DP, Sauer MJ (2001) Phys Chem A 105:7989
17. Cognet L, Harms Blab GA, GS, Schmidt T (2000) Appl Phys Lett 77:4052
18. Tellinghuisen J, Goodwin PM, Ambrose WP, Martin JC, Keller RA (1994) Anal Chem 66:64

19. Hofkens J, Schroeyers W, Loos D, Cotlet M, Köhn F, Vosch T, Maus M, Herrmann A, Müllen K, Gensch T, De Schryver FC (2001) Sp Acta A 57:2093
20. Morgenroth F, Kubel C, Mullen KJ (1997) Mater Chem 7:1207
21. Maus M, Rousseau E, Cotlet M, Hofkens J, Schweitzer G, Van der Auweraer M, De Schryver FC (2001) Rev Sci Instrum 72(1):36
22. Baker S, Cousin RD (1984) Nucl Instrum Methods 221:437
23. Hall P, Selinger BJ (1981) Phys Chem 85:2941
24. Maus M, Cotlet M, Hofkens J, Gensch T, De Schryver FC, Schaffer J, Seidel CAM (2001) Anal Chem 73(9):2078
25. Janssens LD, Boens N, Ameloot M, De Schryver FC (1990) J Phys Chem 94:3564
26. Stelzer EHK (1995) in Pawley JB (ed) Handbook of biological confocal microscopy. Plemum Press, New, p 142
27. Lakowicz JR (1986) Principles of fluorescence spectroscopy. Plenum Press, New York
28. Yip WT, Hu D, Yu J, Vanden Bout DA, Barbara PF (998) J Phys Chem A 102: 7564
29. Garcia-Parajo M.F, Segers-Nolten GMJ, Veerman JA, Greve J, van Hulst NF (2000) Proc Natl Acad Sci USA 97:7237
30. Novikov E, Boens N, Hofkens J (2001) Chem Phys Lett 338:151
31. Novikov E, Hofkens J, Cotlet M, Maus M, De Schryver FC, Boens N (2001) Sp Acta A 57:2109
32. Saleh B (1978) Photoelectron statistics. Springer, Berlin
33. Cox DR, Isham V (1980) Point processes. Chapman & Hall, London
34. Daley DJ, Vere-Jones D (1988) An introduction to the theory of random point processes. Springer, New York
35. Hofkens J, Latterini L, De Belder G, Gensch T, Maus M, Vosch T, Karni Y, Schweitzer G, De Schryver FC, Hermann A, Mullen K (1999) Chem Phys Lett 304:1
36. Bopp MA, Jia YW, Li LQ, Cogdell RJ, Hochstrasser RM (1997) Proc Natl Acad Sci USA 94:10630
37. Vanden Bout DA, Yip WT, Hu DH, Fu DK, Swager TM, Barbara PF (1997) Science 277:1074
38. A similar dendritic molecule, but bearing only one chromophore (model compound in [15]) was found to display an intersystem crossing quantum yield of about 10^{-7}.
39. Cotlet M, Hofkens J, Köhn F, Michiels J, Dirix G, Van Guisc M, Vanderleyden J, De Schryver FC (2001) Chem Phys Lett 336:328
40. Schroeyers W, Hofkens J, Cotlet M, De Schryver FC (unpublished results)
41. Habuchi S, Hofkens J, Cotlet M, De Schryver FC (submitted)

Fluorescence Intensity Distribution Analysis (FIDA) and related fluorescence fluctuation techniques: theory and practice

P. KASK, C. EGGELING, K. PALO, Ü. METS, M. COLE, AND K. GALL

Fluorescence fluctuation methods are a versatile means of probing molecular interactions with single molecule sensitivity. We review a family of such techniques that are centrally concerned with the monitoring of the fluorescence intensity of dilute, heterogeneous samples through histogramming photon counts in consecutive time intervals. The subsequent fitting of a theoretical model to this intensity data resolves and quantifies different molecular species present in the sample. Each of these methods shares a common confocal experimental set-up with fluorescence detection using avalanche photodiodes. Using the "parent" method, fluorescence intensity distribution analysis (FIDA), it is possible to determine both the concentration and specific brightness values of a number of fluorescent species in solution. FIDA may be extended to enable simultaneous monitoring of two different polarization states or spectral bands (2D-FIDA). Further pertinent molecular information can be revealed through measurement of the fluorescence lifetime or the diffusion coefficient of the sample under investigation. It is possible to extend FIDA to incorporate measurement of either of these properties. In the case of fluorescence intensity and lifetime distribution analysis (FILDA) simultaneous determinations of concentration, brightness and fluorescence lifetime are made. Fluorescence intensity multiple distribution analysis (FIMDA) enables simultaneous measurement of the concentration, brightness and diffusion coefficient. We discuss both the mathematical techniques and experimental considerations that are necessary for successful realization of these methods. The superior statistical accuracy of FIDA and its sister techniques, combined with their fast data acquisition and analysis make them particularly well suited for use in demanding industrial applications. To this end, we present several examples that demonstrate the ability of such methods to fulfill the demands of high throughput screening for drug development.

9.1
Introduction

Since the first reported observation of the phenomenon of fluorescence almost 150 years ago [1], the versatility and relative simplicity of fluorescence measurements have led to an explosive and concomitant growth in both measurement techniques and applications. All of these techniques and applications exploit one or more of the multitude of parameters associated with fluorescence emission (intensity, polarization, wavelength, lifetime, etc.).

Advances in fluorescence microscopy paved the way for the introduction of one such fluorescence-based measurement technique, namely fluorescence correlation spectroscopy (FCS), in 1972 [2], which in turn laid the foundations for a whole new family of methods that are now collectively referred to using the term fluorescence fluctuation spectroscopy (FFS). These techniques exploit fluctuations in the fluorescence signal intensity and, with judicious design of the experimental set-up (e.g., inclusion of confocal optics), enable the detection of individual fluorescent molecules. Fluorescence fluctuation methods are therefore particularly well suited to the probing of biophysical processes, in which the number of biologically significant molecules may well approach the single molecule level [3].

FCS and related FFS techniques rely on the detection of fluctuations in the fluorescence signal due to the passage of single molecules through the detection volume, a task usually rendered impossible by the presence of a high background signal due to the presence of other scattering or fluorescing molecules in the observation volume. This problem can successfully be addressed through the use of a confocal detection geometry which drastically reduces the detection volume (to the femtoliter level), thus reducing the average number of scattering or fluorescing molecules present and consequently suppressing the background signal. Assuming use of a low (~ nanomolar) concentration of fluorophores, the fluorescence peaks that result from diffusion of individual molecules through the detection volume will be clearly visible above the remaining (much reduced) background signal. Analysis of such fluctuations can yield detailed information about both the molecule under investigation and any reactions it may undergo [4, 5].

In the case of FCS, the fluorescence fluctuations are analyzed by calculating their autocorrelation function [2, 6–9]. The autocorrelation function characterizes the average residence time of the molecules in the observation volume. Assuming that the size of the observation volume is known, it may be used together with the residence time to calculate the diffusion coefficient of the species under investigation. More recently, the functionality of FCS has been extended by the inclusion of a second detector. In this technique, known as fluorescence cross correlation spectroscopy (FCCS), the two detectors are configured to monitor and cross correlate two different spectral bands of the fluorescence [10].

Although FCS has evolved into a user-friendly procedure for sensitive analysis of molecular species in solution and molecular processes in and on living cells and tissues, it has limitations that preclude its use for several important biological applications. First, most biologically significant reactions involve the interaction of two or more different species. In order to probe the behavior of multiple fluorescent species using FCS their molecular weights must differ significantly (e.g., typically a difference in molecular weight of at least a factor of 8 is necessary, corresponding to a factor of 2 difference in diffusion coefficient), a condition that is not fulfilled by a number of important processes. Second, FCS is not capable of accurately evaluating cases in which the species have different fluorescence emission rates (i.e., count rate per particle or specific brightness), a case that commonly arises in aggregation studies. Attempts to quantify aggregation using FCS have only been successfully realized through the use of higher order correlation functions of fluorescence intensity fluctuations [11, 12].

In response to the shortcomings of the autocorrelation approach, a number of other FFS methods have been introduced that separate species based on their molecular brightness rather than their molecular weight. An initial solution was proposed that makes use of the moments of photon count number distributions [13, 14]. This approach may be considered as a simplified version of the higher order FCS cited above, since the moments calculated and used in the analysis have the same physical meaning as the amplitudes of the corresponding higher order correlation functions. Subsequently, two research groups independently reported theoretical models capable of directly fitting photon count number histograms [15, 16] and resolving species of different brightness [16, 17]. These methods were given two equivalent names: photon counting histogram analysis (PCH) [15] and fluorescence intensity distribution analysis (FIDA) [16].

As in other FFS techniques, the photon count number histogram-based methods are also centrally concerned with the detection of individual fluorescence photons (single photon counting). In a typical measurement the acquisition time is divided into time windows of equal width (the counting time interval, T) and the number of photon counts detected in each counting time interval is monitored. In FIDA and PCH a frequency histogram of these numbers is constructed and fitted directly. Thus, such theories express the probability of detecting a certain number of photon counts within a time interval T.

Further powerful histogram methods, based on the detection of single-molecule transits, have also been proposed [18, 19]. These methods gather photon count numbers solely from fluorescence bursts due to single molecules. Thus, the disadvantage of these techniques when compared with FIDA and PCH is the prerequisite of an extremely low fluorophore concentration (~pM) and also the need for lengthy data processing that lacks the ability to fit the photon count number histogram of the whole data stream.

The FIDA theory has formed the basis for the development of a series of related methods, namely two-dimensional FIDA (2D-FIDA); [20], fluorescence intensity multiple distribution analysis [21] and fluorescence intensity and lifetime distribution analysis (FILDA); [22], which yield extended molecular information in addition to the molecular brightness. It should also be noted that both 2D-FIDA and FILDA provide extremely high statistical accuracy. The robustness of FIDA data, together with its short data acquisition and processing time make it extremely well suited for application in demanding industrial applications such as high-throughput drug screening (HTS) [23].

This chapter will review the theory of FIDA and its sister theories (Sect. 9.6–9.8). Experimental prerequisites will be discussed (Sect. 9.2), along with mathematical tools (Sect. 9.4) that form the backbone for the FIDA fitting procedure. A brief description of an experimental application of each technique will also be given.

9.2
Instrumentation for Fluorescence Fluctuation Spectroscopy

FFS experiments generally require that there should only be a small number of fluorescent molecules present in the detection volume at any one time, a condition that can be most easily fulfilled through use of a confocal geometry in conjunction with low fluorophore concentrations. A number of other advances have also contributed to the extensive application of FFS methods today, including the increased availability of stable, user-friendly laser sources with a variety of wavelengths (primarily diode lasers and all-solid-state laser systems), the development of efficient correlator electronics along with improvements in computer processing speed and the introduction of high quantum efficiency detectors with broad spectral sensitivity (primarily avalanche photodiodes (e.g., SPCM-AQ-131, EG&G Optoelectronics, Vaudreuil, Quebec, Canada) and, more recently, a new generation of photomultiplier tubes (e.g., H7421-series, Hamamatsu, Japan)). This section will discuss the various considerations necessary for the successful experimental realization of FFS methods. We discuss the generalized set up of a confocal geometry with laser illumination and fluorescence detection using an avalanche photodiode (APD), since this corresponds to the instrumentation used for the cases discussed in later sections.

The use of confocal optics for FFS has been described several times in detail elsewhere [20, 24–26]. The basic principle of confocal fluorescence microscopy is to monitor only a small volume of the sample (the observation volume) in the focal plane of the microscope at any one time. For conventional, single photon excitation, this is achieved by illuminating the sample with a focussed laser beam. Fluorescence emitted by the sample is imaged back to a pinhole by the objective and is subsequently detected. The pinhole may be adjusted to limit the observation volume to a size of the order of one femtoliter. Multiphoton excitation (MPE) greatly facilitates the realization of confocal excitation, since the nonlinearity of the absorption process (i.e., the quadratic dependence of absorption on excitation intensity) confines the fluorescence emission to the focal region of the excitation beam [27], thus a confocal pinhole is no longer required. Additionally, MPE minimizes sample photobleaching and heating effects and also permits imaging of UV-absorbing samples without the need for expensive quartz optics. These features combined with the reduced phototoxicity of near infrared light [28] make MPE particularly well suited for measurement of live cells [29].

It should be noted that the use of a confocal set-up induces a highly inhomogeneous spatial excitation and detection profile. For accurate evaluation of FFS data, this inhomogeneity must be taken into account. The problem may be addressed by considering the "spatial brightness profile", referred to in some cases as the point spread function, which takes such inhomogeneity into account. Considering the case of excitation with a tightly focussed laser beam, we note that the spatial intensity distribution of the excitation light is well described by a Gaussian-Lorentzian beam profile. This case was shown to represent data taken in a photon counting histogram (PCH) measurement with 2-photon excitation accurately, where the square of the Gaussian-Lorentzian illumination intensity profile was used, since this is the expected brightness profile for 2-photon excitation [15]. In the case of conventional confocal spectroscopy, the presence of the confocal detection pinhole must also be taken into account. In this case the spatial brightness function is calculated by multiplying the inhomogeneity of the excitation profile (Gaussian-Lorentzian) with that of the detection profile (defined by the size of the pinhole) [26, 30]. Since inhomogeneity in the spatial brightness profile leads to a distribution of fluorescence intensities that are dependent on the position of the fluorophore within the observation volume, it is imperative that it is correctly accounted for in any FFS-related theory. We choose to define the spatial brightness profile as $B(\mathbf{r})$ which is a product of the excitation intensity and the detection efficiency and is a function of the spatial coordinates, $\mathbf{r}$, of the fluorescent particle in the observation volume.

One possible approach for selecting an appropriate brightness profile is to look for an empirical relationship between the brightness profile and volume, V, which combines spatial sections, dV, of equal B-values. A sufficiently flexible model of $B(\mathbf{r})$ is represented by:

$$\frac{dV}{du} \propto (1 + a_1 u + a_2 u^2)u^{a_3} \tag{9.1}$$

with the variable $u - \ln[B(0)/B(\mathbf{r})]$ and adjustment parameters a_1, a_2 and a_3. The adjustment parameters are equipment dependent and (in our case) typically take values $a_1 \approx -0.4$, $a_2 \approx 0.08$, $a_3 = 1$. It should be noted that Eq. 9.1 was introduced as a means of obtaining good agreement between measured and calculated data and is able to account for both variations in optical design parameters (e.g., pinhole diameter or convergence angle of the laser beam) and also any systematic error that may arise in the adjustment procedure (e.g., beam or pinhole misalignment). Indeed the fit quality is significantly better than that obtained using a Gaussian model [31]. The Gaussian case that represents MPE can however be accounted for using Eq. 9.1 by setting $a_1 = a_2 = 0$, $a_3 = 0.5$.

Further considerations are necessary to account for noise and artifacts that may arise during the photodetection process. Background counts, mainly due to dark counts and Raman scattering of the solvent, can easily be accounted for by a simple control experiment prior to data acquisition. A second problem, namely the dead-time of the APDs, must also be considered and is somewhat more difficult to correct.

Typically, semiconductor photodetectors are unable to detect photons in a time interval of a certain duration (the dead-time) after the previous detection event. In the case of a photodiode with active quenching, the dead-time is approximately 35 ns, imposing a maximum photon detection rate limit of ~30 MHz. The count--rate per molecule of a typical, high quantum yield species in the focus of the microscope is usually around 200–300 kHz (assuming no saturation of the triplet state). Assuming the presence of only a single molecule in the observation volume the count-rate remains well below the 30 MHz limit and the dead-time will not lead to severe artifacts. In the case where samples of high concentration or multiple fluorophore species are studied, resulting in high, ~MHz, count rates, dead-time effects may lead to artifacts in the acquired data. Such artifacts arise from a combination of factors, but manifest themselves most visibly in an artificial lowering of the observed count-rate. Dead-time correction algorithms have been studied in the general context of photon statistics [32]. Although an exact correction of dead-time is possible, the calculation is too lengthy to be viable for industrial applications. We have therefore developed a faster, approximate solution that increases the count-rate range and thus the concentration range that can be used by a factor of 5 or more.

9.3
Assumptions and Conventions

There are a number of important assumptions and conventions, that are used throughout FFS. Before proceeding with a detailed description of FIDA and its related techniques (Sect. 9.6–9.8), this section gives an overview of these common assumptions.

1. Throughout different versions of FFS theory (with only a few exceptions, e.g., [33, 34] different particles are assumed to have independent coordinates, and thus to contribute independently to the fluorescence signal. This assumption permits the use of convolutions for combining the contributions to the photon count number from different particles, species or volume elements. Direct calculation of convolutions is a mathematically clumsy and time-consuming process, however this problem may be circumvented (as described in detail in Sect. 9.4) through the use of generating functions.

2. The duration of the photon-counting time interval is assumed to be short compared to the typical diffusion time of the fluorescent particles through the detection volume. Changes in brightness during a counting time interval are neglected. This assumption has been used (explicitly or otherwise) in all earlier studies including moment analysis [14], FIDA and PCH [16, 35] and 2D-FIDA [20]. This condition was relaxed only for the FIMDA case [21].

3. The light intensity emitted by a particle at a certain position, $\mathbf{r}$, of the inhomogeneous observation volume is expressed as a product of its specific brightness, q (mean count rate per particle), and the spatial brightness function, $B(\mathbf{r})$, (see

Sect. 9.2) only. Other effects such as saturation and triplet effects are neglected, or assumed to be similar for all species. This assumption was explicitly introduced in FCS by Koppel et al. in 1976 [24]. Since then, it has been used for histogram analysis by Qian and Elson [14] and has often been used thereafter. There is a tendency towards modifying this assumption both in correlation and histogram analysis in order to include cases that are not easily described using translational coordinates. Rotational motion [36] and electronic transitions have been accounted for in the theory of FCS [37, 38] and to some extent in the theory of FIMDA [21].

Conventionally, the distribution of count numbers due to the fluorophores present in the observation volume is described using Poisson statistics. Assuming that we divide the observation volume into a number of spatial sections with sizes dV_i, each having an approximately constant value of spatial brightness, B_i, then the distribution of the number of photon counts (i.e., probability of detecting n photon counts within the counting time interval, T) emitted by molecules in the i-th section $P_i(n)$ is expressed as:

$$P_i(n) = \sum_{m=0}^{\infty} P(m)P(n|m) \tag{9.2}$$

where $P(m)$ is the Poissonian distribution of the number of molecules with the mean value cdV_i, and c is the concentration of the molecules. $P(n|m)$ denotes the conditional distribution of the number of photon counts, providing that there are m molecules inside the confocal volume element, dV_i, and is also Poissonian in nature with a mean value mqB_iT where q is the specific brightness, and T is the width of the counting time interval. Thus, $P(m)$ expresses the probability of finding m molecules in the volume section cdV_i, when the overall concentration is c. $P(n|m)$ represents the probability of detecting n photon counts in the time interval T, providing m molecules are present inside dV_i. Therefore the distribution $P_i(n)$ is double Poissonian with two parameters, cdV_i and qB_iT, it can be expressed in full as follows:

$$P_i(n) = \sum_{m=0}^{\infty} \frac{(cdV_i)^m}{m!} e^{-cdV_i} \frac{(mqB_iT)^n}{n!} e^{-mqB_iT} \tag{9.3}$$

Eq. 9.3 describes the ideal case of noninteracting molecules that remain at fixed positions during the counting interval, T. Thus, Eq. 9.3 may only be used for low values of T during which the molecule does not change its brightness significantly as a result of diffusion through the volume element. To account for the full observation volume it is necessary to convolve the contributions P_i from each individual volume element dV_i (see assumption 1, Sect. 9.3).

Having described the aspects of instrumentation and theory that are common to all FFS techniques, we now proceed to examine the theory and practice of FIDA and its related methods in more detail. We begin by addressing the theory of generating functions, a mathematical tool that forms a cornerstone of the FIDA theory.

9.4
Generating Functions for FIDA and Related Methods

The central mathematical task in the FIDA process is the fitting of a theoretical count number distribution to the measured histogram of photon count numbers. The calculation of the theoretical count number distribution must take into account all contributions from each independent source in the observation volume. As mentioned in Sect. 9.3, the individual distributions corresponding to each independent source may be combined, albeit in a mathematically inefficient fashion, using convolutions. A much more efficient approach is offered, however, by the theory of generating functions.

The generating function principle is a convenient mathematical representation that is widely used in addressing problems in mathematical statistics. It has the drawback of lacking a simply understandable physical meaning, but conversely has the advantage of enabling the formulation of theories that would otherwise be prohibitively complex. The use of generating functions in calculating a theoretical photon count number distribution is outlined below.

Let us assume a sequence of probabilities, P, of detecting different numbers of photons under defined conditions (e.g., apparatus configuration, width of counting time interval and sample type):

$$P = (P(0), P(1), ...) \tag{9.4}$$

The generating function of the probabilities is defined as:

$$G(\xi) = \sum_{n=0}^{\infty} P(n)\xi^n \tag{9.5}$$

where ξ is a complex parameter. It is, of course, important that the probabilities can be recovered from $G(\xi)$ of Eq. 9.5. This can be achieved by making the substitution $\xi \rightarrow \exp(i\phi)$ (i.e., effectively restricting ξ to a complex unit circle) and reinterpreting Eq. 9.5 as a Fourier series. The generating function is then expressed as:

$$G(\phi) = \sum_{n=0}^{\infty} P(n)e^{in\phi} \tag{9.6}$$

The probabilities can then be retrieved via the inverse Fourier transform of Eq. 9.6:

$$P(n) = \frac{1}{2\pi} \oint G(\phi)e^{-in\phi}d\phi \tag{9.7}$$

This is particularly convenient for computational purposes, due to the existence of fast Fourier transform (FFT) algorithms [39].

The multidimensional generalization of Eq. 9.5 is straightforward:

$$G(\vec{\xi}) = \sum_{n=0}^{\infty} P(\vec{n})(\vec{\xi})^{\vec{n}} \tag{9.8}$$

where $\vec{n} = (n_1, n_2, ..., n_N)$, $\vec{\xi} = (\xi_1, \xi_2, ..., \xi_N)$ and $(\vec{\xi})^{\vec{n}} = \xi_1^{n_1} \xi_2^{n_{21}} ... \xi_N^{n_N}$.

The most useful property of generating functions for our purposes is that they are able to map a convolution into a product. Consider the case of two distributions, $P^{(a)}(n)$ and $P^{(b)}(n)$, their convolution $P^{(ab)}(n)$ is defined as:

$$P^{(ab)}(n) = (P^{(a)} \otimes P^{(b)})(n) = \sum_{k=0}^{n} P^{(b)}(k)P^{(b)}(n-k) \tag{9.9}$$

where, for example, if $P^{(a)}(n)$ is the probability of collecting n photons from source a and $P^{(b)}(n)$ the probability of collecting n photons from source b, then $P^{(ab)}(n)$ is the probability of collecting the total of n photons from both sources combined. Using the representation of generating functions, this operation is greatly simplified. It can be verified that

$$G^{(ab)}(\xi) = G^{(a)}(\xi)G^{(b)}(\xi) \tag{9.10}$$

The same relationship holds for the multidimensional case with $\xi \to \vec{\xi}$.

The second key property of generating functions is their linearity. Considering an array of conditional probabilities $P(0|\alpha), P(1|\alpha), ...$ where α is an additional variable that the probabilities depend on, the unconditional probabilities may be expressed through $P(\alpha)$ as:

$$P(n) = \sum_{\alpha} P(\alpha)P(n|\alpha) \tag{9.11}$$

A similar formula holds in the generating functions case:

$$G(\xi) = \sum_{\alpha} P(\alpha)G(\xi|\alpha) \tag{9.12}$$

Having demonstrated the linearity of generating functions, and also their ability to conveniently map a convolution into a product, we will now proceed to show how this is of particular use when applied to the problem of calculating theoretical count number distributions in FIDA (Sect. 9.5) and FIDA related methods (Sect. 9.6–9.8).

9.5
FIDA

As mentioned in Sect. 9.4, the central task in the evaluation of FIDA data is the fitting of a theoretical count number distribution to the measured histogram of photon count numbers. The histogram of photon count numbers is acquired by monitoring the signal intensity over time and plotting the frequencies of different numbers of photon counts detected in consecutive time intervals of a given width, T, typically 40 μs. Typical FIDA histograms for simple solutions of two individual dyes and also their mixture are presented in Fig. 9.1.

The theoretical count number distribution shown in Fig. 9.1 for Rh6G was calculated using generating functions as follows. Recall that Eq. 9.5 defines the generating function, $G(\xi)$ of a probability distribution $P(n)$. Furthermore, recall that the distribution of the number of photon counts arising from an ensemble of non-interacting molecules of concentration, c, and specific brightness, q, in a volume element dV with spatial brightness $B(\mathbf{r})$, that are assumed to remain stationary over a counting interval, T, is described by a double Poissonian (Eq. 9.3). By substituting Eq. 9.3 into Eq. 9.5, the generating function $G(\xi; dV)$ for this case is obtained:

$$G(\xi; dV) = e^{-cdV} \sum_{m=0}^{\infty} \frac{(cdV)^m}{m!} e^{-mqBT} \sum_{n=0}^{\infty} \frac{(m\xi qBT)^n}{n!}$$

$$= e^{-cdV} \sum_{m=0}^{\infty} \frac{\{cdV\exp[(\xi-1)qBT]\}^m}{m!} \tag{9.13}$$

$$= \exp\left[cdV\left(e^{(\xi-1)qBT} - 1\right)\right]$$

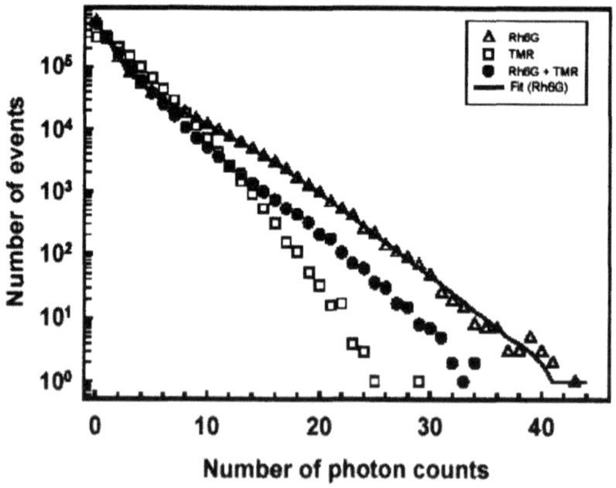

Fig. 9.1. Representative experimental FIDA histograms for Rhodamine 6G (Rh6G) ([Rh6G]=0.5 nM), Tetramethylrhodamine (TMR) ([TMR]=1.5 nM), and a mixture of Rh6G+TMR ([TMR]=0.8 nM, [Rh6G]=0.1 nM) in deionized water. The counting time interval was $T = 40$ μs. A fit to the data (for the case of Rh6G) is performed using Eq. 9.15. The estimated fit parameters are $a_1 = -0.380$, $a_2 = 0.077$, $a_3 = 1$, $c = 0.461$, $q = 107.2$ kHz

where the identity $\sum_n \dfrac{x^n}{n!} = e^x$ is used twice.

$G(\xi)$ for the whole ensemble is calculated by a simple multiplication (rather than the convolution that would otherwise be necessary) of the contributions from different volume elements, dV, and different species (denoted by index j).

$$G(\xi) = \exp\left(\sum_j c_j \int_V \left\{ \exp\left[(\xi-1)q_j TB(\mathbf{r})\right] - 1 \right\} dV \right) \tag{9.14}$$

Using Eq. 9.1, the integral on the right-hand side of Eq. 9.14 can be calculated numerically. Photon counts due to background signal are also Poissonian distributed with a mean background count number; λT; corresponding to the mean count rate; λ; and can be accounted for by inclusion in Eq. 9.14 of a factor $\exp[(\xi-1)\lambda T]$; which is the generating function of the Poisson distribution.

$$G(\xi) = \exp\left((\xi-1)\lambda T + \sum_j c_j \int_V \left\{ \exp\left[(\xi-1)q_j TB(\mathbf{r})\right] - 1 \right\} dV \right) \tag{9.15}$$

After calculation of $G(\xi)$ using Eq. 9.15, the probability distribution is retrieved by carrying out an inverse Fourier transform (see Eq. 9.7).

Fitting to the experimental data is carried out by calculating theoretical photon count number distributions with varying concentration and specific brightness parameters, c_j and q_j, (and in exceptional cases also a_1, a_2, a_3 and λ) and minimizing the deviations between theory and experiment. In most photon histogramming techniques, the quantification of the deviations is usually carried out using one of two different methods [40]. When the number of detected photons is sufficiently high, either the least squares, χ^2, fitting method or the maximum likelihood criterion may be applied. In the case of lower photon count numbers, due to either short data acquisition times or low signal intensities, it is only appropriate to use the maximum likelihood method [41, 42, 43]. The actual fitting is carried out using a Marquardt algorithm with an appropriate weighting as described by [20]. This algorithm also returns the theoretical statistical errors of the estimated parameters. Thus by fitting the photon count distribution in this manner, the concentration and specific brightness of the fluorophores under observation can be determined.

The estimates of the spatial adjustment parameters, a_1, a_2, and a_3, and the background count rate, λ, are determined by control experiments. First, the background is determined from a measurement using a sample of pure solvent. A second control experiment using a single dye solution is then required for characterization of the adjustment parameters. Their values are then fixed for subsequent studies when multi-component samples are analyzed.

An example of an assay ideally suited for FIDA analysis is presented in Fig. 9.2. The binding of a labeled ligand to an antibody expressed on the surface of a bacterium is characterized. Although the accumulation of the ligand on the bacte-

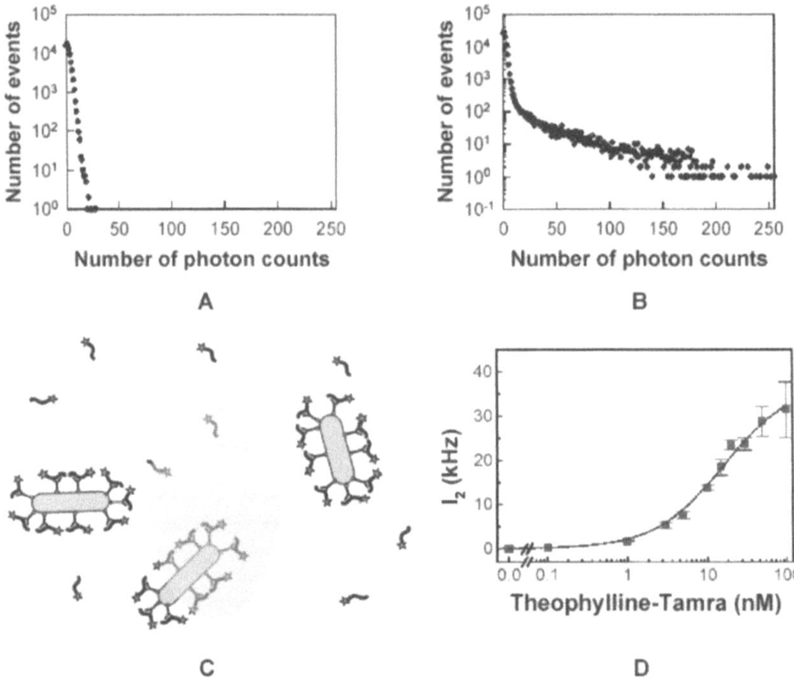

Fig. 9.2. Example assay for FIDA. Binding of Tamra-labeled theophylline to polyclonal antibody linked to the surface of a bacterium by exposed protein A derivative. The principle of the assay is shown in **C**. Upon binding the bacteria accumulate labeled theophylline. The corresponding FIDA data is shown in **A** and **B**. **A** Sample without bacteria solely containing free theophylline. **B** Sample with bacteria containing both bound and free theophylline. The accumulation of theophylline on the bacterial surface is indicated by the increased frequency (in **B**) of events with high photon numbers. A FIDA fit to these curves yields the brightness values q_x of free $(x = 1)$ and bacteria accumulated $(x = 2)$ theophylline: $q_1 = 7.6 \pm 0.3$ kHz and $q_2 = 1620 \pm 27$ kHz. In the subsequent titration series of labeled theophylline these brightness values were both fixed and the readout calculated as light flux from the complex; $I_2 = q_2 \times c_2$, which is shown in **D**. The binding is characterized by a binding constant of $K_D = 16 \pm 2$ nM, determined by a hyperbolic fit to the data in **D**. For further details, see [44]

rial surface does not induce a significant change in the overall fluorescence intensity of the sample. However, since there are on average 100–1000 ligand binding receptors per bacterium, the count rate per particle is enormously increased upon binding, thus monitoring of the reaction is possible using FIDA [44].

Having introduced the basic concept of FIDA, we now proceed to discuss extensions to this method that enable more detailed probing of the sample under investigation.

9.6
2D-FIDA

Many fluorescence applications necessitate experiments using two detectors that monitor two different spectral or polarization components of the fluorescence emission from a single detection volume. This can be realized by inclusion of a second detector in the experimental set up along with an appropriate dichroic or polarization beamsplitter. The extension of the basic FIDA theory to the two detector case is referred to as 2D-FIDA [20]. Pairs of count numbers are collected for each counting time interval, and from this data a two-dimensional histogram of photon count numbers is created.

A typical 2D-FIDA histogram is shown in Fig. 9.3 along with the raw photon count data, i.e., the instantaneous photon count numbers in consecutive time bins, that forms the basis of the 2D-FIDA histogram. In this case, the histogram results from the detection of photons in two different fluorescence emission wavelength ranges (rather than two different polarization states). The analysis is a simple extension of the FIDA procedure described in Sect. 9.5. In this case a theoretical joint distribution of photon count numbers from both detectors must be calculated. Once again values of the concentration, c, and mean count rate per particle of each species for each detector, q_1, and q_2, are estimated. The equipment is characterized by the same brightness profile parameters as in FIDA, but two background count rates (one for each detector) must now be included in the calculation.

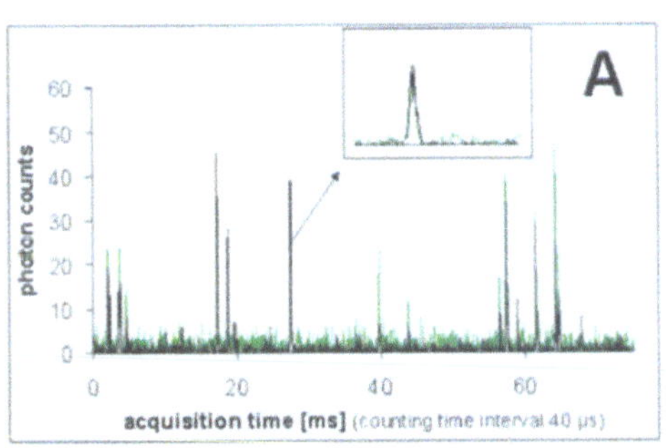

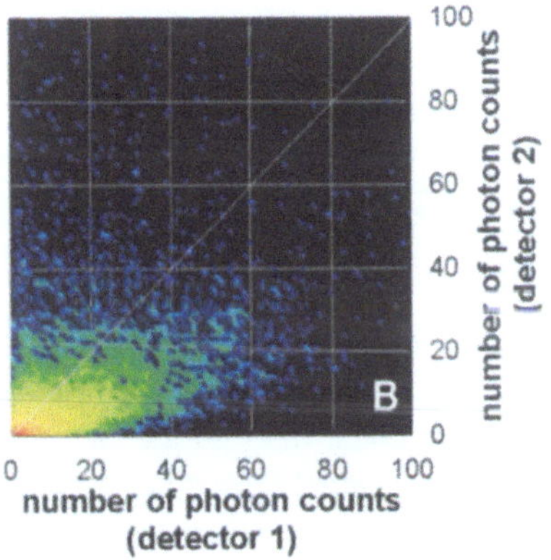

Fig. 9.3. Exemplary 2D-FIDA data **B** with corresponding raw data (i.e., instantaneous count numbers in successive time bins) **A**. The number of photon counts of two detectors (green trace detector 1, black trace detector 2) are registered in consecutive counting time intervals of constant width (40 µs). In the 2D-FIDA histogram **B** the frequency of pairs of photon count numbers registered in the same counting time interval is plotted (color scale). The sample under investigation contained vesicles, which were colored with a membrane dye and accumulated multiple dye-labeled ligands. While the fluorescence emission of the dye-labels was detected on detector 1, the emission of the membrane dye was detected on detector 2. For details, see [20]

In order to generalize Eqs. 9.13–9.15 for the 2D-FIDA case, we now include two photon count numbers, n_1 and n_2, which in turn can be used to obtain the joint photon count number distribution, $P(n_1,n_2)$, i.e., the probability of detecting n_1 photons on detector 1 and n_2 photons on detector 2 within the same counting time interval, T. According to Eq. 9.8 the two-dimensional generating function $G(\xi_1,\xi_2)$ of $P(n_1,n_2)$ now reads:

$$G(\xi_1,\xi_2) = \sum_{n_1=0}^{\infty} \sum_{n_2=0}^{\infty} \xi_1^{n_1} \xi_2^{n_2} P(n_1,n_2) \tag{9.16}$$

Once again, we derive the generating function of the count number distribution from a particular species and a selected spatial section. The contributions from different species and all sections are then multiplied. The distribution of the numbers of photon counts from each species within a volume element, $P_{dV}(n_1,n_2)$ and its generating function $G_{dv}(\xi_1,\xi_2)$ can be expressed in a similar way to Eqs. 9.3 and 9.13.

$$P_{dV}(n_1,n_2) = \sum_m \frac{(cdV)^m}{m!} e^{-cdV} \frac{(mq_1 TB(\mathbf{r}))^{n_1}}{n_1!} e^{-mq_1 TB(\mathbf{r})} \frac{(mq_2 TB(\mathbf{r}))^{n_2}}{n_2!} e^{-mq_2 TB(\mathbf{r})}$$

$$\begin{aligned}
G_{dv}(\xi_1,\xi_2) &= e^{-cdV} \sum_m \frac{(cdV)^m}{m!} e^{-mq_1 B(\mathbf{r})T} e^{-mq_2 B(\mathbf{r})T} \sum_{n_1} \frac{(m\xi_1 q_1 B(\mathbf{r})T)^{n_1}}{n_1!} \sum_{n_2} \frac{(m\xi_2 q_2 B(\mathbf{r})T)^{n_2}}{n_2!} \\
&= e^{-cdV} \sum_m \frac{\{cdV \exp[(\xi_1-1)q_1 B(\mathbf{r})T] \exp[(\xi_2-1)q_2 B(\mathbf{r})T]\}^m}{m!} \\
&= \exp\left[cdV\left(e^{(\xi_1-1)q_1 B(\mathbf{r})T} e^{(\xi_2-1)q_2 B(\mathbf{r})T} - 1\right)\right]
\end{aligned} \tag{9.17}$$

and thus the generating function of the distribution from the whole sample, $G(\xi_1,\xi_2)$ can be written as

$$G(\xi_1,\xi_2) = \exp\left[(\xi_1-1)\lambda_1 T + (\xi_2-1)\lambda_2 T + \sum_j c_j \int_V \left\{ e^{(\xi_1-1)q_{1j}TB(\mathbf{r})} e^{(\xi_2-1)q_{2j}TB(\mathbf{r})} - 1 \right\} dV \right] \tag{9.18}$$

where different species are denoted by the subscript j and λ_1 and λ_2 are the background count rates on the two detectors. The joint photon count distribution is finally recovered through a two dimensional inverse FFT of Eq. 9.18.

2D-FIDA directly determines three specific quantities per fluorescent species in one measurement, namely their absolute concentrations and two fluorescence intensities per molecule. When configured for detection of two polarization states, for example, the anisotropy of the individual molecules under investigation may directly be determined. Using this supplementary information, all participating species may be discriminated and the binding behavior may be quantified. Polarization 2D-FIDA is an ideal tool for the quantitative description of systems exhibiting multiple binding steps, aggregation and multimerization phenomena.

When configured for detection of two different spectral bands of the fluorescence, 2D-FIDA may be used to probe molecular coincidence, for example in vesicle-based ligand binding assays where it is necessary to discriminate fluoro-

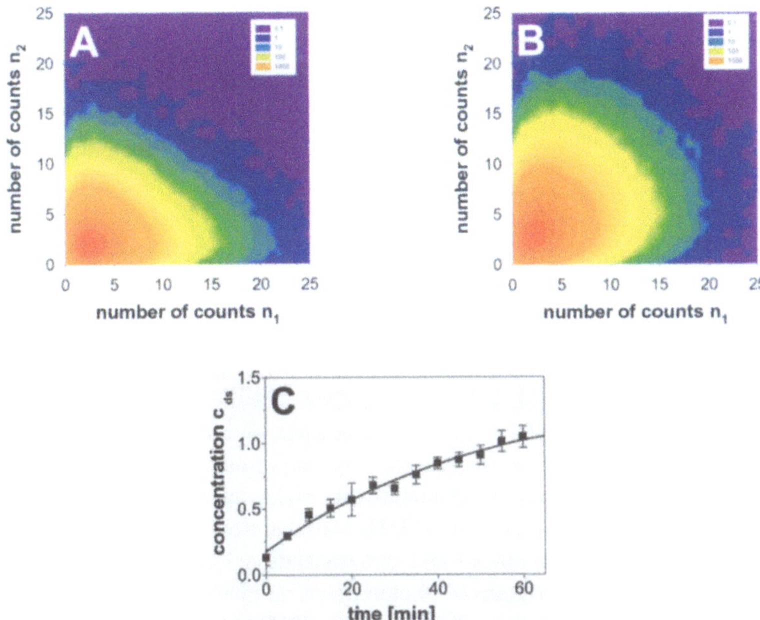

Fig. 9.4. Example assay for 2D-FIDA. Hybridization of two dye-labeled single-stranded oligonucleotides measured over time. Two complementary oligonucleotides (66mer) are labeled with Rhodamine Green (RhGr) and MR121, respectively. Excitation was performed with two lasers simultaneously at 488 nm and 632 nm (corresponding to the absorption peaks of the two dyes. While the RhGr fluorescence is monitored on detector 1 (photon counts n_1), the MR121 fluorescence is registered on detector 2 (photon counts n_2). In the two FIDA histograms presented the colors represent the number of events with a joint photon count number pair (n_1, n_2). The counting time interval was 40 μs and the measurement time 2 seconds. In case **A**, the measurement was performed directly after mixing of the two single strands (each at ~2.5 nM concentration). In case **B**, sample consisted solely of the hybridized double strand (2.5 nM). The amount of hybridization is apparent from the distribution of the counts between the two axes. In **B** the distribution expands between the two axes, corresponding to a simultaneous, coincident fluorescence emission in both wavelength ranges. Conversely, in **A**, the distribution extends predominantly along the axes, corresponding to an independent fluorescence emission in the two wavelength bands. For further details, see [20]. 2D-FIDA histograms were recorded at times 0–60 mins after mixing of the two single strands. The 2D-FIDA fit directly yields the concentration, c_{ds}, of the double-stranded DNA (**C**). A 3-component analysis is applied with the brightness values fixed to those obtained from control samples; MR121-single strand ($q_1 = 0$, $q_2 = 17.9$ kHz), RhGr-single strand ($q_1 = 29.0$ kHz, $q_2 = 0.6$ kHz), double-strand ($q_1 = 20.8$ kHz, $q_2 = 21.6$ kHz). The hybridization is characterized by a time-constant of 51.8 ± 6.8 min as determined by an exponential fit to the data

phore bearing vesicles from free ligand (see Fig. 9.3), or alternatively interactions involving fluorescence resonance energy transfer (FRET), e.g., in cleavage assays.

A typical assay-based application of 2D-FIDA is depicted in Fig. 9.4. Here, 2D-FIDA is used to monitor the hybridization of two differently labeled complementary oligonucleotides. 2D-FIDA provides a clear means of distinguishing the coincident fluorescence that arises in the case of a hybridized double strand. In this way 2D-FIDA enables the direct determination of the concentration of the double stranded DNA.

9.7
FIMDA

As demonstrated in the 2D-FIDA case, the power of FIDA can be increased by introducing a further parameter to jointly characterize and distinguish different fluorescent species. In the case of fluorescence intensity multiple distribution analysis, this is achieved by combining the features of FIDA and FCS [21], using only a single detector for data acquisition. FIMDA is therefore sensitive to changes in both the fluorescence brightness and the translational diffusion. In contrast to FIDA, where a single histogram of photon count numbers is acquired using a single width, T, of the counting time interval, in FIMDA a series of count number histograms with different counting time interval widths are collected in parallel and jointly fitted. Movement of the molecules during a counting time interval, that was explicitly neglected as a central assumption in FIDA and 2D-FIDA (Sect. 9.3, assumption 2), can no longer be ignored in FIMDA. Indeed, quantifying this particle motion is one of the key tasks of the FIMDA theory. Thus for a single acquisition, FIMDA yields the concentration, c, and brightness, q, in addition to the translational diffusion coefficient, D, for each species contributing to the fluorescence. A typical set of FIMDA histograms is shown in Fig. 9.5. Note that with increasing counting interval width, T, higher photon count numbers occur and the count number distribution becomes broader.

Even though the movement of particles during any of the counting intervals is no longer neglected in FIMDA, the assumption that the fluorescence from different particles is independent (assumption 1, Sect. 9.3) is still assumed to be valid. As before, this assumption alone reduces the problem of several particles to that of a single particle, since the individual contributions are easily combined (using the generating function principle).

The movement of a molecule in each counting time interval can be fully characterized by a spatial brightness integral over the random path of the particle. Thus the static spatial brightness function used in Eq. 9.15 must be replaced by a dynamic probability distribution of the path integrated brightness, after which the photon count number distributions may be calculated using the FIDA algorithms, even for time windows exceeding the characteristic diffusion time. A simple approach for accounting for the movement of particles was proposed when FIMDA was introduced [21]. Here, variations in the shape of the distribution of the brightness path integral with changes in the duration of the counting time interval were ignored. Using this approach, the FIDA theory can simply be applied to individual

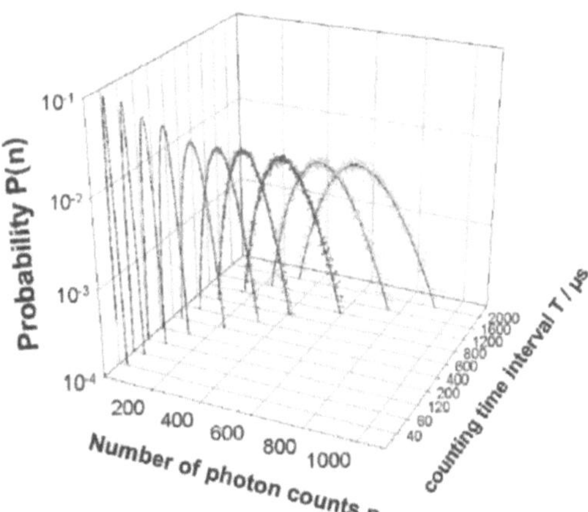

Fig. 9.5. Typical FIMDA data: Series of photon count number distributions, $P(n)$ (scattered plots), recorded at different counting interval widths, T, together with the resulting FIMDA fits. The sample consisted of an aqueous 3.8 nM Cyanine 5 solution (measurement time 2 seconds). The parameters yielded by the FIMDA fit are $c = 3.1 \pm 0.1$, $q = 38.2 \pm 0.8$ kHz, and $D = (3.9 \pm 0.2) \times 10^{-6}$ cm²/s (assuming $\omega_x = 0.75$ µm), while $\lambda = 0.5$ kHz, $a_1 = 0.38$, $a_2 = -0.075$, and $a_3 = 1$ were fixed to values predetermined from control measurements. For details, see [21]

histograms with an apparent concentration, c_{app}, and apparent brightness, q_{app}, whose values vary with the width of the counting time interval.

The individual photon count number distributions, $P(n; T)$, of a FIMDA series, with counting intervals, T, are fitted using Eq. 9.15 with the apparent concentrations, $c_{app,j}(T)$, and brightness, $q_{app,j}(T)$, for different species j.

$$P(n;T) = \mathrm{FFT}^{-1}\, G(\xi;T)$$

$$G(\xi;T) = \exp\left((\xi - 1)\lambda T + \sum_j c_{app,j}(T) \int_V \left\{ \exp\left[(\xi - 1)\, q_{app,j}(T)\, T\, B(\mathbf{r})\right] - 1 \right\} dV \right) \tag{9.19}$$

In order to express the dependency of $c_{app,j}(T)$ and $q_{app,j}(T)$ on the width of the counting time interval, T, we introduce the concepts of factorial moments and factorial cumulants of a probability distribution. Both factorial moments and cumulants have a direct relationship to the generating functions described in Sect. 9.4. The m-th factorial moment, M_m, and factorial cumulant, K_m, are defined in terms of the generating function $G(\xi)$ as:

$$M_m = \sum_n n(n-1)...(n-m+1) P(n) = \left(\frac{\partial}{\partial \xi}\right)^m G(\xi)\Big|_{\xi=1} \tag{9.20a}$$

$$K_m = \left(\frac{\partial}{\partial \xi}\right)^m \ln(G(\xi))\Big|_{\xi=1} = M_m - \sum_{k=1}^{m-1} \binom{m-1}{k} K_k\, M_{m-k} \tag{9.20b}$$

As a result of their relationship to the generating functions, contributions from independent sources to cumulants are additive. Therefore, we will, as before, proceed by analyzing the single species case with generalized values of $c_{app}(T)$ and $q_{app}(T)$ and extend this later to the multi-species case. Eq. 9.20 applied to a single species yields:

$$K_1(T) = M_1(T) = \langle n \rangle_T = c_{app}(T) q_{app}(T) T \tag{9.21a}$$

$$K_2(T) = M_2(T) - M_1(T)^2 = \langle n(n-1) \rangle_T - \langle n \rangle_T^2$$
$$= c_{app}(T) q_{app}^2(T) T^2 \tag{9.21b}$$

where the bracket, $\langle ... \rangle$, denotes the ensemble average. Thus, $c_{app}(T)$ and $q_{app}(T)$ can be expressed in terms of K_1 and K_2:

$$c_{app}(T) = \frac{K_1(T)^2}{K_2(T)} \qquad q_{app}(T) = \frac{K_2(T)}{K_1(T)T} \tag{9.22}$$

Using standard photon statistics identities, we can relate the moments of the photon count numbers, n, to the integrated light intensity $W = \int_0^T I(t)\,dt$

$$\langle n \rangle = \langle W \rangle \tag{9.23a}$$

$$\langle n(n-1) \rangle = \langle W^2 \rangle \tag{9.23b}$$

Assuming a stationary process, one can easily deduce from Eqs. 21a and 23a that

$$c_{app}(T) q_{app}(T) = \langle I \rangle \tag{9.24}$$

where the right-hand-side of Eq. 9.24 is the mean count rate and is independent of T. Under the same assumption we can substitute Eq. 9.23b into Eq. 9.21b and deduce that:

$$K_2(T) = \left\langle \left(\int_0^T I(t)\,dt \right)^2 \right\rangle - \left\langle \int_0^T I(t)\,dt \right\rangle^2 = \left\langle \int_0^T I(t)\,dt \int_0^T I(t')\,dt' \right\rangle - T^2 \langle I \rangle^2$$

$$= \int_0^T dt \int_0^T dt' \langle I(t)I(t') \rangle - T^2 \langle I \rangle^2 = \int_{-T}^T \langle I(0)I(s) \rangle (T - |s|)\,ds - T^2 \langle I \rangle^2 \tag{9.25}$$

$$= \int_{-T}^T \left(\langle I(0)I(s) \rangle - \langle I \rangle^2 \right) (T - |s|)\,ds$$

The results of the theory of FCS express the auto-correlation function of the count rate, $\langle I(0)I(t_c) \rangle - \langle I \rangle^2$, as a product of the (true) mean particle number, c, in the detection volume, the square of the (true) mean count rate per particle, q^2, and a shape function, $G(t_c)$, of the correlation delay time, t_c, which generally depends on different dynamic properties of the particle (translational and rotational diffusion coefficients, singlet-triplet transition rates) as well as on some parameters of the equipment (such as the beam waist radius) [6, 34, 45]:

$$\langle I(0)I(t_c)\rangle - \langle I\rangle^2 = c\,q^2\,G(t_c)\tag{9.26}$$

For example, when accounting only for translational diffusion, $G(t_c)$ for a three--dimensional Gaussian brightness profile with equal beam waist radii in x and y-directions, $w_y = w_x$, but an expanded radius in z-direction, $w_z = a_{\mathrm{axial}}w_x$, is expressed as [45]

$$G_{\mathrm{diff}}(t_c) = \left(1 + \frac{4Dt_c}{w_x^{\,2}}\right)^{-1}\left(1 + \frac{4Dt_c}{a_{\mathrm{axial}}^{\,2}w_x^{\,2}}\right)^{-1/2}\tag{9.27}$$

where D denotes the diffusion coefficient of the observed molecules. When accounting also for triplet trapping, this is extended to:

$$G(t_c) = \left(1 + A_t e^{-t_c/\tau_t}\right)\left(1 + \frac{4Dt_c}{w_x^{\,2}}\right)^{-1}\left(1 + \frac{4Dt_c}{a_{\mathrm{axial}}^{\,2}w_x^{\,2}}\right)^{-1/2}\tag{9.28}$$

where A_t is the "amplitude" and τ_t the relaxation time of the triplet state, as determined by FCS. Substituting Eq. 9.25 into Eq. 9.24 results in the definition of a function, $\Gamma(T)$, which combines the diffusion [38] properties and the counting interval width, T.

$$K_2 = c\,q^2 \int_{-T}^{T} G(s)(T - |s|)\,\mathrm{d}s = c(qT)^2\,\Gamma(T) \quad\text{with}$$

$$\Gamma(T) = \frac{1}{T^2}\int_{-T}^{T} G(s)(T - |s|)\,\mathrm{d}s\tag{9.29}$$

Using Eq. 9.21 and requiring that $cq = c_{\mathrm{app}}(T)q_{\mathrm{app}}(T) = \langle I\rangle$, we are able to relate the true to the apparent brightness and particle numbers, $q_{\mathrm{app}}(T) = q\,\Gamma(T)$ and $c_{\mathrm{app}}(T) = \dfrac{c}{\Gamma(T)}$. These relationships are explained by the following observation.

With increasing time interval width, T, an enhanced number of particles move through the detection volume and thus the apparent particle number increases. In contrast, each particle spends a decreasing fraction of the time T in the detection volume, i.e., the apparent brightness decreases. Note that different species are characterized by different function values of $\Gamma_j(T)$, due to different diffusion coefficients, D_j.

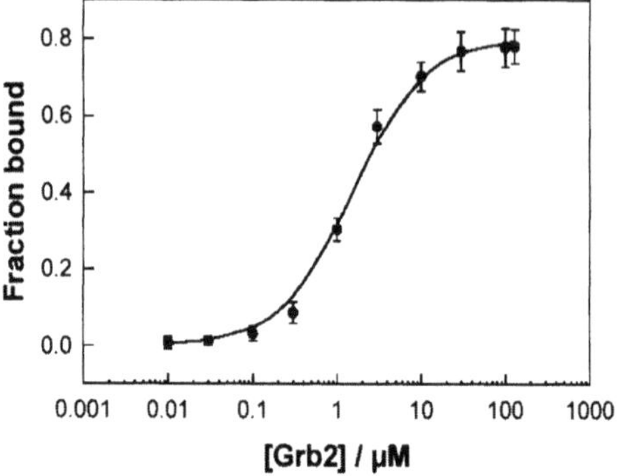

Fig. 9.6. Binding of a small Cyanine 5-labeled peptide to the SH2 domain of the adapter protein Grb2 as measured by FIMDA. Upon titration of Grb2 a series of FIMDA data was recorded and the fraction of bound peptide determined. The FIMDA fit enables direct determination of the fraction of bound peptides, since the brightness (q) and diffusion times ($\tau_{diff} = \omega_x^2/4D$) of both free and bound peptide were pre-determined from control measurements and subsequently fixed in a two-component fit; free peptide $q = 31.7$ kHz, $\tau_{diff} = 407$ μs and bound peptide $q = 39.5$ kHz, $\tau_{diff} = 913$ μs. For details, see [21]

It is worth mentioning that the dependence of photon count number distributions on the dwell time, T, is of interest not only for FIMDA, but also for other related histogram methods. The theory of FIMDA provides us with a correction algorithm for estimating true particle numbers and brightness values rather than just apparent ones. However, the correction requires knowledge of the diffusion times of different species which are not automatically provided when only a single counting time interval width is used, as is typical in FIDA, 2D-FIDA, and FILDA.

An assay that is well suited for analysis using FIMDA is presented in Fig. 9.6. The binding of a small labeled peptide to the SH2 domain of a Grb2 protein is monitored by directly determining the fraction of bound peptide present in the sample. The accuracy of this measurement can be significantly improved using FIMDA, especially when compared with the results obtainable using either FCS or FIDA alone, since each of the bound and unbound ligands are characterized by both different values of specific brightness and the diffusion coefficient.

9.8
FILDA

Fluorescence lifetime is an intrinsic molecular property and, as such, is largely independent of features such as experimental setup and excitation intensity. It is therefore an extremely robust parameter, well suited for resolving multiple fluorescent species. Having discussed the combination of FCS with FIDA in Sect. 9.7, we now proceed to show how fluorescence lifetime analysis (FLA) may be combined with FIDA, a technique that we refer to as fluorescence intensity and lifetime distribution analysis (FILDA).

For our time-domain FILDA acquisition, two main experimental modifications are necessary. First, a pulsed laser source is used to excite the sample. Ideally, the pulse duration should be short in comparison with the lifetime values that one expects to measure (typically, a pulse length of 100ps is sufficiently short). A short excitation pulse results in both improved statistics and also improved time resolution. Second, a modification to the detection electronics is necessary. For every photon that is detected, the excitation-to-detection time delay is also recorded with the aid of a time-to-digital converter (TDC), each delay time being recorded as a bin number. The delay time bins form a discrete array characterized by a given bin width (i.e., resolution time), Δ. In standard FLA, the histogram of the bin numbers (of all detected fluorescence photons) are fitted to a corresponding theoretical distribution from which the fluorescence lifetime values may be retrieved.

The equipment characterization process borrows from that carried out for FIDA with additional considerations to ensure artifact free determination of the lifetime. As in FIDA, the spatial brightness function must be determined. For FLA it is necessary to determine the instrument response function, IRF. The IRF represents the time profile of the laser excitation pulse as recorded by the detector. Both the pulse width and the random error of the photon detection time contribute to the IRF. Additionally, the pulse repetition rate of the excitation laser and the bin width of the TDC are required. Usually, the time-to-digital conversion is not distortion--free, due to electronic imperfections. Thus, the quality of analysis can be improved by accounting for the individual width value of each bin. This is carried out by recording a FLA histogram at constant illumination. Last, two background count rates must be determined, one that is directly related to excitation pulses (scattered component) and a second background that has a fully random detection time not related to excitation pulses (i.e., dark counts of the detector).

After the equipment characterization has been completed a FILDA measurement can be made. A histogram of two stochastic variables, $P(n, \theta)$, is collected and fitted. Here n denotes the number of photon counts in time intervals of a given width, T, (as in standard FIDA measurements) and θ denotes the sum of excitation-to-detection delay time bin numbers over these n photons. The use of the sum of delay bin numbers (rather than each individual delay bin number) for histogram building may at first seem to be an unusual choice. This choice of variable is used since it has been verified that the accuracy of FILDA in resolving multiple fluorescent species is improved when the histogram is built on an integrated rather than a single delay time basis. This fact is illustrated in Fig. 9.7, where the presence of two species can be resolved visually in the integrated distributions, although this is not possible in the case of a single-photon delay bin distribution.

As in FIDA, assumptions 1–3 of Sect. 9.3 are used. Additionally, each fluorescent species is assumed to have a single characteristic excitation-to-detection delay time distribution, which is independent of the delay time recorded for the previously detected photon. Generally, this delay time distribution need not be a mono-exponential function. Nevertheless, for the sake of simplicity we will use a model with mono-exponential decays for each species in the examples presented here. As in FIDA, the central task in FILDA is the fitting of a theoretical probability distribution to the measured histogram. A typical FILDA histogram is pre-

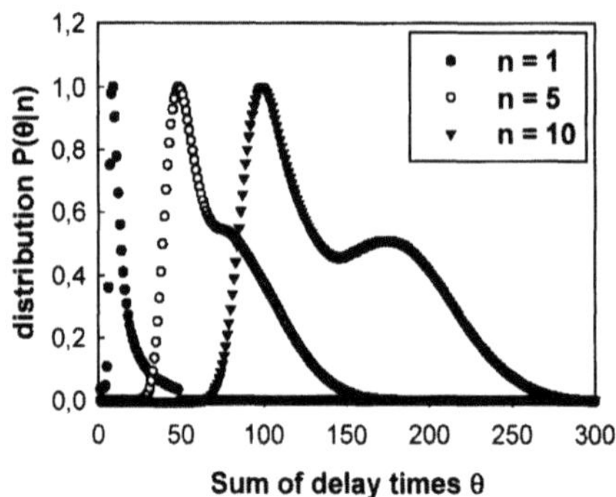

Fig. 9.7. Calculated distributions of the integrated excitation-to-detection delay time numbers, $P(\theta \mid n)$, for different photon numbers, n. The assumed sample comprised a mixture of two species with lifetimes of 1 ns and 4 ns and equal fractional fluorescence count rates. The case of $n = 1$ corresponds to the case of the ordinary FLA distribution, $P^{\mathrm{FLA}}(k)$, of single delay time numbers, k. Integrated delay bin distributions calculated for 5 and 10 photons are also presented. The presence of two species can be resolved visually in the integrated distributions, however this is not possible in the case of the single-photon delay bin distribution

sented in Fig. 9.8. Even rough visual inspection of this histogram confirms the ability of FILDA to successfully resolve multiple fluorescent species.

In order to calculate the theoretical distribution $P(n,\theta)$, we use a method similar to that applied in Sect. 9.4–9.6. First, we express $P(n,\theta)$ as a product of two factors, $P^{\mathrm{FIDA}}(n)$ is the FIDA distribution, i.e., the probability of detecting n photon counts in a certain time-window, and $P(\theta \mid n)$ is the distribution of the sum of delay time bin numbers provided there are n photon counts. The probability $P(n,\theta)$ can be represented as the simple product of both distributions:

$$P(n,\theta) = P^{\mathrm{FIDA}}(n)P(\theta \mid n) \tag{9.30}$$

From this point onwards, the probability distribution of the number of counts and its generating function will be denoted by the superscript FIDA and the distribution of delay time bin numbers and its generating function by the superscript FLA. Thus, the generating function, $G(\xi,\eta)$ of the FILDA probability distribution is defined as:

$$G(\xi,\eta) = \sum_{n=0}^{\infty} \sum_{\theta=0}^{\infty} P(n,\theta)\xi^{n}\eta^{\theta} \tag{9.31}$$

The theory for calculating the FIDA distribution for a single species, $P^{\mathrm{FIDA}}(n)$ was explained in Sect. 9.5, it is given by the inverse Fourier transform of Eq. 9.15. In order to calculate $P(\theta \mid n)$, we must first calculate the expected distribution of delay time bin numbers of single photons, $P^{\mathrm{FLA}}(k)$, for a single species. This has been addressed in other publications [46, 47], nevertheless, we shall give a brief description here. The expected distribution of delay bin numbers is related to the

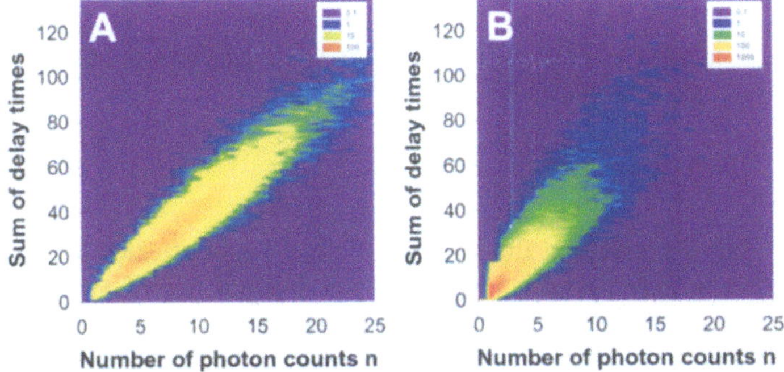

Fig. 9.8. Representative FILDA histograms mapping the joint distribution of the number of photon counts, n, and the integrated excitation-to-detection delay time bins, θ, of these n photons. The colors represent the number of events of a joint photon number pair (n, θ). A pulsed diode laser (repetition rate of 80 MHz and a pulse width of 300 ps) was used to excite aqueous solutions of **A** Cyanine 5 (5 nM), and **B** a mixture of Cyanine 5 (0.7 nM) and Bodipy 650 (0.3 nM). FILDA histograms were recorded for a measurement time of 2 seconds, with counting time intervals of $T = 100$ µs, and using a delay time bin width of $\Delta = 0.56$ ns. The two species may be distinguished through the perturbation they cause to the shape of the FILDA distribution. A FILDA fit to the data resulted in the following values: **A** $c = 5.15 \pm 0.05$, $q = 19.0 \pm 0.2$ kHz, $\tau = 0.73 \pm 0.002$ ns, $a_1 = 0.41 \pm 0.01$, $a_2 = 0.085 \pm 0.002$, $a_3 = 1$ (fixed) and **B** $c_1 = 0.72 \pm 0.008$, $q_1 = 19.5 \pm 0.3$ kHz, $\tau_1 = 0.61 \pm 0.003$ ns and $c_2 = 0.27 \pm 0.005$, $q_2 = 23.3 \pm 0.3$ kHz, $\tau_2 = 3.1 \pm 0.02$ ns, $a_1 - a_3$ fixed to the above values. The background count rates were fixed to predetermined values, $\lambda_{scat} = 0.4$ kHz and $\lambda_{dark} = 0.15$ kHz

photon detection function, $P^{FLA}(t)$, which in turn is the convolution of the experimentally recorded instrument response function, P^{FLA}_{IRF}, and a theoretical fluorescence decay function, $P^{FLA}_{decay}(t)$:

$$P^{FLA}(t) = \left(P^{FLA}_{IRF} \otimes P^{FLA}_{decay} \right)(t) \tag{9.32}$$

In the most simple case, the decay function is monoexponential, with a fluorescence lifetime, τ:

$$P^{FLA}_{decay}(t) = \frac{1}{\tau} e^{-\frac{t}{\tau}} \tag{9.33}$$

Note that the argument, t, of the functions in Eq. 9.32 denotes real time rather than bin number. For the numeric calculation of the convolution, it is recommended to artificially reduce the bin width (i.e., use an increased number of delay time bins) and then, after convolution, to revert back to the cruder ("true") time-bin axis, k. This procedure improves the accuracy of the convolution, since the error of the discrete convolution calculation increases with the width of bin used. The typical bin width value selected in FILDA is a few times higher than that used for FLA in order to circumvent excessive values of the integrated bin number, θ, and to minimize the number of data points to be fitted. The generating function $G^{FLA}(\eta)$ of $P^{FLA}(k)$ is given in the usual way:

$$G^{FLA}(\eta) = \sum_k P^{FLA}(k)\eta^k \tag{9.34}$$

At this point we note that fluorescence lifetime is not affected by variation in the excitation or detection efficiency, i.e., it is independent of variations in the spatial brightness profile. Thus $P^{FLA}(k)$ is dependent on a single parameter, the fluorescence lifetime, τ. Since $P^{FLA}(k)$ for a single species is characteristic for the given species, depending neither on the number of photons emitted or detected previously, not on the coordinates of the molecule emitting the photon, $P(\theta\,|n)$ can be calculated from $P^{FLA}(k)$ by an n-fold convolution, or alternatively in the generating function representation, from the n-th power of the generating function, $G^{FLA}(\eta)$, of $P^{FLA}(k)$

$$P(\theta\,|n) = \left(P^{FLA} \otimes ...(n\text{ times})... \otimes P^{FLA}\right)\theta)$$

$$G(\eta\,|n) = \sum_\theta P(\theta\,|n)\eta^\theta = \left[G^{FLA}(\eta)\right]^n \tag{9.35}$$

The combination of Eqs. 9.30, 9.31, and 9.35 leads to the following expression[1]:

$$G^{FLA}(\xi,\eta) = \sum_n P^{FIDA}(n)\left[G^{FLA}(\eta)\right]^n \xi^n \tag{9.36}$$

$$= G^{FIDA}\left(\xi\,G^{FLA}(\eta)\right) \tag{9.37}$$

Finally, we use Eqs. 9.15 and 9.37 to express the generating function $G(\xi,\eta)$ of $P(n,\theta)$:

$$G(\xi,\eta) = \exp\left[c\int dV\left(\exp\left\{\left[\xi G^{FLA}(\eta)-1\right]qB(\mathbf{r})T\right\}-1\right)\right] \tag{9.38}$$

As mentioned previously, two different "backgrounds", due to dark and scattered counts, must also be accounted for. They may be considered as two additional "species", the count rates of which are Poissonian in nature, with mean count numbers $\lambda_{dark}T$ and $\lambda_{scat}T$ respectively and a generating function as described in Sect. 9.5. The contribution of these species may be calculated by using Eq. 9.35, the distribution of delay times of dark counts $P_{dark}^{FLA}(k)$, which is in

---

[1] According to Eq. 9.36, each column of the $G(\xi,\eta)$-matrix corresponding to a given value is a one-dimensional Fourier transform of the function $P^{FIDA}(n)[G^{FLA}(\eta)]^n$, while according to Eq. 9.37, each element of the $G(\xi,\eta)$-matrix can also be expressed as a Fourier image of $P^{FIDA}(n)$ at the point $\xi\,G^{FLA}(\eta)$.

principle a constant value over all delay time bins, and of scatter counts $P_{\mathrm{IRF}}^{\mathrm{FLA}}(k)$, which is the instrument response function, and their respective generating functions or Fourier images, $G_{\mathrm{dark}}^{\mathrm{FLA}}(\eta)$ and $G_{\mathrm{IRF}}^{\mathrm{FLA}}(\eta)$

$$G_{\mathrm{dark}}(\xi,\eta) = \exp\left[\left(\xi G_{\mathrm{dark}}^{\mathrm{FLA}}(\eta)-1\right)\lambda_{\mathrm{dark}}T\right] \tag{9.39}$$

$$G_{\mathrm{scat}}(\xi,\eta) = \exp\left[\left(\xi G_{\mathrm{scat}}^{\mathrm{FLA}}(\eta)-1\right)\lambda_{\mathrm{scat}}T\right] \tag{9.40}$$

For multiple species including both types of background counts, Eqs. 9.38–9.40 combined yield the product of all contributions:

$$G(\xi,\eta) = \exp\left[\begin{array}{l}\left(\xi G_{\mathrm{dark}}^{\mathrm{FLA}}(\eta)-1\right)\lambda_{\mathrm{dark}}T + \left(\xi G_{\mathrm{scat}}^{\mathrm{FLA}}(\eta)-1\right)\lambda_{\mathrm{scat}}T \\ + \sum_j c_j \int dV\left(\exp\left\{\left[\xi G_j^{\mathrm{FLA}}(\eta)-1\right]q_j B(\mathbf{r})T\right\}-1\right)\end{array}\right] \tag{9.41}$$

where the subscript, j, denotes contributions from different species. Once again, calculation of the Fourier transform of Eq. 9.41 completes the calculation of the theoretical distribution function, $P(n,\theta)$. Thus by fitting the FILDA distribution calculated in this manner, the concentration, specific brightness and the fluorescence lifetime of the fluorophores under observation can be determined from a single, one-detector experimental set-up.

In order to demonstrate the features of FILDA, a binding assay which causes a change in lifetime and brightness is shown in Fig. 9.9. The ability to distinguish a bound and unbound state through two complementary parameters enables a much

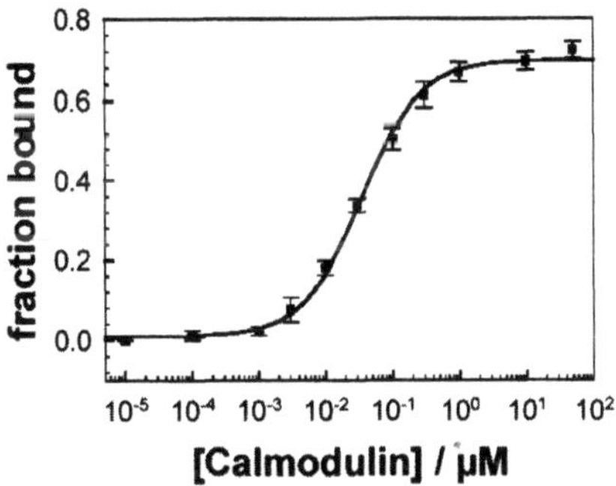

Fig. 9.9. Binding of a MR121-labeled peptide to the calmodulin protein as monitored by FILDA. Upon binding the fluorescence brightness as well as the lifetime of the label changes as determined from control measurements of unbound ($q_1 = 6.5$ kHz, $\tau_1 = 1.9$ ns) and bound ligand ($q_1 = 16.7$ kHz, $\tau_1 = 3.29$ ns). A subsequent two-component fit to FILDA data obtained from a titration series of calmodulin (with brightness and lifetime fixed to the above values) directly resulted in the fraction of bound peptide. A hyperbolic fit to the data (*solid line*) yields a binding constant for the calmodulin–peptide interaction of $K_{\mathrm{D}} = 34 \pm 3$ nM. For further details, see [22]

more robust and accurate determination of the bound fraction. This improved robustness enables us to apply FILDA at acquisition times as short as 100 ms, an aspect that is clearly of great importance for HTS applications.

9.9
Conclusions

This paper has reviewed the technique of FIDA, a fluorescence fluctuation measurement and analysis technique that due to its speed of acquisition and analysis, its versatility and also its robustness, is extremely well suited for application not only in basic research but also in demanding industrial applications. The mathematical foundations of FIDA have been described in detail, along with associated experimental considerations. We have described the extension of FIDA to acquisition with two detectors (2D-FIDA), which enables probing of fluorescence polarization, two color coincidence and FRET phenomena. Combination of FIDA with the measurement principles of FCS and fluorescence lifetime has also been discussed, yielding the techniques FIMDA and FILDA, respectively. As confirmation of the applicability of these techniques outside of specialist research laboratories, examples from the field of assay development for high throughput drug screening have been discussed in each case.

The future progress of FIDA and its related methods is likely to be dominated by associated experimental developments and also the requirements of its users. The ongoing developments in solid-state pulsed laser development, will hopefully enable the application of FILDA across a wider range of wavelengths than is currently commercially viable. Such developments are also likely to facilitate the use of two-photon excitation for FIDA and its sister techniques. This would be highly desirable for live-cell measurement, due to the associated reduction in photodamage. In cellular screening applications, the use of imaging is becoming increasingly important for visualization of important biological processes, suggesting a new challenge that FIDA and its sister theories are likely to have to address in the near future.

Acknowledgement. The authors gratefully acknowledge Drs Leif Brand, Stefan Jäger, Sabine Schärtl, and Joachim Krämer for valuable contributions.

References

1.   Stokes GG (1852) On the change of refrangibility of light. Phil Trans R Soc London 142:463–562
2.   Magde D, Elson EL, Webb WW (1972) Thermodynamic fluctuations in a reacting system-measurement by fluorescence correlation spectroscopy. Phys Rev Lett 29:704–708
3.   Eigen M, Rigler R (1994) Sorting single molecules: application to diagnostics and evolutionary biotechnology. Proc Natl Acad Sci USA 91:5740–5747

4. Keller RA, Ambrose WP, Goodwin PM, Jett JH, Martin JC, Wu M (1996) Single-molecule fluorescence analysis in solutions. Applied Spectroscopy 50:12A–32A
5. Xie XS, Trautman JK (1998) Optical studies of single molecules at room temperature. Annu Rev Phys Chem 49:441–480
6. Elson EL, Magde D (1974) Fluorescence correlation spectroscopy. I. Conceptual basis and theory. Biopolymers 13:1–27
7. Thompson NL (1991) Fluorescence correlation spectroscopy In: Lakowicz JR (ed) Topics in fluorescence spectroscopy. Vol. 1: Techniques. Plenum Press, New York, pp 337–378
8. Widengren J, Rigler R (1998) Fluorescence correlation spectroscopy as a tool to investigate chemical reactions in solutions and on cell surfaces. Cell Mol Biol 44:857–879
9. Visser AJWG, Hink MA (1999) New perspectives of fluorescence correlation spectroscopy. J Fluoresc 9:81–87
10. Schwille P, Meyer-Almes FJ, Rigler R (1997) Dual-color fluorescence cross-correlation spectroscopy for multicomponent diffusional analysis in solution. Biophys J 72:1878–1886
11. Palmer AG, Thompson NL (1987) Molecular aggregation characterized by high order autocorrelation in fluorescence correlation spectroscopy. Biophys J 52:257–270
12. Palmer AG, Thompson NL (1989) High-order fluorescence fluctuation analysis of model protein clusters. Proc Natl Acad Sci USA 86:6148–6152
13. Qian H, Elson EL (1990) Distribution of molecular aggregation by analysis of fluctuation moments. Proc Natl Acad Sci USA 87:5479–5483
14. Qian H, Elson EL (1990) On the analysis of high order moments of fluorescence fluctuations. Biophys J 57:375–380
15. Chen Y, Müller JD, So PT, Gratton E (1999) The photon counting histogram in fluorescence fluctuation spectroscopy. Biophys J 77:553–567
16. Kask P, Palo K, Ullmann D, Gall K (1999) Fluorescence-intensity distribution analysis and its application in biomolecular detection technology. Proc Natl Acad Sci USA 96:13756–13761
17. Müller JD, Chen Y, Gratton E (2000) Resolving heterogeneity on the single molecular level with the photon-counting histogram. Biophys J 78:474–486
18. Fries JR, Brand L, Eggeling C, Köllner M, Seidel CAM (1998) Quantitative identification of different single molecules by selective time-resolved confocal fluorescence spectroscopy. J Phys Chem 102:6601–6613
19. Eggeling C, Berger S, Brand L, Fries JR, Schaffer J, Volkmer A, Seidel CA (2001) Data registration and selective single-molecule analysis using multi- parameter fluorescence detection. J Biotechnol 86:163–180
20. Kask P, Palo K, Fay N, Brand L, Mets Ü, Ullmann D, Jungmann J, Pschorr J, Gall K (2000) Two-dimensional fluorescence intensity distribution analysis: Theory and applications. Biophys J 78:1703–1713
21. Palo K, Mets Ü, Jäger S, Kask P, Gall K (2000) Fluorescence intensity multiple distributions analysis: concurrent determination of diffusion times and molecular brightness. Biophys J 79:2858–2866
22. Palo K, Brand L, Eggeling C, Kask P, Gall K (2002) Fluorescence intensity and lifetime distribution analysis: Towards higher accuracy in fluorescence fluctuation spectroscopy. Biophys J (submitted)
23. Ullmann D, Busch M, Mander T (1999) Fluorescence correlation spectroscopy-based screening technology. Inn Pharm Tech 30–40

24. Koppel DE, Axelrod D, Schlessinger J, Elson EL, Webb WW (1976) Dynamics of fluorescence marker concentration as a probe of mobility. Biophys J 16:1315–1329
25. Rigler R, Widengren J (1990) Ultrasensitive detection of single molecules by fluorescence correlation spectroscopy. BioScience 40:180–183
26. Rigler R, Mets Ü, Widengren J, Kask P (1993) Fluorescence correlation spectroscopy with high count rate and low background: analysis of translational diffusion. Eur Biophys J 22:169–175
27. Denk W, Strickler JH, Webb WH (1990) Two-photon laser scanning fluorescence microscopy. Science 248:73–76
28. Hockberger PE, Skimina TA, Centonze VE, Lavin C, Chu S, Dadras S, Reddy JK, White JG (1999) Activation of flavin-containing oxidases underlie slight-induced production of H_2O_2 in mammalian cells. Proc Natl Acad Sci USA 96:6255–6260
29. Piston DW (1999) Imaging living cells and tissues by two-photon excitation microscopy. Trends Cell Biol 9:66–69
30. Qian H, Elson EL (1991) Analysis of confocal laser-microscope optics for 3-D fluorescence correlation spectroscopy. Appl Opt 30:1185–1195
31. Kask P, Palo K (2001) Introduction to the theory of fluorescence intensity distribution analysis. In: Rigler R, Elson EL (eds) Fluorescence correlation spectroscopy: Theory and applications. Springer, Berlin, Heidelberg, New York, pp 396–409
32. Saleh B (1978) In: MacAdam DL (ed) Photonelectron statistics. Springer, Berlin, Heidelberg, New York
33. Phillies GDJ (1975) Fluorescence correlation spectroscopy and nonideal solutions. Biopolymers 14:490–508
34. Abney JR, Scalettar BA, Hackenbrock CR (1990) On the measurement of particle number and mobility in nonideal solutions by fluorescence correlation spectroscopy. Biophys J 58:261–265
35. Chen Y, Müller JD, Tetin SY, Tyner JD, Gratton E (2000) Probing ligand protein binding equilibria with fluorescence fluctuation spectroscopy. Biophys J 79:1074–1084
36. Ehrenberg M, Rigler R (1974) Rotational brownian motion and fluorescence intensity fluctuations. Chem Phys 4:390–401
37. Kask P, Piksarv P, Mets Ü (1985) Fluorescence correlation spectroscopy in the nanosecond time range: photon antibunching in dye fluorescence. Eur Biophys J 12:163–166
38. Widengren J, Mets Ü, Rigler R (1995) Fluorescence correlation spectroscopy of triplet states in solution – a theoretical and experimental study. J Phys Chem 99:13368–13379
39. Brigham EO (1974) The fast fourier transform. Prentice-Hall, Englewood Cliffs
40. Baker S and Cousins RD (1984) Clarification of the use of chi-square and likelihood functions in fits to histograms. Nuclear Instruments & Methods in Physics Research 221:437–442
41. Hall P, Selinger B (1981) Better estimates of exponential decay parameters. J Phys Chem 85:2941–2946
42. Brand L (1998) Zeitaufgelöster Nachweis einzelner Moleküle in Lösung. Cuvillier verlag, Göttingen
43. Köllner M, Wolfrum J. (1992) How many photons are necessary for fluorescence-lifetime measurements? Chem Phys Letters 200:199–204

44. Schaertl S, Meyer-Almes FJ, Lopez-Calle E, Siemers A, Krämer J (2000) A novel and robust homogeneous fluorescence-based assay using nanoparticles for pharmaceutical screening and diagnostics. J Biomol Screen 5:227–238
45. Aragón SR, Pecora R (1976) Fluorescence correlation spectroscopy as a probe of molecular dynamics. J Chem Phys 64:1791–1803
46. Grinvald A, Steinberg IZ (1974) On the analysis of fluorescence decay kinetics by the method of least-squares. Anal Biochem 59:583–598
47. Lakowicz JR (1983) Principles of fluorescence spectroscopy. Plenum Press, New York

Single Molecule Reactions of the Enzyme LDH and of Restriction Endonucleases in the Fluorescence Microscope

B. NASANSHARGAL, B. SCHÄFER, AND K. O. GREULICH

Two types of single molecule enzyme reactions can be directly observed in the fluorescence microscope: reactions, which convert nonfluorescing small substrate molecules into fluorescing products (or vice versa) and reactions of enzymes on macromolecules stained by a fluorescence dye or visualized otherwise. As an example of the first type of reaction, the conversion of nonfluorescent NAD^+ into fluorescing NADH, or vice versa, by a few molecules of lactate dehydrogenase in femtodroplets is described. The femtodroplet-pipetting method is essentially a subattomol technique with high accuracy. Lineweaver Burk plots are obtained with approximately the kinetic constants of the enzyme known from conventional biochemistry. On the other hand, the femtodroplet-in-substrate method allows the observation of the action of individual enzyme molecules. The second type of single molecule enzyme reactions is the sequence-specific cutting of individual DNA molecules held by optical tweezers. It is shown that such molecules can be characterized by the cutting (restriction) pattern generated by the restriction endonucleases *ApaI*, *SmaI* and *EcoRI*.

10.1
Introduction

With the advent of single molecule detection an old dream of spectroscopists, observation of molecular spectra without inhomogeneous line broadening, has become reality. Analytical (bio-)chemistry is now on the verge also to make use of the enormous chances that are offered by single molecule techniques. The direct observation of single molecule reactions is becoming possible, i.e., the ultimate limit of analytical (bio-)chemistry is going to be reached. What this means may be illustrated in a simple example: provided one has a single molecule sensor that can distinguish a specific type of molecule from all other molecules in the environment (human taste and smell sensors are probably of that quality), one can detect a source of molecules that emits a few liters of material at a distance of hundreds of kilometers (for quantitative details, see[1]), for example a person exhaling air with specifically smelling molecules. This can explain why some people are believed to "smell" friends over large distances, and it also may explain part of the enormous abilities of some animals to find their way over long distances. So far poorly understood observations may thus be transferred from the realm of parapsychology to strict single molecule biochemistry. To achieve this, knowledge of single molecule reactions of taste or smell sensors would be required, which is so far not yet available. However, a first step towards this goal, the single molecule study of more commonly known enzyme reactions, is now possible and one of these studies will be reported in the present contribution.

A second type of reaction aims even more at molecular individuality: when fluorescently labeled DNA is cut by sequence-specific enzymes (restriction endonucleases), a specific cutting pattern will be generated. Modifications (mutations) of the recognition site will result in a different cutting (restriction) pattern, by which two individual molecules can be distinguished. As a bulk technique, using a large number of molecules, such a restriction analysis is the basis for the identification of individuals, for example in paternity tests or in crime cases. As a single molecule technique the true individuality of DNA molecules can be addressed. How important it is to regard DNA as a molecular individual becomes clear when one realizes that in order to synthesize only one copy of each possible short DNA molecule of 120 base pairs in length, the visible mass of the universe would not suffice (for quantitative details, see ref. [1], page 162). This shows that

---

[1]   A person exhales every three seconds approx. 3 L of air. Using for this approximate discussion noble gas data (22.4 L = 1 mol = 6×10^{23} molecules) this corresponds to 2.7×10^{22} molecules per second. When these molecules are evenly distributed by convection into a cylinder of 100 km in radius and 10 km (thickness of Earth's atmosphere) in height (3.14×10^{17} liter) this corresponds to a flow of 0.9×10^5 molecules per second of the exhaled air into each liter of air at a distance of 100 km. If only 10 ppm of these molecules have a specific smell, a smell receptor will detect a flow of one molecule per second. If the person starts to move in direction of the detector, a flow gradient will be noticed which detects this motion at a distance of 100 km, i.e., the person's future arrival can be "anticipated".

sequencing the human genome is only a first step for understanding inter-individual differences in the human genome. In order to understand diseases throughout large populations single molecule techniques are badly needed. The second part of the present contribution presents first steps towards the goal of single molecule DNA restriction analysis.

10.2
Femtodroplets and the Poisson Statistics

Single molecule enzyme reactions can be best interpreted, when the experiments are performed at conditions as close as possible to those in bulk reactions, i.e., when the experiments just represent a down scaling of conventional biochemistry. Particularly then, for example single molecule kinetic parameters such as reaction rates or "Michaelis Menten constants" can be directly compared with values known from conventional biochemistry. The most straightforward strategy to do that is to perform the experiments at low enzyme concentrations and in minute reaction volumes so that only one or a few enzyme molecules are in the observed volume. For example, a 1 nanomolar (6×10^{14} molecules per liter) enzyme concentration corresponds to somewhat less than one molecule per femtoliter. The latter is a droplet with linear dimensions of approximately one micrometer and can be generated by pipetting with commercially available equipment well known from patch clamp or microinjection techniques.

The technique has some disadvantages, which, however, can be alleviated by suitable data evaluation. If for determination of the kinetic constants of an enzyme a substrate concentration in the low micromolar range is required, only a few thousand substrate molecules per enzyme molecule are available, i.e., substrate consumption during the reaction has to be taken into consideration. A second disadvantage of such femtoliter droplet techniques is, that the true number of enzyme molecules in a femtodroplet is given by Poisson statistics. If, on average, one molecule is expected to be present in a droplet, the true distribution is:

$$P(n) = 1^n / n! \times e^{-1} \tag{10.1}$$

where n is the number of molecules in the droplet, $P(n)$ the probabilty of occurrence of such a number in a droplet. The percentage is $100 \times P(n)$. e $= 2.71$, $n! = 1 \times 2 \times \ldots \times n$, with 0! defined as 1. The probability of finding no or one molecule in the respective droplet is 37% each ($1/e = 0.369$). In 26% of the droplets, more than one molecule will be found.

Similarly, if on average two molecules are expected, the corresponding distribution is:

$$P(n) = 2^n / n! \times e^{-2} \tag{10.2}$$

For a mean value of two molecules per droplet the probability of "empty" droplets is 13.6%. Droplets with one and two molecules are found with 27.2% each.

Three molecules are found in 18.1%, four molecules still in 9% of the droplets. In approximately 5% of all droplets, five or more molecules are expected.

This Poisson statistics on the one hand complicates the evaluation of single molecule experiments. In turn, if one measures reaction rates for, say, ten reactions under identical conditions and finds Poisson statistics verified, one is on the safe side to assume true single molecule conditions. This has not been considered in a number of single molecule enzyme reactions published so far, and enormous differences in individual reaction rates have been invoked. Probably, at least part of these large differences in individual reaction rates can be attributed to Poisson statistics.

10.3
From Concentrations to Intermolecular Distances

For discussions on single molecule reactions it is often more convenient to use the microscopic intermolecular distance "d" for a concentration instead of the macroscopic concentration value c. The equation for conversion can be derived as follows:

$$c' = c \text{ [mol/L]} \times 0.001 \text{ [L/cm}^3\text{]} \times 6 \times 10^{23} \text{ [molecules/mol]} = 6 \times 10^{20} \times c \qquad (10.3)$$

where c' is now expressed in molecules/cm^3 when c is given in mol/L.

The intermolecular distance can then be calculated as the inverse cubic root and becomes:

$$d = 1.185/c^{1/3} \qquad (10.4)$$

where d is the average intermolecular distance in nanometer when c is given in mol/L. For example, the intermolecular distance at a concentration of 1 mol/L is 1.185 nm. For a concentration of 1 μmol/L it is 118.5 nm.

If a molecule of radius r diffuses through space, the probability to hit a point-like object at the distance R is:

$$P = \frac{\text{cross section of the molecule}}{\text{surface of a sphere with radius R}} = \frac{r^2\pi}{4R^2\pi} = \left(\frac{r}{2R}\right)^2 \qquad (10.5)$$

If an enzyme molecule with a diameter of 8 nm searches for DNA by diffusion from a distance of 20 nm, the probability to find it is 16% (actually, it is somewhat better since the DNA has a diameter of 2 nm and is, for this calculation, infinitely long).

10.4
Why Fluorescence Microscopy?

There are several methods available for single molecule detection (SMD). Essentially, the most classical one is electron microscopy. AFM (atomic force microscopy) and SNOM (scanning near-field optical microscopy) require less sample preparation. SNOM allows for multicolor observation. In spite of the advantage of very high resolution, all these methods have one disadvantage: their temporal resolution is low and sample preparation is different from that of classical biochemical preparation techniques. This is the reason, why classical far field microscopy has found its firm place in single molecule detection, and particularly in studies of single molecule reactions. The temporal resolution of far field microscopy is, with a simple CCD camera, 40 (PAL) or 33 (NTSC) milliseconds and can be increased to better than 1 millisecond with specialized cameras. It is no real problem to achieve single molecule sensitivity similar to the competitive techniques above. The apparent disadvantage of lower resolution is often not as severe as it may appear: in most cases either single molecules are observed in absence of other molecules in close vicinity. Then, the only price one has to pay is that the molecules appear to be too thick. Or, one wants to observe many molecules in a given space, for example the reaction products of an enzyme reaction catalyzed by one or a few enzyme molecules: then spatial resolution is not too important. On the other hand, in all these cases the high temporal resolution is of unbeatable value.

10.5
Single Molecule Enzyme Reactions with Small Substrates

One class of reactions which can be observed down to the single molecule level are those with small molecules as substrate as well as product. One of them has to be fluorescent and one does not fluoresce, or has at least a considerably smaller fluorescence emission. In such cases the emergence or reduction of fluorescence is a measure for the progress of the reaction. There are a number of reactions in biochemistry where, among other molecules, NADH, which fluoresces at 440 nm when excited at 360 nm, is converted into almost nonfluorescent NAD^+. One example is the reaction of lactate dehydrogenase:

$$\text{pyruvate} + \text{NADH} \leftrightarrows \text{lactate} + \text{NAD}^+ \tag{10.6}$$

There are a number of general strategies to observe such reactions. In the "femtodroplet pipetting" method the reaction proceeds in a capillary. In equidistant time intervals, typically every few minutes, femtodroplets (a few µm in diameter, i.e., 10–20 femtoliters) are pipetted onto a glass surface and the fluorescence is measured. In reactions, where fluorescence is reduced, the fluorescence intensity of the droplets decreases with time. Fig. 10.1 shows such a result.

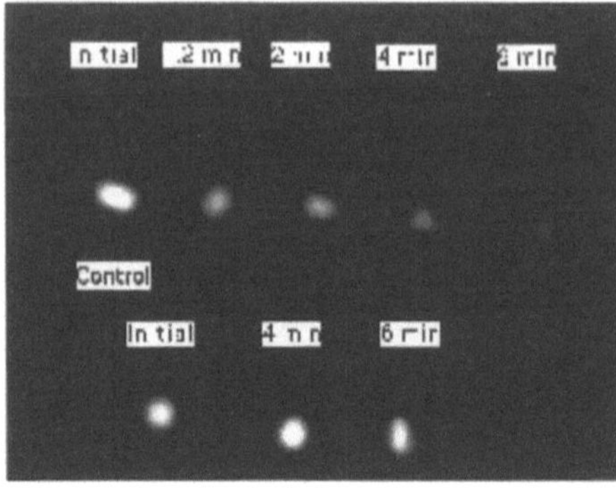

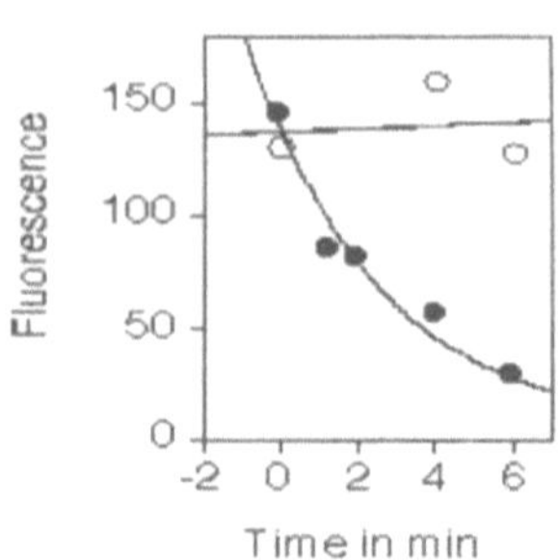

Fig. 10.1. *Left*: Micrograph of fluorescent femtodroplet pipetted every two minutes onto the coverslide. *Upper row*: Reaction, *lower row*: control. *Right*: Graphic representation of the fluorescence in these droplets

The fluorescence can be measured by image analysis with a suitable program, here NIH image. In Fig. 10.1(*right*) the fluorescence of the spots of Fig. 10.1(*left*) is plotted as a function of time. The initial slope of the curve through the solid symbols is proportional to the reaction rate v. Also shown, with open symbols, is the time course when no enzyme is present, indicating that the decrease of fluorescence is not caused by photobleaching.

When the reaction rate v, thus determined at different substrate concentrations [S] is plotted according to Eq. 10.7 (i.e., $1/v$ versus $1/[S]$, see Fig. 10.2 below) two typical parameters of biochemistry, the maximum rate v_{max} and the Michaelis constant K_m can be determined:

$$1/v = 1/v_{max} + K_m/(v_{max} \times [S]) \tag{10.7}$$

The plot in Fig. 10.2 follows from the Michaelis-Menten equation and is called the Lineweaver Burk plot. Its slope gives K_m/v_{max}, its intercept with the x-axis gives $-1/K_m$, from which K_m can be immediately calculated.

The result of this experiment is that the Michaelis constant can be determined for single molecules with high accuracy. Even the slopes are quite similar. However, the pattern seen in Fig. 10.2 indicates some not yet understood noncompetitive inhibition of the single molecule reaction by some so far unidentified impurity. Another result could have been, that all lines have a common intersection point on the y axis instead the x axis (competitive inhibition) or that all lines are parallel (uncompetitive inhibition).

The femtodroplet pipetting method, though yielding single molecule sensitivity, is not exactly a genuine single molecule method for the following reason: a given reaction volume, which finally is pipetted onto the glass surface, is open. That means that molecules can diffuse in and out of this volume. On average, each volume contains one or a few enzyme molecules, but it is not always the same one. Thus, the femtodroplet pipetting method cannot uncover differences between individual molecules. On the other hand, as just discussed, it gives enzyme kinetic parameters with an accuracy which is close to the accuracy of bulk measurements

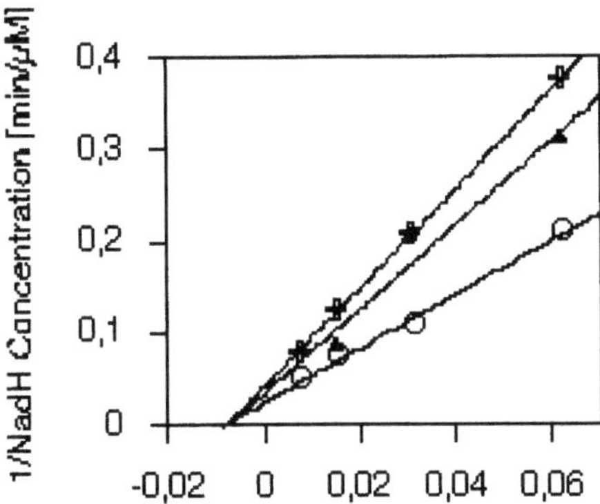

Fig. 10.2. Lineweaver Burk plot according to Eq. 10.7. The straight line with the open circles is the result of a bulk experiment. The lines with the larger slopes result from femtodroplet pipetting

with 10 to 15 orders of magnitude more material. Thus, femtodroplet pipetting is the method of choice, when extremely small amounts of material are available, for example in mass screening of disease, where only blood droplets and not milliliter samples can be taken from the person to be tested.

A method better suited to test individual enzyme molecules is the femtodroplet--in-substrate technique. Here, a femtodroplet of enzyme solution with a higher viscosity due to the addition of glucose, is pipetted into a substrate solution with lower viscosity. The droplet with the enzyme remains stable even inside the substrate solution for more than an hour. At the interface between enzyme femtodroplet and substrate solution enzyme and substrate come into contact and the reaction, here one which generates fluorescence, can proceed. Due to the high viscosity of the enzyme-containing droplet, the position of single enzyme molecules is comparatively stable. Thus, progress of reaction results in generation of fluorescent spots at the surface of the enzyme femtodroplets. Fig. 10.3 shows the emergence of such localized fluorescence spots at the surface of a femtodroplet.

In early phases of the reaction, single spots can be easily separated and evaluated, in later stages the spots flow into each other due to slow, but not neglible diffusion. A surprise in this reaction (the inverse reaction of Eq. 10.6) was that at the position of the fluorescence spots nanocrystals could be observed, when the microscope was switched from fluorescence to bright field mode. So far it is not yet clear what the nanocrystals are. One explanation might be that the reaction induces growth of crystals from the sugar in the solution. Nevertheless, whatever the crystals are made from, the effect is very quantitative and definitely reflects the progress of the reaction (Fig. 10.4).

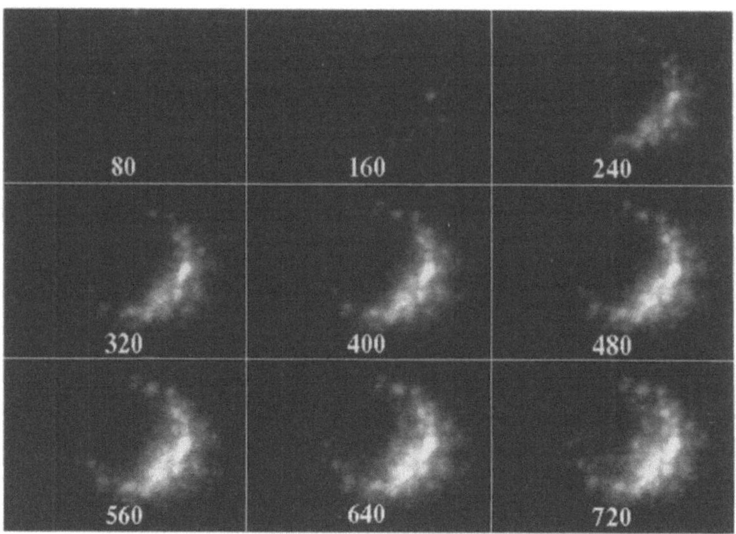

Fig. 10.3. Emergence of fluorescent spots at the surface of a high viscosity femtodroplet of enzyme solution embedded in low viscosity substrate solution. After 80 seconds the first spots become visible, after 720 seconds the have merged with each other. (From [6])

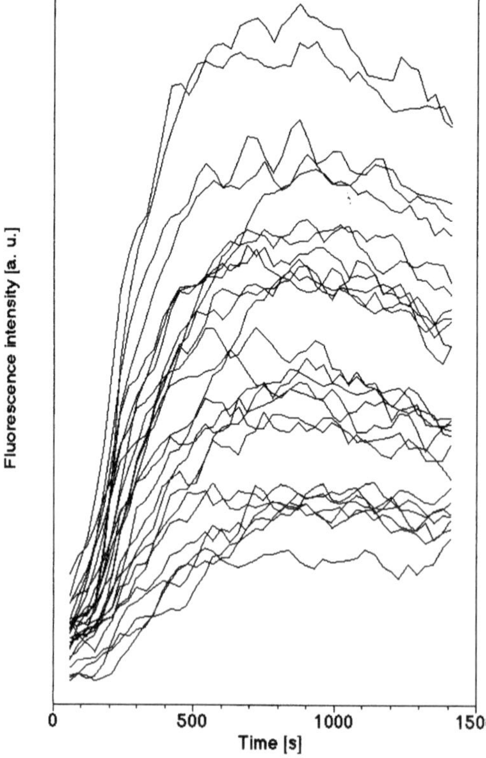

Fig. 10.4. Time course of the fluorescence for 28 randomly selected spots of Fig. 10.3. (From [6])

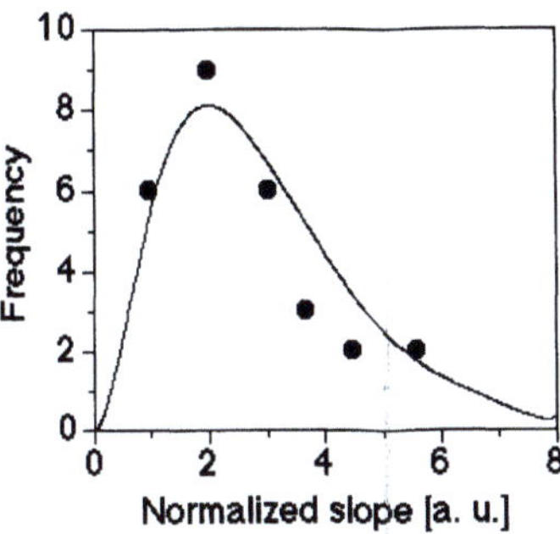

Fig. 10.5. Number of curves in Fig. 10.4 with a given initial slope plotted versus this slope. The curve drawn through the data points is the smoothened Poisson distribution according to Eq. 10.2. (From [6])

Different groups of curves are observed. When the initial slopes of these curves are plotted with the assumption, that they represent integer multiples of a basic slope, a straight line is obtained (not shown). This suggests that fluorescent spots generated by one, two, three ... up to six enzyme molecules in their center, are observed. Such an assumption gets additional confirmation, when the number of curves with a given initial slope in Fig. 10.4 is plotted against the initial slope (Fig. 10.5 below), i.e. the putative number of enzyme molecules in the center of the spot, the data can be fitted by a Poisson distribution (see Eq. 10.2) with a mean value of 2.

Taken together, these observations are consistent with the assumption that here indeed the reactions of individual enzyme molecules are seen.

The method is not as error proof as the femtodroplet pipetting method and it suffers from the fact, that diffusion is highly impeded, i.e., that kinetic constants do not represent the constants of classical biochemistry. Nevertheless, it is the first technique available to study reactions of single enzyme molecules in a quite direct and highly parallel way.

10.6
Restriction Endonuclease Reactions

A second type of single molecule reactions which can be well studied in the fluorescence microscope are reactions of enzymes acting on macromolecules. Here, the macromolecule is labeled with a fluorescence dye and can be directly observed. Examples are the reactions of polymerases or restriction endonucleases on DNA molecules. The latter are labeled with an unspecific fluorescence dye that binds to every tenth to thirtieth base. Only a small fraction of all possible dyes are suitable for these studies, since care has to be taken that the reaction to be studied is at most affected moderately by the staining procedure. DAPI and SYBRGreen have been found suitable. The DNA molecules (here DNA from the bacteriophage λ, a virus that infects bacteria) are coupled to polystyrene microbeads (1–2 micro-

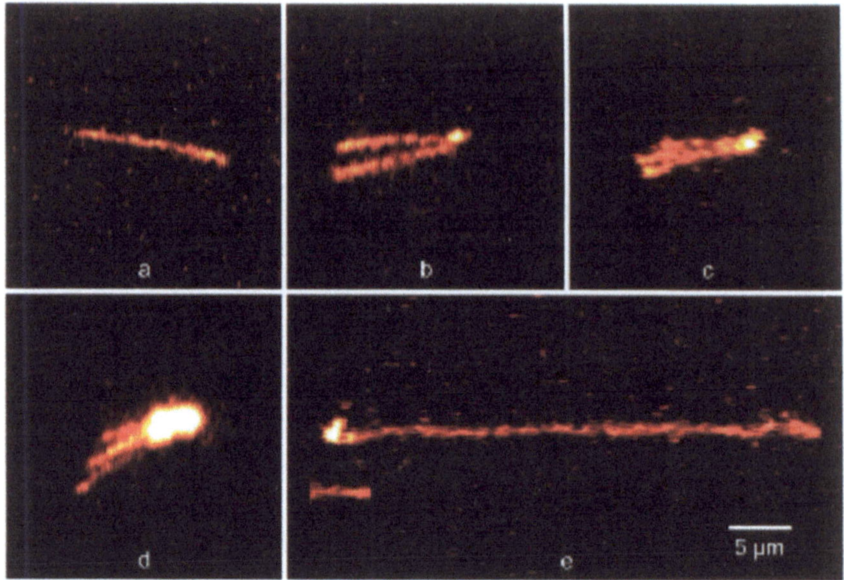

Fig. 10.6. One, two and three fluorescently stained phage λ DNA molecules (**a–c**), a cluster of DNA molecules (**d**) and a DNA concatemer (**e**) consisting of several DNA molecules

meter in diameter) by end-labeling (using a commercial kit) the DNA molecule with biotin and coating the beads with avidin [2]. Due to the high binding constant between avidin and biotin (of the order of 10^{16}) biotin end-labeled DNA molecules bind to beads upon simple mixing and stirring. If stirring is not done very carefully, the DNA molecules wrap around the otherwise nonfluorescing beads which then appear to be fluorescent. The beads can be held with optical tweezers [1, 3] and moved along complex trajectories in the visual field of a microscope. Fig. 10.6a–d [3] shows a number of different outcomes from such a binding procedure, from a single DNA molecule without wrapping, via two and three molecules wrapped around the bead, up to a whole cluster of DNA molecules. Fig. 10.6e shows a longer DNA molecule.

All DNA molecules are held in a hydrodynamic flow. This is necessary, since at physiological conditions at which the enzyme reactions typically proceed, DNA molecules tend to collapse into globular structures. The enzyme is added at the site of the bead by a micropipette (for details, see [3, 4]). While enzyme reactions usually proceed even on such collapsed DNA molecules, direct observation is only possible with the extended DNA filaments shown in Fig. 10.6.

In Fig. 10.7, the action of the enzymes *Apa*I, *Sma*I and *Eco*RI on a phage λ DNA molecule is shown (see also [5, 6]). In the sequence of this DNA molecule there are five sequence elements of the type 5'...G↓AATTC...3' (the recognition sequence of *Eco*RI) and this means that one expects five cuts in a single phage λ DNA molecule. Similarly, three cuts are expected by *Sma*I and one cut by *Apa*I. The expected cutting pattern is given schematically in the lower right part of Fig. 10.7.

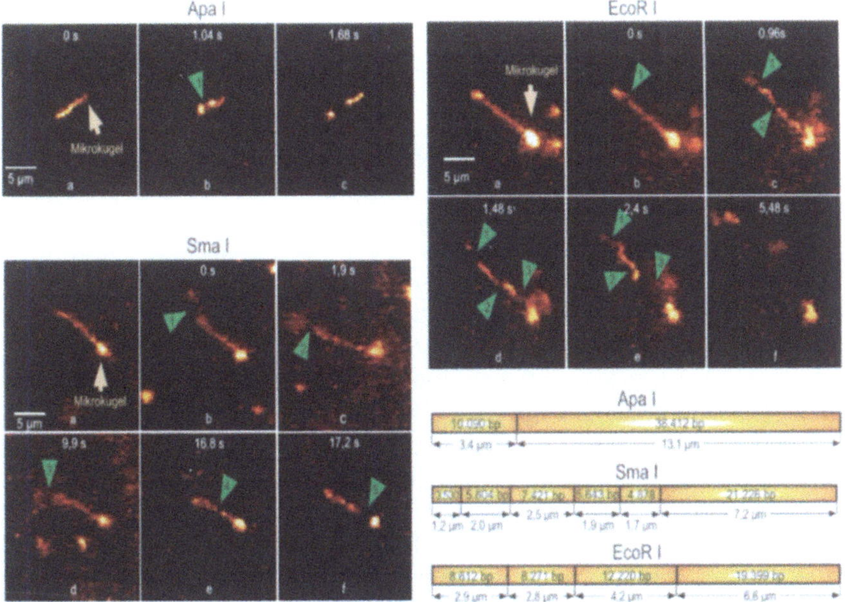

Fig. 10.7. Restriction of individual phage λ molecules by the restriction endonucleases *Apa*I (with one expected cutting site), *Sma*I (with three expected cutting sites) and *Eco*RI (with five expected sites). Note that for *Sma*I and *Eco*RI the cuts are consecutive and that no cutting site is omitted

In the micrographs of Fig. 10.7 three details are notable:

1. the reactions of *Eco*RI and *Apa*I start at the DNA end distant from the bead

2. the reaction proceeds against the hydrodynamic flow, and

3. no expected cutting site is skipped.

Binding to the distant end can be explained by the "shadow" of the 1 micrometer diameter bead. In order to bind within 5 μm from the bead, only those enzyme molecules are available, which arrive at an angle determined by tan α > 1:5. In order to bind at a distance of 16 μm from the bead, i.e., at the distant end, all enzyme molecules arriving from an angle determined only by more than tan α > 1:16 can find the DNA molecule. These are much more enzyme molecules, i.e., the probability of reaction start at the distant end is highest.

The fact that immediately after the reaction start only reactions against the direction of hydrodynamic flow are seen is trivial. Cases where the enzyme molecule moves in the direction of the flow would not be recognized, since in that case the enzyme molecule, which has bound close to the distant end of the DNA molecule, falls off again and no cut is observed. The interesting observation is not that the enzyme proceeds against hydrodynamic flow in all cases, but that such cases are possible at all. Finally, the fact that no cutting site is omitted gives important information on the reaction mechanism and will be discussed as follows.

There are several models discussed on the way an enzyme molecule proceeds along the DNA molecule. In the sliding model, the enzyme remains permanently in contact with the DNA molecule. In the hopping model the strength of binding varies, but there remains always some contact. In the jumping model, there are instants, where there is no longer contact and the enzyme molecule can diffuse freely through space. The jumping model, though invoked frequently, is highly improbable for the following reason. In order to be completely detached an enzyme molecule has to be at a distance from the DNA molecule of more than its diameter (which is of the order of 8 nm). Since it is not known at what distance an enzyme molecule is really detached from the DNA molecule, we make the assumption that this is the case at 2.5 times of its radius. From this point, the enzyme molecule can diffuse into all space directions. The probability of re-finding the DNA molecule is then 16%. (see Eq. 10.5). Correspondingly, the chance to find it two times is below 1% and to find it more often is vanishingly small. Thus, any jumping model should be ruled out for restriction endonucleases. For the same reason which excludes the jumping model, it is also highly improbable that several different enzyme molecules interact in a way that would result in the reaction pattern seen in Fig. 10.7. The experiments described here cannot distinguish between the sliding or the hopping model. They confirm however, that the enzyme molecule remains at least in some contact with the DNA molecule, as long as the reaction proceeds.

10.7
Conclusions

Fluorescence techniques are particularly useful for single molecule studies, since they allow the use of conventional far field microscopy with its high temporal resolution and with the possibility to use sample preparation techniques known from conventional biochemistry. The low spatial resolution, as compared to electron microscopy, AFM or SNOM, is not as disadvantageous as it might appear at a first glance.

References

1. Greulich KO (1999) Micromanipulation by light in biology and medicine: The laser microbeam and optical tweezers. Birkhäuser, Basel
2. Hoyer C, Monajembashi S, Greulich KO (1996) Laser manipulation and UV induced single molecule reactions of individual DNA molecules. J Biotechnol 52:65–73
3. Ashkin A (1997) Optical trapping and manipulation of neutral particles using lasers. Proc Natl Acad Sci USA 94:4853–4860
4. Schäfer B, Gemeinhardt H, Uhl V, Greulich KO (2000) Single molecule DNA restriction analysis in the light microscope. Single Molecules 1:33–40
5. Schäfer B, Gemeinhardt H, Greulich KO (2001) Direct microscopic observation of the time course of single molecule DNA restriction reactions. Angew Chem (in press)
6. Uhl V, Pilarczyk G, Greulich KO (1998) Fluorescence microscopic observation of catalysis by single or few LDH1 enzyme molecules. Biol Chem 379:1175–1180

Monitoring γ-Subunit Movement in Reconstituted Single EF$_o$F$_1$ ATP Synthase by Fluorescence Resonance Energy Transfer

M. Börsch, M. Diez, B. Zimmermann, R. Reuter, and P. Gräber

The membrane-bound enzymes H$^+$ ATP synthases contain two coupled rotary motors that drive catalysis. We applied a single molecule spectroscopy approach to monitor the internal rotation of the γ-subunit of the F$_1$ part against its static counterpart, the b-subunits of the F$_o$ part. We specifically attached two fluorophores to H$^+$ ATP synthase from *E. coli*, namely Cy5 at the γ-subunit and tetramethylrhodamine at one b-subunit. After reconstitution into liposomes, these enzymes regained their full catalytic activity as measured by ATP synthesis rates. Fluorescence resonance energy transfer (FRET) was monitored in photon bursts of freely diffusing proteoliposomes using a confocal setup for single molecule detection. Incubation with non-hydrolyzable AMPPNP resulted in stable intensity ratios within a photon burst. This corresponds to a fixed γ-subunit orientation. We detected three different FRET efficiencies, i.e., γ-subunit orientations. After addition of ATP a consecutive order of three distinguishable FRET efficiencies was observed within the bursts, indicating a stepwise unidirectional γ-subunit movement against the b-subunits.

11.1
Introduction

H^+ ATP synthases ($F_o F_1$ ATP synthases) catalyze the synthesis of ATP from ADP and inorganic phosphate in the membranes of mitochondria, chloroplasts and bacteria. Endergonic ATP synthesis is coupled to proton translocation across the membrane due to a difference in the electrochemical potential of protons. This large enzyme consists of two parts. The hydrophobic, membrane-integrated F_o part is involved in proton transport and contains subunits a, b_2 and c_{10-12} for the *Escherichia coli* enzyme. The catalytic binding sites are located on the three β-subunits in the hydrophilic F_1 part with subunit composition $\alpha_3\beta_3\gamma\delta\varepsilon$.

Currently, H^+ ATP synthase is thought to contain two rotary motors converting electrochemical energy via a mechanical form of energy to chemical energy [1]. In the F_o part proton transport across the membrane causes rotation of a ring of 10 to 12 c-subunits with respect to the nonrotating a and b-subunits. The γ and ε-subunits of the F_1 part are connected to the c-ring and thus are forced to rotate within the $\alpha_3\beta_3$ hexamer. The stepwise rotation of the γ-subunit induces conformational changes in the β-subunits, which leads to synthesis and release of ATP. According to their ADP and ATP binding affinities three different conformational states of the subunits are distinguished. The relative γ-subunit position determines the actual state of each β-subunit. Turning the γ-subunit in the catalytic cycle initiates cooperative sequential changes in the β-subunit conformations.

This enzyme can also work as a proton pump by hydrolyzing ATP. During ATP hydrolysis conformational changes of the β-subunits in the F_1 part force the γ-subunit and also the ring of c-subunits in the F_o part to rotate in reverse direction. The rotating domain of the ATP synthase, $F_1\gamma\varepsilon$-$F_o c_{10-14}$, is called "rotor". The central stalk in electronmicroscopic images of the *E. coli* enzyme (Fig. 11.1a [2]) is therefore identified as a part of this "rotor". All other subunits belong to the nonrotating counterpart called "stator". A second stalk on the right hand side of this electronmicroscopic image in Fig. 11.1a probably consists of the b-subunit dimer and is part of the "stator".

11.2
Visualizing Intersubunit Rotation

ATP-driven rotation of the γ-subunit in isolated F_1 parts of ATPases from a thermophilic bacterium was convincingly visualized by Noji et al. [3]. For this purpose, the F_1 parts were attached with three His-tags onto the glass surface of microscopic cover slides. Connecting a highly fluorescent actin filament to the γ-subunit as a pointer, the actual orientation of this subunit with respect to the $\alpha_3\beta_3$ hexamer was measurable by videomicroscopy at millisecond time resolution. The maximum speed of rotation depended on the length of the actin filament at high ATP concentrations. At very low ATP concentrations, rotation occured in discrete

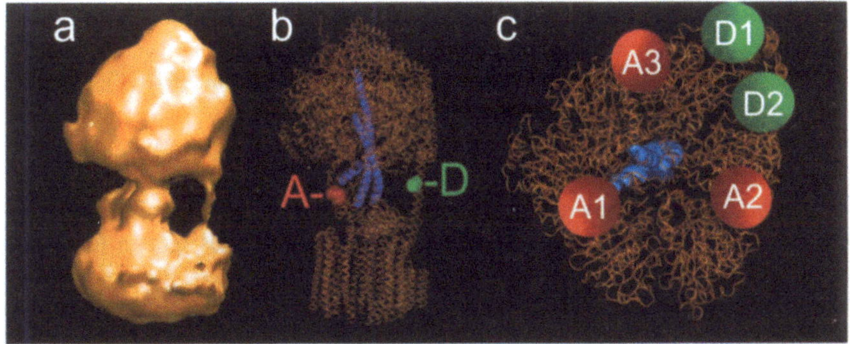

Fig. 11.1. a Surface representation of electron density after threedimensional reconstruction of electronmicroscopic images of F$_o$F$_1$ ATP synthase from *E. coli* [2]. **b** Modified model of EF$_o$F$_1$ combining alignment of structual data for the γ and ε-subunit [18] with a homology model by S. Engelbrecht [19]. Amino acid position for the FRET donor on the b-subunit dimer is indicated with <u>D</u>, FRET acceptor position <u>A</u> is located at the γ-subunit. **c** Cross--section view at the fluorophore level from the membrane side (F$_o$) to the top of F$_1$, showing the γT106C cysteine position as <u>A1</u> and the two bQ64C positions <u>D1</u> and <u>D2</u>. The distance between <u>A1</u> and <u>D1</u> is 7.3 nm. Counterclockwise 120° stepped rotation of the γ-subunit during ATP hydrolysis will approximately result in <u>A2</u> and <u>A3</u> positions for the γT106C cysteine, respectively, in a sequence of A1 → A2 → A3 → A1 → transitions

120° steps. The fluorescent actin filament method has been applied also to the F$_1$ parts from chloroplasts and *E. coli* (see review [1]). In all cases the γ-subunit rotated counterclockwise during ATP hydrolysis when viewed from the membrane side to the top of F$_1$ (see Fig. 11.1c, the expected sequence of γ-subunit orientations is → A1 → A2 → A3 → A1 →).

To demonstrate the rotational motion of the membrane-integrated F$_o$ motor, a fluorescent actin filament was attached to the c-ring. F$_o$F$_1$ was bound to the glass surface with His-tags. Addition of ATP resulted in rotational movement of the fluorescent actin filament [4]. High concentrations of detergent and BSA were required to prevent sticking of the filaments to the surface. However, these enzymes were not sensitive for the inhibitors generally used to show that the F$_o$F$_1$ ATP synthases are coupled [5]. At enzyme concentrations in the nanomolar range used for single molecule studies, treatment with detergent leads to a loss of subunits. In the case of the chloroplast F$_1$ part, the dissociation of the δ-subunit within minutes was detected with fluorescence correlation spectroscopy [6] (also for EF$_1$, M.B., unpublished results). Crystallization studies of the yeast F$_o$F$_1$ enzyme resulted in the loss of subunits, including the a-subunit, by the use of detergent [7].

In solution, single protein dynamics can be investigated by single fluorophore detection techniques. Freely diffusing enzymes are expected not to be disturbed by interactions with surfaces. Confocal microscopy is used in combination with sensitive detectors to measure photon bursts from fluorescently labeled enzymes as they diffuse through the focal volume. If the enzymes are tagged with two different fluorophores, the conformational dynamics during catalysis can be observed by distance-dependent fluorescence resonance energy transfer (FRET). The photon bursts are detected in two spectral channels and ratiometric data analysis allows for calculation of FRET efficiencies and Förster distances at a millisecond time resolution (see review [8]).

Membrane-bound enzymes like the H^+ ATP synthases can be reintegrated into the membrane of lipid vesicles. These reconstituted enzymes remain fully functional for several hours even at nanomolar concentrations [9], if stored at room temperature. Determing enzyme activity as the rate of ATP synthesis is the crucial test, which requires a fully intact, coupled F_oF_1 ATP synthase. These proteoliposomes diffuse freely and the detection time in the focal volume is increased by at least one order of magnitude due to slow diffusion of the vesicles (100 nm size). With a confocal detection volume of about 5.5 fL, the characteristic time of diffusion for the proteoliposomes is in the range of 30 ms, as determined by fluorescence correlation spectroscopy. In photon bursts of single H^+ ATP synthases with more than 500 ms duration, several turnover steps are expected during ATP hydrolysis. To monitor intersubunit rotation during catalysis, we measured relative distance changes between two amino acids – one at the rotor and one at the stator domain of F_oF_1 – by FRET, with single enzymes reconstituted into liposomes.

11.3
FRET-labeled F_oF_1 ATP Synthase EF_o-b64-TMR-F_1-γ106-Cy5

For the FRET experiments two fluorescent labels were attached to cysteine residues in different subunits of H^+ ATP synthase from *E. coli* (Fig. 11.1b). The two cysteine residues were introduced by site-directed mutagenesis. We labeled EF_oF_1 with Cy5 at the rotating γ-subunit and tetramethylrhodamine at the b-subunits. To enhance the specificity we labeled the γ-subunit of EF_1 separately.

The cysteine mutation at the γ-subunit (γT106C) in the F_1 part is described in [10] and has been used previously for single molecule studies [11]. From this mutant only the F_1 parts of the enzyme (EF_1) were isolated [12]. The γ-subunit was labeled with Cy5-maleimide according to [13]. Cy5-maleimide was kindly provided by E. Schweinberger and C.A.M. Seidel (MPI für Biophysikalische Chemie, Göttingen, Germany) and H.-D. Martin (Institut für Organische Chemie und Makromolekulare Chemie, Universität Düsseldorf, Germany). The degree of labeling was determined by the corrected UV-absorption of the enzyme at 280 nm and dye absorption at 641 nm. Approximately 50 percent of the F_1 parts were labeled with the FRET acceptor Cy5. EF_1 was stored in liquid nitrogen.

We separately introduced cysteines in the two b-subunits belonging to the stator part of the enzyme (bQ64C, to be published elsewhere). This enzyme did not carry the γT106C cysteine mutation and was genetically engineered to have no other cysteines in the F$_o$ part. EF$_o$F$_1$ was isolated according to [9] and solubilized using dodecylmaltoside. The b-subunits were labeled with tetramethylrhodamine-maleimide (TMR, Molecular Probes) as FRET donor. The degree of labeling was adjusted to be in the 20 per cent range to avoid double labeling, which might occur since the b-subunit in the F$_o$ part of *E. coli* exists as a dimer (Fig. 11.1c). TMR-labeled EF$_o$F$_1$ was reconstituted into liposomes [9] and the F$_1$ parts were removed using a procedure adapted from [14]. TMR-labeled F$_o$ parts, one per liposome, were reassembled with Cy5-labeled EF$_1$ according to [15].

11.3.1
Synthezising ATP with Reconstituted EF$_o$-b64-TMR-F$_1$-γ106-Cy5

Enzyme activity was measured by the rate of ATP synthesis in an acid–base transition [17]. Labeling of the b-subunits with TMR did not affect enzyme activity. The reconstituded EF$_o$-b64-TMR-F$_1$ catalyzed ATP synthesis with a rate of 60 s^{-1} at ΔpH = 4. After stripping off the F$_1$ parts, no ATP synthesis was measurable, i.e., F$_1$ parts had been removed quantitatively. Addition of Cy5-labeled EF$_1$ yielded a reassembled enzyme EF$_o$-b64-TMR-F$_1$-γ106-Cy5, which synthesized ATP with rate of 31 s^{-1} at ΔpH = 4. For comparison, reassembling EF$_o$-b64-TMR with unlabeled EF$_1$ resulted in a rate of 29 s^{-1}. During the reassembling procedure several centrifugation steps were required to remove unbound EF$_1$. Centrifugation and subsequent resuspension of the proteoliposome pellet were shown to reduce the rates of ATP synthesis by up to 45 percent [17] (from 65 s^{-1} to 35 s^{-1}). We conclude that the FRET-labeled H$^+$ ATP synthases EF$_o$-b64-TMR-F$_1$-γ106-Cy5 are fully functional after reconstitution into liposomes.

11.3.2
Set-up for Single Enzyme FRET Analysis

FRET experiments with freely diffusing single H$^+$ ATP synthases in liposomes were performed using a confocal set-up of local design. A frequency-doubled Nd:YAG laser (532 nm, 50 mW, Coherent, Germany) was used for excitation. The circularly polarized laser beam was attenuated to 120 μW and focussed into the liposomes, containing buffer solution, by a water immersion objective (UAPO 40x, N.A. 1.15, Olympus). For epi-illumination a dichroic mirror DCLP 545 nm (AHF, Tübingen, Germany) was used. Out-of-focus fluorescence was filtered out by a 100 μm pinhole (OWIS, Staufen, Germany). Fluorescence was separated into two spectral regions by a dichroic mirror DCLP 630 nm (AHF, Germany). Single photons were detected with avalanche photodiodes (SPCM AQR 151, EG&G, Canada), after passing an interference filter HQ 575 nm/65 nm for TMR and a HQ 665 nm LP for Cy5 (AHF, Germany). Photons in two channels were registered simultaneously with a PC-card (PMS 300, Becker & Hickl, Berlin, Germany).

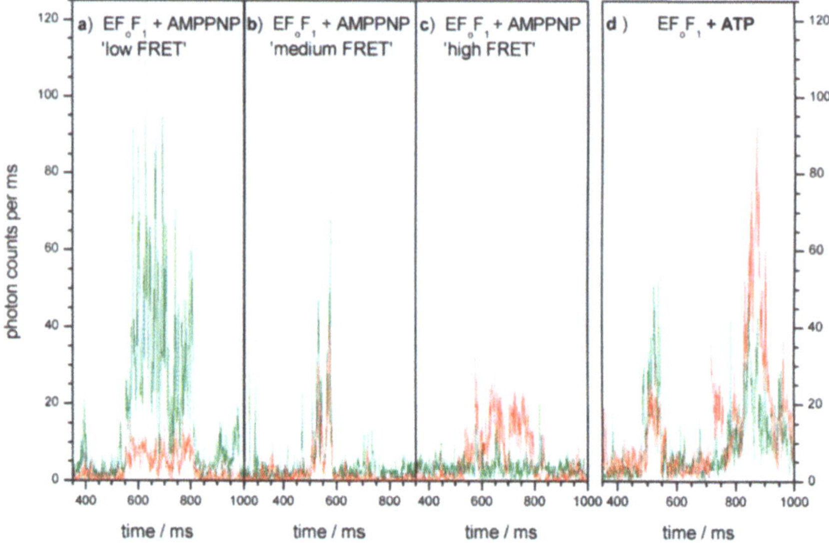

Fig. 11.2. Photon bursts of single EF$_o$F$_1$ in liposomes. Green trace is the TMR count rate, red trace is the Cy5 count rate with 1 ms time resolution. Addition of AMPPNP (**a–c**) leads to three populations with different FRET efficiencies. The FRET donor–acceptor distance remains constant within one photon burst. **d** Addition of ATP results in fluctuating FRET efficiencies within one photon burst

From the fluorescence intensities I of FRET donor (*green traces* in Fig. 11.2, $I_{(Donor)}$) and acceptor (*red traces* in Fig. 11.2, $I_{(Acceptor)}$) we calculate a so-called proximity factor P :

$$P = I_{(Acceptor)}/[\, I_{(Donor)} + I_{(Acceptor)}\,]$$

This proximity factor P allows for simple estimation of the FRET efficiency and the distance between FRET donor and acceptor. However, the value of $P = 0.5$ corresponds to an approximate Förster radius $R_0 = 6.5$ nm for the FRET pair TMR $\rightarrow$ Cy5 only in an ideal case, where the correction factor is 1 for the detection efficiencies of the set-up and the fluorescence quantum yields [8].

For fluorescence correlation spectroscopy (FCS), the multiplexed signals were fed in parallel to a hardware autocorrelator PC-card (ALV 5000/E FAST, ALV, Langen, Germany). The actual detection volume V was calculated by FCS [11]. With a measured mean diffusion time of $\tau_D = 350$ μs for Rhodamine 6G in water and a translational diffusion coefficient of $D = 2.8 \times 10^{-6}$ cm^2 s^{-1} according to [16], the computed radial and axial $1/e^2$ radii are $\omega_0 = 0.63$ μm and $z_0 = 2.5$ μm. The confocal volume is therefore $V = \pi^{1.5} \times \omega_0^2 z_0 = 5.5$ fL.

Single enzyme FRET measurements were performed in 50 mM Hepes/NaOH buffer (pH 8.0, 2.5 mM MgCl$_2$). Fluorescent impurities in the buffer solutions were removed by activated charcoal granula (Merck, Germany) with subsequent sterile filtration (200 nm pore size). ATP and AMPPNP were obtained from Boehringer (Mannheim, Germany). The ATP concentrations were held constant by a biochemical ATP regeneration kit containing 2.5 mM phosphoenolpyruvate, 10 mM KCl and 18 units/mL pyruvate kinase.

11.3.3
Discrimination of Three γ-Subunit Positions with AMPPNP

The orientation of the γ-subunit with respect to the b-subunits in the reassembled holoenzyme is not known. During the isolation procedure of EF$_1$ most of the bound nucleotides – ATP or ADP – are lost. The conformations of all three β-subunits under these conditions are presumably in the "open state" and the γ-subunit can wobble within the α$_3$β$_3$ hexamer due to Brownian motion. By adding AMPPNP – a non-hydrolyzable ATP derivative – the position of the γ-subunit is fixed. The mitochondrial F$_1$ part [20] and the yeast F$_o$F$_1$ enzyme, which were crystallized in the presence of AMPPNP, showed well-defined γ-subunit orientations [7].

We incubated the reconstituted FRET-labeled H$^+$ ATP synthases with 1 mM AMPPNP and diluted the proteoliposomes to a final concentration below 1 nM. In the well separated photon bursts of EF$_o$-b64-TMR-F$_1$-γ106-Cy5 three different classes of FRET efficiencies were found (Fig. 11.2a–c). Within one photon burst the mean proximity factor P remained constant.

In Fig. 11.2a the proximity factor $P = 0.19$ was calculated for the photon burst starting at $t = 550$ ms. The standard deviation sd was $= \pm 0.11$ within this burst. This was a so-called "low FRET" case and most probably represented the A1 position of the γ-subunit in Fig. 11.1c. To ensure the existence of FRET in this case, we compared this proximity factor with the appropriate intensity ratio for a reconstituted EF$_o$-b64-TMR without the Cy5 label. The "donor only"-labeled EF$_o$F$_1$ exhibited an intensity ratio $I_{(Acceptor)}/(I_{(Donor)} + I_{(Acceptor)})$ of 0.05 (data not shown), which is significantly smaller than the lowest FRET efficiency detected.

In Fig. 11.2b the proximity factor was $P = 0.45$ ($sd = \pm 0.15$) for the burst starting at $t = 510$ ms. This was a "medium FRET" case and presumably corresponded to the A2 position of the γ-subunit. The signals on both detector channels decreased in parallel within the middle of the burst. Rotational correlation times of EF$_o$F$_1$ within the lipid membranes have previously been determined to be in the range of 100–200 μs [21]. The intensity fluctuations in the millisecond time range were therefore caused by translational diffusion of the proteoliposome in and out of the focal volume or were due to rotational movement of the proteoliposome.

For the photon burst starting at $t = 540$ ms in Fig. 11.2c, the calculated proximity factor was $P = 0.75$ ($sd = \pm 0.15$). This was a "high FRET" case, probably due to the A3 position with respect to a D1 position of the FRET donor.

We calculated the proximity factors of 541 F$_o$F$_1$ ATP synthases in the presence of AMPPNP (Fig. 11.3) to analyze the distribution of the three distinct FRET efficiencies observed in single photon bursts. Data registration was automated by the "event mode" of the multichannel scaler PC-card. The enzyme concentration was further decreased and the binning time interval enlarged to 100 ms, which is the 3-

fold value of the mean diffusion time of the proteoliposomes, to get the mean value of the proximity factor for every single F_0F_1 ATP synthase. A minimum threshold of 500 counts per channel per 100 ms interval was applied. The proximity factor histogram showed a broad distribution from $P = 0.1$ to 0.9. Whereas the "low FRET" and the "medium FRET" efficiencies seemed to be equally distributed, the "high FRET" efficiency with $P > 0.65$ was significantly under-represented. This could indicate a preferred orientation of the γ-subunit with respect to the b-subunits in the reassembled enzyme.

The FRET efficiencies were not well separated in this histogram. Intensity fluctuations due to photophysical properties of protein-bound TMR [22, 11] and photoinduced isomerization of Cy5 [23] were expected to occur in the sub-millisecond range. Also polarization effects due to rotational motion of the enzyme within the lipid membrane and the rotation of the proteoliposome itself should be averaged in a 100 ms binning time interval. In the model of EF_0-b64-TMR-F_1-γ106-Cy5 (Fig. 11.1c) the two possible FRET donor positions on the b-subunit dimer are indicated. Binding of TMR to D1 or D2 can result in two independent sets of Förster distances, which might slightly differ in all of the three FRET efficiencies. In principle, we should find six FRET efficiencies, but these were not resolved in the histogram. However, identification of three classes of FRET efficiencies after addition of AMPPNP was the basic prerequisite for a discrimination of all γ-subunit orientations during the catalytic cycle, i.e., ATP hydrolysis or synthesis.

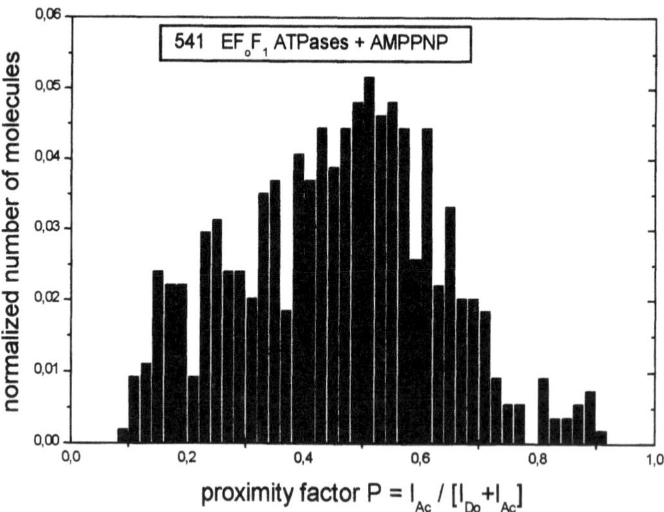

Fig. 11.3. Histogram of proximity factors P calculated from photon bursts of 541 EF_0F_1 in liposomes in the presence of AMPPNP. Photons were counted in 100 ms bins. Only events with more than 500 counts on both detector channels were taken for the histogram

11.3.4
Unidirectional Motion of the γ-Subunit with ATP

After addition of 1 mM ATP to reconstituted EF_o-b64-TMR-F_1-γ106-Cy5 strong fluctuations of the fluorescence intensities within the photon bursts of single enzymes were measured (see Fig. 11.2d). This behavior was not observed in the absence of ATP. A biochemical ATP-regenerating kit with phosphoenolpyruvate and pyruvate kinase was used to avoid the ADP-induced inhibition of ATP hydrolysis. Due to additional fluorescent impurities the background signals were increased.

Within the photon burst starting at $t = 482$ ms three different steps of proximity factors were distinguishable. For the first 12 ms, a mean value of $P = 0.22 \pm 0.1$ was calculated from the FRET acceptor and donor count rates. A different value of $P = 0.45 \pm 0.1$ was calculated starting from $t = 494$ ms. This decreased slightly to $P = 0.35 \pm 0.1$ after $t = 521$ ms. In the time interval from 547 ms to 567 ms the calculated proximity factor was $P = 0.65 \pm 0.15$. In this photon burst the different levels of P were separated by similar decreases in intensity down to levels near the background. The sequence of proximity factors (low → medium → high FRET efficiency) might correspond to a sequence of γ-subunit positions (A1 → A2 → A3) with respect to a donor position D1 in Fig. 11.1c and could indicate counterclockwise rotation.

A second enzyme was observed by a photon burst starting at $t = 715$ ms, clearly identified by the increase in the FRET acceptor count rate (*red trace*). In a detailed analysis we calculated $P = 0.8 \pm 0.1$ for the first 45 ms, then the FRET acceptor count rate dropped suddenly. For the next 6 ms donor and acceptor count rates were similar at low intensities and $P = 0.5$. From $t = 767$ to 786 ms the donor count rate increased and $P = 0.3 \pm 0.1$. The acceptor intensity rised at $t = 790$ ms and for the next 21 ms the proximity factor fluctuated around $P = 0.7 \pm 0.15$. At $t = 812$ ms the acceptor count rate dropped to the donor count rate resulting in $P = 0.45 \pm 0.15$ for 7 ms. From $t = 823$ to 858 ms the calculated proximity factor was $P = 0.6 + 0.1$; from $t = 859$ to 921 ms it was $P = 0.75 \pm 0.1$ and from $t = 922$ to 939 ms it was $P = 0.5 \pm 0.1$. At $t = 940$ ms the proximity factor dropped to $P = 0.3 \pm 0.2$ for 20 ms. For the last 40 ms we calculated $P = 0.65 \pm 0.15$.

In this photon burst the expected sequence of proximity factors for the counterclockwise rotation of the γ-subunit was not clearly seen. The mean values for P were centered around $P = 0.8$ for "high FRET", $P = 0.3$ for "low FRET" and $P = 0.6$ for "medium FRET". Between the A3 → A1 transition we always found a proximity factor around $P = 0.6$, which is the same as for a A2 position. It remains uncertain, whether we observed a conformational substep [26] or whether the FRET donor of the enzyme was labeled at the D2 position, resulting in a reversed order of proximity factors during counterclockwise rotation.

The maximum ATP hydrolysis rates of reconstituted EF_oF_1 range from 200 s^{-1} to about 2 s^{-1} depending on produced ADP concentration and the ΔpH across the lipid membrane, which is due to proton pumping during catalysis [17]. We also found several photon bursts of other enzymes in an "inhibited state" or with stable γ-subunit orientation for more than 200 ms (data not shown). The second enzyme in Fig. 11.2d was highly active. We identified the A3 position of the γ-subunit by the count rate ratio as the starting point for the catalytic cycle of the enzyme. Reaching the proximity factor $P \approx 0.8$ in the time trace indicated three turnovers or three hydrolyzed ATP. This was observed at $t = 790$, 859 and 960 ms. Therefore, nine transitions were observed within 300 ms. This corresponds to a rate between 30 s^{-1} and 50 s^{-1}, which is the maximum rate measured in bulk (40 s^{-1} without added valinomycin and nigericin).

All time intervals observed for the transition of the γ-subunit positions, as identified by the changes of acceptor count rate, range from 2 to 5 ms, which is near the detection limit imposed by the 1 ms time resolution. We conclude that the γ-subunit in EF_oF_1 rotates stepwise during ATP hydrolysis, at least those regions of the γ-subunit around residue 106 with respect to the residue position 64 in the b-subunits of F_o.

11.4
Conclusions

Labeling F_oF_1 ATP synthases with two fluorophores on different subunits in combination with ratiometric single molecule FRET analysis [8] is a promising new approach to determine the orientation of the rotating γ-subunit. Addition of AMPPNP fixes the γ-subunit position and results in stable proximity factors within photon bursts of single enzymes, as they traverse the focal volume. Three different FRET efficiencies are found, which correspond to three possible orientations of the γ-subunit with respect to the static counterpart, the b-subunits of EF_o-b64-TMR-F_1-γ106-Cy5.

Polarization measurements with a single fluorophore as a reporter of the actual γ-subunit orientation in the F_1 part [24, 25] require attachment of the enzyme to a surface and run into the problem of subunit dissociation. In contrast, the reconstituted EF_oF_1 ATP synthases are stable at nanomolar concentrations and fully functional. Conformational dynamics are undisturbed as shown by the high ATP synthesis rates. Therefore, the fluctuations of the proximity factor P within the photon burst of one EF_oF_1 ATP synthase after addition of ATP directly indicate the transitions of the γ-subunit orientation.

To identify conformational substeps [26] during ATP hydrolysis the time resolution has to be improved. The excitation power has to be adjusted to the photophysical limits of both fluorophores to reduce the fluctuations of the proximity factor. By analyzing the FRET donor fluorescence lifetime changes independently within the photon burst, it will be possible to reduce the time resolution to 100 μs. The time interval for the transition between two orientations of the γ-subunit is resolvable even for highly active enzymes. The concept of elastic power transmission between the F_1 and F_o part [21] can be probed directly, because the distances of the $\gamma T106C$ position at the γ-subunit and the b-subunits are measured by the FRET experiment. Finally, the ambiguitiy of the FRET donor position on the b-subunit dimer (D1 or D2) might be overcome by the use of a bisfunctional crosslinking fluorophore like Cy3-bis-maleimide so that the direction of γ-subunit rotation during ATP synthesis can be monitored.

References

1. Yoshida M, Muneyuki E, Hisabori T (2001) Nature Rev Mol Cell Biol 2:669
2. Böttcher B, Bertsche I, Reuter R, Gräber P (2000) J Mol Biol 296:449
3. Noji H, Yasuda R, Yoshida M, Kinosita K (1997) Nature 386:299
4. Wada Y, Sambongi Y, Futai M (2000) Biochim Biophys Acta 1459:499
5. Tsunoda S, Aggeler R, Yoshida M, Capaldi RA (2001) Proc Natl Acad Sci USA 98:898
6. Häsler K, Pänke O, Junge W (1999) Biochemistry 38:13759
7. Stock D, Leslie AGW, Walker JE (1999) Science 286:1700
8. Deniz AA, Laurence TA, Dahan M, Chemla DS, Schultz PG, Weiss S (2001) Annu Rev Phys Chem 52:233
9. Fischer S, Gräber P (1999) FEBS Lett 457:327
10. Aggeler R, Capaldi RA (1992) J Biol Chem 267:21355
11. Börsch M, Turina P, Eggeling C, Fries JR, Seidel CAM, Labahn A, Gräber P (1998) FEBS Lett 437:251
12. Gogol EJ, Luecken U, Bork T, Capaldi RA (1989) Biochemistry 28:4709
13. Turina P, Capaldi RA (1994) J Biol Chem 269:13465
14. Lötscher HR, de Jong C, Capaldi RA (1984) Biochemistry 23:4128
15. Perlin DS, Cox DN, Senior AE (1983) J Biol Chem 258:9793
16. Widengren J, Mets Ü, Rigler R (1995) J Phys Chem 99:13368
17. Fischer S, Gräber P, Turina P (2000) J Biol Chem 275:30157
18. Rodgers A, Wilce M (2000) Nature Struct Biol 7:1051
19. Engelbrecht S, see: http://131.173.26.96/se/se.html
20. Abrahams JP, Leslie AGW, Lutter R, Walker JE (1994) Nature 370:621
21. Junge W, Pänke O, Cherepanov DA, Gumbiowski K, Müller M, Engelbrecht S (2001) FEBS Lett 504:152
22. Wazawa T, Ishii Y, Funatsu T, Yanagida T (2000) Biophys J 78:1561
23. Widengren J, Schwille P (2000) J Phys Chem A 104:6416
24. Häsler K, Engelbrecht S, Junge W (1998) FEBS Lett 426:301
25. Adachi K, Yasuda R, Noji H, Itoh H, Harada Y, Kinosita K (2000) Proc Natl Acad Sci USA 97:7243
26. Yasuda R, Noji H, Yoshida M, Kinosita K, Itoh H (2001) Nature 410:898

Part 3
Application of Fluorescence in Biological Membrane and Enzyme Studies

Application of the Wavelength-selective Fluorescence Approach to Monitor Membrane Organization and Dynamics

A. CHATTOPADHYAY

Wavelength-selective fluorescence comprises a set of approaches based on the red edge effect in fluorescence spectroscopy, which can be used to monitor directly the environment and dynamics around a fluorophore in a complex biological system. A shift in the wavelength of maximum fluorescence emission toward higher wavelengths, caused by a shift in the excitation wavelength toward the red edge of the absorption band, is termed red edge excitation shift (REES). This effect is mostly observed with polar fluorophores in motionally restricted media such as very viscous solutions or condensed phases where the dipolar relaxation time for the solvent shell around a fluorophore is comparable to or longer than its fluorescence lifetime. REES arises from slow rates of solvent relaxation (reorientation) around an excited state fluorophore, which is a function of the motional restriction imposed on the solvent molecules in the immediate vicinity of the fluorophore. Utilizing this approach, it becomes possible to probe the mobility parameters of the environment itself (which is represented by the relaxing solvent molecules) using the fluorophore merely as a reporter group. Furthermore, since the ubiquitous solvent for biological systems is water, the information obtained in such cases will come from the otherwise "optically silent" water molecules. This makes REES and related techniques extremely useful since hydration plays a crucial modulatory role in a large number of important cellular events including lipid–protein interactions and ion transport. The application of REES and related techniques (wavelength-selective fluorescence approach) as a powerful tool to monitor organization and dynamics of probes and peptides bound to membranes and membrane-mimetic medium such as micelles is discussed.

12.1
Introduction

Biological membranes are complex assemblies of lipids and proteins that allow cellular compartmentalization and act as the interface through which cells communicate with each other and with the external milieu. The biological membrane constitutes the site of many important cellular functions including transfer of information from outside to the interior of the cell. However, our understanding of these processes at the molecular level is limited by the lack of high resolution three-dimensional structures of membrane-bound molecules. It is extremely difficult to crystallize membrane-bound molecules for diffraction studies. Only a few years back was the first complete X-ray crystallographic analysis of an integral membrane protein successfully carried out [1]. Even high resolution NMR methods have limited applications for membrane-bound molecules because of slow reorientation times in membranes [2].

Due to the inherent difficulty in crystallizing membrane-bound molecules, most structural analyses of membranes have utilized other biophysical techniques with an emphasis on spectroscopic approaches. Fluorescence spectroscopy has been one of the principal techniques to study organization and dynamics of biological and model membranes because of its suitable time scale, minimal perturbation, noninvasive nature and intrinsic sensitivity [3–7]. This review is focussed on the application of a novel approach, the wavelength-selective fluorescence approach, as a powerful tool to monitor organization and dynamics of probes and peptides bound to membranes and membrane-mimetic systems such as micelles.

12.2
Red Edge Excitation Shift (REES)

In general, fluorescence emission is governed by Kasha's rule which states that fluorescence normally occurs from the zero vibrational level of the first excited electronic state of a molecule [8, 9]. It is obvious from this rule that fluorescence should be independent of wavelength of excitation. In fact, such a lack of dependence of fluorescence emission parameters on excitation wavelength is often taken as a criterion for purity and homogeneity of a molecule. Thus, for a fluorophore in a bulk nonviscous solvent, the fluorescence decay rates and the wavelength of maximum emission are usually independent of the excitation wavelength.

However, this generalization breaks down in case of polar fluorophores in motionally restricted media such as very viscous solutions or condensed phases, that is, when the mobility of the surrounding matrix relative to the fluorophore is considerably reduced. This situation arises because of the importance of the solvent shell and its dynamics around the fluorophore during the process of absorption of a photon and its subsequent emission as fluorescence. Under such conditions, when the excitation wavelength is gradually shifted to the red edge of the absorption band, the maximum of fluorescence emission exhibits a concomitant shift toward higher wavelengths. Such a shift in the wavelength of maximum emission toward higher wavelengths, caused by a corresponding shift in the excitation wavelength toward the red edge of the absorption band, is termed the red edge excitation shift (REES) [7, 10–18]. Since REES is observed only under conditions of restricted mobility, it serves as a reliable indicator of the dynamics of fluorophore environment.

The genesis of REES lies in the change in fluorophore–solvent interactions in the ground and excited states brought about by a change in the dipole moment of the fluorophore upon excitation, and the rate at which solvent molecules reorient around the excited state fluorophore [12, 15–21]. For a polar fluorophore, there exists a statistical distribution of solvation states based on their dipolar interactions with the solvent molecules both in the ground and excited states. Since the dipole moment of a molecule changes upon excitation, the solvent dipoles have to reorient around this new excited state dipole moment of the fluorophore, so as to attain an energetically favorable orientation. This readjustment of the dipolar interaction of the solvent molecules with the fluorophore essentially consists of two components. First, the redistribution of electrons in the surrounding solvent molecules because of the altered dipole moment of the excited state fluorophore, and then, the physical reorientation of the solvent molecules around the excited state fluorophore. The former process is almost instantaneous, i.e., electron redistribution in solvent molecules occurs on about the same time scale as the process of excitation of the fluorophore itself (10^{-15} s). The reorientation of the solvent dipoles, however, requires a net physical displacement. It is thus a much slower process and is dependent on the restriction to their mobility as offered by the surrounding matrix. More precisely, for a polar fluorophore in a bulk non-viscous solvent, this reorientation occurs at a time scale of the order of 10^{-12} s, so that all the solvent molecules completely reorient around the excited state dipole of the fluorophore well within its excited state lifetime, which is typically of the order of 10^{-9} s. Hence, irrespective of the excitation wavelength used, all emission is observed only from the solvent-relaxed state. However, if the same fluorophore is now placed in a viscous medium, this reorientation process is slowed down to 10^{-9} s or longer. Under these conditions, excitation by progressively lower energy quanta, i.e., excitation wavelength being gradually shifted towards the red edge of the absorption band, selectively excites those fluorophores which interact more strongly with the solvent molecules in the excited state. These are the fluorophores around which the solvent molecules are oriented in such a way as to be more similar to that found in the solvent-relaxed state. Thus, the necessary condition for giving rise to REES is that a different average population is excited at each excitation wavelength and, more importantly, that the difference is maintained in the time-scale of fluorescence lifetime. As discussed above, this requires that the dipolar relaxation time for the solvent shell be comparable to or longer than the fluorescence lifetime, so that fluorescence occurs from various partially relaxed states. This implies a reduced mobility of the surrounding matrix with respect to the fluorophore.

The essential criteria for the observation of the red edge effect can thus be summarized as follows: (i) the fluorophore should normally be polar so as to be able to suitably orient the neighboring solvent molecules in the ground state, (ii) the solvent molecules surrounding the fluorophore should be polar, (iii) the solvent reorientation time around the excited state dipole moment of the fluorophore should be comparable to or longer than the fluorescence lifetime, and (iv) there should be a relatively large change in the dipole moment of the fluorophore upon excitation. The observed spectral shifts thus depend both on the properties of the fluorophore itself (i.e., the vectorial difference between the dipole moments in the ground and excited states), and also on properties of the environment interacting with it (which is a function of the solvent reorientation time). It has previously been shown for 7-nitrobenz-2-oxa-1,3-diazol-4-yl (NBD)-labeled phospholipids incorporated into model membranes, that a dipole moment change of ~4 D upon excitation is enough to give rise to significant red edge effects [22]. A recent comprehensive review on the red edge effect is provided in [23].

12.3
The Wavelength-selective Fluorescence Approach

In addition to the dependence of fluorescence emission maxima on the excitation wavelength (REES), fluorescence polarization and lifetime are also known to depend on the excitation and emission wavelengths in viscous solutions and in otherwise motionally restricted media. Taken together, these constitute the wavelength-selective fluorescence approach which consists of a set of approaches based on the red edge effect in fluorescence spectroscopy, which can be used to directly monitor the environment and dynamics around a fluorophore in a complex biological system [7].

Early applications of REES and wavelength-selective fluorescence to systems of biological relevance has been restricted mainly to indole, tryptophan, and other fluorescent probes in viscous solvents, and when present in proteins. The application of REES and related techniques to elucidate organization and dynamics in proteins have previously been reviewed [20, 24, 25] and will not be addressed here since it is beyond the scope of this review. The application of the wavelength-selective fluorescence approach as a powerful, yet sensitive tool to monitor organization and dynamics of probes and peptides bound to membranes and micelles constitutes the subject of this review.

12.4
The Wavelength-selective Fluorescence Approach:
A Novel Tool to Monitor Organization and Dynamics of
the Membrane Interfacial Region

Organized molecular assemblies such as membranes can be considered as large cooperative units with characteristics very different from the individual structural units that constitute them. A direct consequence of such highly organized systems is the restriction imposed on the mobility of their constituent structural units. It is well known that interiors of biological membranes are viscous, with the effective viscosity comparable to that of light oil [34, 35]. The biological membrane, with its viscous interior, and distinct motional gradient along its vertical axis, thus provides an ideal system for the utilization of REES in particular and wavelength-selective fluorescence in general to study various membrane phenomena. The use of this technique becomes all the more relevant in view of the fact that no crystallographic database for membrane-bound probes and proteins exists to date, due to the inherent difficulty in crystallizing such molecules.

Among the three major regions in the membrane, the interfacial region is characterized by unique motional and dielectric characteristics [27] different from the bulk aqueous phase (experienced by charged aqueous probes such as ANS and TNS) and the more isotropic hydrocarbon-like deeper regions of the membrane and plays an important role in functional aspects such as substrate recognition and activity of lipolytic enzymes [36]. This specific region of the membrane exhibits slow rates of solvent relaxation and is also known to participate in intermolecular charge interactions [37] and hydrogen bonding through the polar headgroup [38–40]. These structural features which slow down the rate of solvent reorientation have previously been recognized as typical features of solvents giving rise to significant red edge effects [15]. It is therefore the membrane interface which is most likely to display red edge effects and is sensitive to wavelength-selective fluorescence measurements (see Fig. 12.1).

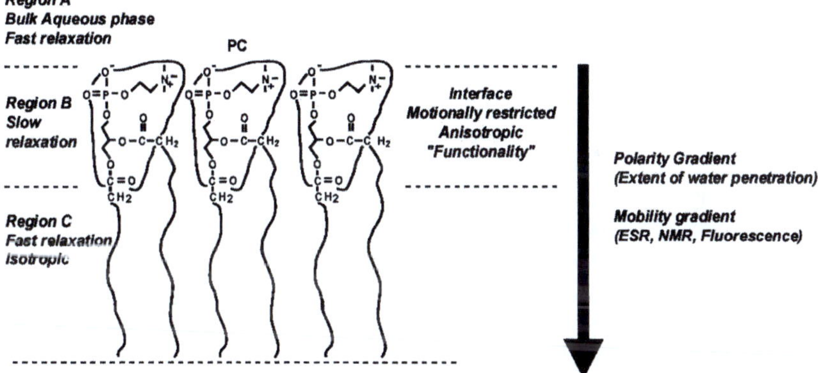

Fig. 12.1. A schematic diagram of half of the membrane bilayer showing the motional anisotropy of the bilayer. The dotted horizontal line at the bottom indicates the center of the bilayer. The membrane anisotropy along the axis perpendicular to the plane of the bilayer divides the membrane leaflet into three broad regions exhibiting very different dynamics (as revealed by various spectroscopic techniques such as ESR, NMR and fluorescence measurements). *Region A*: bulk aqueous phase, fast solvent relaxation; *Region B*: slow (restricted) solvent relaxation, hydrogen bonding (important for functionality), water penetration (interfacial water), highly anisotropic medium; *Region C*: bulk hydrocarbon-like environment, isotropic, fast solvent relaxation. A polarity gradient is also set up along this axis. Fluorescent probes and peptides localized in the interfacial region B are most likely to display red edge effects and are sensitive to wavelength-selective fluorescence measurements

Initial reports, in which excitation wavelength dependence of emission maxima of the fluorescence probes TNS or ANS in model lecithin membranes were investigated, indicated no appreciable red shift [17, 41]. However, REES was reported in case of dioleoyl-*sn*-glycero-3-phosphocholine (DOPC) vesicles labeled with Patman, an amphiphilic phase-sensitive probe [42]. The above results can be rationalized on the basis of the location of these probes in the membrane. On one hand, both TNS and ANS are charged at neutral pH and, therefore, access the phospholipid headgroup from the external aqueous phase. This implies that, for these probes, the immediate environment will be the aqueous phase adjacent to the headgroup, where solvent relaxation is extremely fast, and as such, no red edge effect can be expected. On the other hand, because of the fatty acyl chain in Patman, it partitions well into the membrane, and therefore, experiences a much more motionally restricted environment. In a later report [43], REES of membrane-bound 1-phenylnaphthylamine (1-AN) was monitored. However, this probe has the limitation of not having a unique location in the membrane. Rather, it has a distribution of locations in the membrane. This makes the interpretation of REES data more difficult since the spectral information obtained can no longer be attributed to a unique environment in the membrane.

The choice of a suitable probe is thus of utmost importance in designing membrane-active molecules capable of exhibiting REES. It is desirable that the probe be polar and be able to strongly partition into the membrane and intercalate with its normal components, i.e., the phospholipids. Further, the fluorescent portion of the molecule should be suitably embedded in the membrane. REES is indeed observed when the above criteria are satisfied [44–56]. In addition, it is preferable that the membrane embedded molecule has only one fluorescent group and that it has a unique location in the membrane and not a distribution of locations. Such probes can be used to correlate the extent of REES with a specific fluorophore environment, which in case of such uniquely localized probes, translates to a specific region of the membrane.

One such probe that has been employed to study the phenomenon of REES and related effects in membranes [44, 49] and membrane-mimetic systems [57, 58] is the widely used lipid probe NBD-PE [*N*-(7-nitrobenz-2-oxa-1,3-diazol-4-yl)-1,2--dipalmitoyl-*sn*-glycero-3-phosphoethanolamine. The NBD group is a widely used fluorophore in biophysical, biochemical, and cell biological studies. NBD-labeled lipids are extensively used as fluorescent analogues of native lipids in biological and model membranes to study a variety of processes (for a review on NBD-labeled lipids, see [59]). The NBD moiety possesses some of the most desirable properties for serving as an excellent probe for both spectroscopic and microscopic applications. It is very weakly fluorescent in water. Upon transfer to a hydrophobic medium, it fluoresces brightly in the visible range and exhibits a high degree of environmental sensitivity [22, 44, 60–62]. Fluorescence lifetime of the NBD group is extremely sensitive to the environmental polarity [58, 61]. It is relatively photostable, and lipids labeled with the NBD group mimic endogenous lipids in studies of intracellular lipid transport [63, 64].

In NBD-PE, the fluorescent NBD label is covalently attached to the headgroup of a phosphatidylethanolamine molecule. The precise orientation and location of the NBD group of this molecule in the membrane is known [60, 65–69]. This group has been found to be localized at the membrane interface, which has unique motional and dielectric characteristics distinct from both the bulk aqueous phase and the hydrocarbon-like interior of the membrane [26–33] and this makes it an ideal probe for monitoring red edge effects. Furthermore, previous electrophoretic measurements have shown that the NBD group in NBD-PE is uncharged at neutral pH in the membrane [60]. This would ensure that the NBD group does not project into the external aqueous phase. This is advantageous since the ability of a fluorophore to exhibit red edge effects could very well be dependent on its precise location in the membrane (see later). In addition, the change in dipole moment of the NBD group upon excitation, a necessary condition for a fluorophore to exhibit REES, has been found to be ~4 D [22]. NBD-PE exhibits REES in model membranes of DOPC [44, 49]. Since the precise localization of the fluorescent NBD group in membrane-bound NBD-PE is known to be interfacial [60, 65–69], this result directly implies that the interfacial region of the membrane offers considerable restriction to the reorientational motion of the solvent dipoles around the excited state fluorophore. In another study, REES of membrane-bound NBD-cholesterol, in which the NBD group is covalently attached to the flexible chain of the cholesterol molecule, was reported [49].

12.5
Wavelength-selective Fluorescence as a Membrane Dipstick

The biological membrane is a highly organized molecular assembly, largely confined to two dimensions, and exhibits considerable degree of anisotropy along the axis perpendicular to the membrane plane [26–33]. This not only results in the anisotropic behavior of the constituent lipid molecules, but more importantly, the environment of a probe molecule becomes very much dependent on its precise localization in the membrane. While the center of the bilayer is nearly isotropic, the upper portion, only a few angstroms away toward the membrane surface, is highly ordered [26–33]. As a result, properties such as polarity, fluidity, segmental motion, ability to form hydrogen bonds and extent of solvent (water) penetration would vary in a depth-dependent manner in the membrane (see Fig. 12.1). A direct consequence of such an anisotropic transmembrane environment will be the differential extents to which the mobility of water molecules will be retarded at different depths in the membrane relative to the water molecules in bulk aqueous phase. This offers the possibility of using wavelength-selective fluorescence as a novel approach to investigate the depth of membrane penetration of a reporter fluorophore, i.e., as a membrane dipstick. This was tested by demonstrating that chemically identical fluorophores, varying solely in terms of their localization at different depths in the membrane, experience very different local environments, as judged by wavelength-selective fluorescence parameters [50]. Two anthryoloxy stearic acid derivatives, where the anthroyloxy group has previously been found to be either shallow (2-AS) or deep (12-AS), were used. It was shown that the anthroyloxy moiety of 2 and 12-AS experiences different local membrane microenvironments, as reflected by depth-dependent variation of red edge excitation shift (REES) as well as varying degrees of wavelength dependence of fluorescence polarization and lifetime, and rotational correlation times. These results were attributed to differential rates of solvent reorientation in the immediate vicinity of the anthroyloxy group as a function of its membrane penetration depth. Wavelength-selective fluorescence therefore constitutes a novel approach to probe defined depths in the membrane and can be conveniently used as a dipstick to characterize the depth of penetration of membrane-embedded fluorophores.

12.6
Application of the Wavelength-selective Fluorescence Approach to Membrane Peptides and Proteins

The presence of tryptophan residues as intrinsic fluorophores in most peptides and proteins makes them an obvious choice for fluorescence spectroscopic analyses of such systems. The role of tryptophan residues in the structure and function of membrane proteins has recently attracted a lot of attention [48, 70]. Membrane proteins have been reported to have a significantly higher tryptophan content than soluble proteins [71]. In addition, it is becoming increasingly evident that tryptophan residues in integral membrane proteins and peptides are not uniformly distributed and that they tend to be localized toward the membrane interface, possibly because they are involved in hydrogen bonding [72] with the lipid carbonyl groups or interfacial water molecules. As mentioned earlier, the interfacial region in membranes is characterized by unique motional and dielectric characteristics distinct from both the bulk aqueous phase and the hydrocarbon-like interior of the membrane [26–33]. The tryptophan residue has a large indole side chain that consists of two fused aromatic rings. In molecular terms, tryptophan is a unique amino acid since it is capable of both hydrophobic and polar interactions. In fact, the hydrophobicity of tryptophan, measured by partitioning into bulk solvents, has previously been shown to be dependent on the scale chosen [73]. Tryptophan ranks as one of the most hydrophobic amino acids on the basis of its partitioning into polar solvents such as octanol [74] while scales based on partitioning into nonpolar solvents like cyclohexane [75] rank it as only intermediate in hydrophobicity. This ambiguity results from the fact that while tryptophan has the polar NH group that is capable of forming hydrogen bonds, it also has the largest nonpolar accessible surface area among the naturally occuring amino acids [76]. Wimley and White [77] have recently shown from partitioning of model peptides to membrane interfaces that the experimentally determined interfacial hydrophobicity of tryptophan is highest among the naturally occurring amino acid residues, thus accounting for its specific interfacial localization in membrane-bound peptides and proteins. Due to its aromaticity, the tryptophan residue is capable of π–π interactions and of weakly polar interactions [78]. The amphipathic character of tryptophan gives rise to its hydrogen bonding ability which could account for its orientation in membrane proteins and its function through long-range electrostatic interaction [79]. The amphipathic character of tryptophan also explains its interfacial localization in membranes due to its tendency to be solubilized in this region of the membrane, besides favorable electrostatic interactions and hydrogen bonding. It has already been mentioned (see above) that the membrane interface is most likely to display red edge effects and is sensitive to wavelength-selective fluorescence measurements. This makes study of membrane peptides and proteins by the wavelength-selective fluorescence approach very appropriate. Indeed, tryptophan octyl ester, often used as a simple model for membrane-bound tryptophan residues, exhibits pH-dependent REES when bound to membranes [48].

Melittin, the major toxic component in the venom of the European honey bee, *Apis mellifera*, was one of the earlier membrane peptides studied using the wavelength-selective fluorescence. Results from these studies showed that when bound to zwitterionic membranes, the microenvironment of the sole functionally active tryptophan of melittin was motionally restricted as evident from REES and other results [45]. However, when bound to negatively charged membranes, studies using the wavelength-selective fluorescence approach indicate that the microenvironment of the tryptophan gets modulated and this could be related to the functional difference in the lytic activity of the peptide observed in the two cases [47].

In yet another study, the phenomenon of REES, in conjunction with time-resolved fluorescence spectroscopic parameters such as wavelength-dependent fluorescence lifetimes and time-resolved emission spectra (TRES) were utilized to study the localization and dynamics of the functionally important tryptophan residues in the gramicidin channel [46]. Gramicidin belongs to a family of prototypical channel formers which are naturally fluorescent due to the presence of four tryptophan residues. These interfacially localized tryptophans are known to play a crucial role in the organization and function of the channel. The results from the above study point out the motional restriction experienced by the tryptophans at the peptide-lipid interface of the gramicidin channel. This is consistent with other studies [79, 80] in which such restrictions are thought to be imposed due to hydrogen bonding between the indole rings of the tryptophan residues in the channel conformation and the neighboring lipid carbonyls. The significance of such organization in terms of functioning of the channel is brought out by the fact that substitution, photodamage, or chemical modification of these tryptophan residues are known to give rise to channels with altered conformation and reduced conductivity. Tryptophan residues in another pore-forming toxin, *Staphylococcus aureus* α-toxin, also exhibit REES indicating a restricted and buried environment for these residues [51].

In addition, the red edge effect has also been utilized to study the microconformational heterogeneity of the membrane-binding domain of cytochrome b_5 by comparing the information obtained from the native protein and its mutant which has a single tryptophan residue in this domain [52]. Both these proteins show a red shift in the emission spectrum when excited at the long wavelength edge of the excitation spectrum, indicating thereby that the tryptophan residue(s) in both cases are localized in a region of motional constraint. Very recently, REES of tryptophans in mitochondrial creatine kinase has been reported [53].

12.7
Wavelength-selective Fluorescence in Micelles

Micelles represent yet another type of organized molecular assemblies formed by the hydrophobic effect and are highly cooperative, dynamic assemblies of soluble amphiphiles (detergents). They offer certain inherent advantages in fluorescence studies over membranes since micelles are smaller and optically transparent, have well-defined sizes, and are relatively scatter-free. Furthermore, micelles can be of any desired charge type and can adopt different shapes and internal packing, depending on the chemical structures of the constituent monomers and the ionic strength of the medium. A direct consequence of such organized systems is the restriction imposed on the dynamics and mobility of the constituent structural units. The studies on micellar organization and dynamics assume special significance in view of the fact that the general principles underlying the formation of micelles are common to other related assemblies such as reverse micelles, bilayers, liposomes, and biological membranes. Micelles are extensively used as membrane mimetics in studies of membrane proteins and peptides and as a model for the anesthetic action of pharmacological compounds.

The organization and dynamics of micellar environments, namely, the core, the interface, and the immediately adjacent layers of water near the interface, have been investigated using experimental [81–87] and theoretical [88] approaches. It is fairly well established now that practically all types of molecules have a surface-seeking tendency in micelles (due to very large surface area to volume ratio) and that the interfacial region is the preferred site for solubilization, even for hydrophobic molecules [83, 89–91]. The suitability of micellar systems for studies employing wavelength-selective fluorescence was therefore tested using the interfacial fluorescence probe NBD-PE [57]. NBD-PE exhibits REES when incorporated in micelles formed by a variety of detergents (SDS, Triton X-100, CTAB, and Chaps), which differ in their charge, aggregation number and shape [57, 58]. These results clearly demonstrate that the relaxation rates of micellar interfacial hydration are very different from that of the bulk water and this feature may play an important role in the reactions catalyzed by micelles.

Structural transition can be induced in charged micelles by increasing ionic strength of the medium or amphiphile concentration. Thus, spherical micelles of sodium dodecyl sulfate (SDS) that exist in water at concentrations higher than critical micelle concentration assume an elongated rod-like structure in the presence of increased electrolyte (salt) concentration when interactions among the charged headgroups are attenuated due to the added salt. This is known as sphere-to-rod transition [92]. The change in organization and dynamics that is accompanied with the salt-induced sphere-to-rod transition in SDS micelles was monitored using NBD-labeled lipids utilizing the wavelength-selective fluorescence approach. It thus appears that REES and related parameters are sensitive indicators of the structural transition in micelles induced by salt.

12.8
Conclusions

Water plays a crucial role in the formation and maintenance of both folded protein and membrane architecture in a cellular environment. Knowledge of dynamics of hydration at a molecular level is thus of considerable importance in understanding the cellular structure and function [93–99]. As mentioned earlier, REES is based on the change in fluorophore–solvent interactions in the ground and excited states brought about by a change in the dipole moment of the fluorophore upon excitation, and the rate at which solvent molecules reorient around the excited state fluorophore. Since for most biological systems, the ubiquitous solvent is water, the information obtained in such cases will come from the otherwise optically silent water molecules. The unique feature about REES is that while all other fluorescence techniques (such as fluorescence quenching, energy transfer, polarization measurements) yield information about the fluorophore (either intrinsic or extrinsic) itself, REES provides information about the relative rates of solvent (water in biological systems) relaxation dynamics, which is not possible to obtain by other techniques. This makes the use of REES and the wavelength-selective fluorescence approach extremely useful in membrane biology since hydration plays a crucial modulatory role in a large number of important cellular events involving the membrane such as lipid-protein interactions [95] and ion transport [93–96].

Acknowledgement. Work in my laboratory was supported by the Council of Scientific and Industrial Research and Department of Science and Technology, Government of India. Some of the work described in this article was carried out by former and present members of my laboratory whose contribution I gratefully acknowledge. I thank K. Shanti and H. Raghuraman for critically reading the manuscript and Devaki Kelkar for her help with the figure.

References

1. Deisenhofer J, Epp O, Miki K, Huber R, Michel H (1985) Nature 318:618
2. Opella SJ (1997) Nature Struct Biol 4 (NMR Supplement):845
3. Radda GK (1975) Fluorescent probes in membrane studies. In: Korn ED (ed) Methods in membrane biology. Plenum Press, New York, p 97
4. Lakowicz JR (1981) Fluorescence spectroscopic investigations of dynamic properties of biological membranes. In: Bell JE (ed) Spectroscopy in biochemistry. CRC Press, Boca Raton, Fl, p 195
5. Stubbs CD, Williams BW (1992) Fluorescence in membranes. In: Lakowicz JR (ed) Topics in fluorescence spectroscopy, vol 3, Biochemical applications. Plenum Press, New York, p 231
6. Chattopadhyay A (1992) Membrane penetration depth analysis using fluorescence quenching; a critical review. In: Gaber BP, Easwaran KRK (eds) Biomembrane structure and function: the state of the art. Adenine Press, Schenectady, NY, p 153
7. Mukherjee S, Chattopadhyay A (1995) J Fluorescence 5:237

8.   Birks JB (1970) Photophysics of aromatic molecules. Wiley-Interscience, London
9.   Rohatgi-Mukherjee KK (1978) Fundamentals of photochemistry. Wiley Eastern, New Delhi
10. Chen RF (1967) Anal Biochem 19:374
11. Fletcher AN (1968) J Phys Chem 72:2742
12. Galley WC, Purkey RM (1970) Proc Natl Acad Sci USA 67:1116
13. Rubinov AN, Tomin VI (1970) Opt Spectrosk USSR 29:1082
14. Castelli F, Forster LS (1973) J Amer Chem Soc 95:7223
15. Itoh KI, Azumi T (1975) J Chem Phys 62:3431
16. Demchenko AP (1982) Biophys Chem.15:101
17. Lakowicz JR, Keating-Nakamoto S (1984) Biochemistry 23:3013
18. Macgregor RB, Weber G (1981) Ann NY Acad Sci 366:140
19. Demchenko AP (1986) Ultraviolet Spectroscopy of Proteins. Springer, Heidelberg
20. Demchenko AP (1988) Trends Biochem Sci 13:374
21. Demchenko AP, Ladokhin AS (1988) Eur Biophys J 15:369
Mukherjee S, Chattopadhyay A, Samanta A, Soujanya T (1994) J Phys Chem 98: 2809
23. Demchenko AP (2001) Luminescence (in press)
24. Demchenko AP (1992) Fluorescence and dynamics in proteins. In: Lakowicz JR (ed) Topics in fluorescence spectroscopy, vol 3, Biochemical applications. Plenum Press, New York, p 65
25. Lakowicz JR (2000) Photochem Photobiol 72:421
26. Seelig J (1977) Quart Rev Biophys 10:353
27. Ashcroft RG, Coster HGL, Smith JR (1981) Biochim Biophys Acta 643:191
28. Stubbs CD, Meech SR, Lee AG, Phillips D (1985) Biochim Biophys Acta 815: 351
29. Perochon E, Lopez A, Tocanne JF (1992) Biochemistry 31:7672
30. White SH, Wimley WC (1994) Curr Opinion Str Biol 4:79
31. Slater SJ, Ho C, Taddeo FJ, Kelly MB, Stubbs CD (1993) Biochemistry 32:3714
32. Venable RM, Zhang Y, Hardy BJ, Pastor RW (1993) Science 262:223
33. Gawrisch K, Barry JA, Holte LL, Sinnwell T, Bergelson LD, Ferretti JA (1995) Mol Membr Biol 12:83
34. Cone RA (1972) Nature New Biol 236:39
35. Poo MM, Cone RA (1974) Nature 247:438
36. El-Sayed MY, DeBose CD, Coury LA, Roberts MF (1985) Biochim Biophys Acta 837:325
37. Yeagle P (1987) The membranes of cells. Academic Press, Orlando, Fl, p 89
38. Gennis RB (1989) Biomembranes: molecular structure and function. Springer, New York, p 47
39. Shin TB, Leventis R, Silvius JR (1991) Biochemistry 30:7491
40. Boggs JM (1987) Biochim Biophys Acta 906:353
41. Demchenko AP, Shcherbatska NV (1985) Biophys Chem 22:131
42. Lakowicz JR, Bevan DR, Maliwal BP, Cherek H, Balter A (1983) Biochemistry 22:5714
43. Gakamsky DM, Demchenko AP, Nemkovich NA, Rubinov AN, Tomin VI, Shcherbatska NV (1992) Biophys Chem 42:49
44. Chattopadhyay A, Mukherjee S (1993) Biochemistry 32:3804
45. Chattopadhyay A, Rukmini R (1993) FEBS Lett 335:341
46. Mukherjee S, Chattopadhyay A (1994) Biochemistry 33:5089
47. Ghosh AK, Rukmini R, Chattopadhyay A (1997) Biochemistry 36:14291

48. Chattopadhyay A, Mukherjee S, Rukmini R. Rawat SS, Sudha S (1997) Biophys J 73:839
49. Chattopadhyay A, Mukherjee S (1999) J Phys Chem B 103:8180
50. Chattopadhyay A, Mukherjee S (1999) Langmuir 15:2142
51. Raja SM, Rawat SS, Chattopadhyay A, Lala AK (1999) Biophys J 76:1469
52. Ladokhin AS, Wang L, Steggles AW, Holloway PW (1991) Biochemistry 30:10200
53. Granjon T, Vacheron MJ, Vial C, Buchet R (2001) Biochemistry 40:6016
54. Hutterer R, Schneider FW, Sprinz H, Hof M (1996) Biophys Chem 61:151
55. Santos NC, Prieto M, Castanho, MA (1998) Biochemistry 37:8674
56. MacPhee CE, Howlett GJ, Sawyer WH, Clayton AH (1999) Biochemistry 38:10878
57. Rawat SS, Mukherjee S, Chattopadhyay A (1997) J Phys Chem B 101:1922
58. Rawat SS, Chattopadhyay A (1999) J Fluorescence 9:233
59. Chattopadhyay A (1990) Chem Phys Lipids 53:1
60. Chattopadhyay A, London, E. (1988) Biochim Bioiphys Acta 938:24
61. Lin S, Struve WS (1991) Photochem Photobiol 54:361
62. Fery-Forgues S, Fayet JP, Lopez A (1993) J Photochem Photobiol A 70:229
63. Van Meer GE, Stelzer HK, Wijnaendts-van-Resandt RW, Simons K (1987) J Cell Biol 105:1623
64. Koval M, Pagano RE (1990) J Cell Biol 111:429
65. Chattopadhyay A, London E (1987) Biochemistry 26:39
66. Pagano RE, Martin OC (1988) Biochemistry 27:4439
67. Mitra B, Hammes GG (1990) Biochemistry 29:9879
68. Wolf DE, Winiski AP, Ting AE, Bocian KM, Pagano RE (1992) Biochemistry 31:2865
69. Abrams FS, London E (1993) Biochemistry 32:10826
70. Reithmeier RAF (1995) Curr Opi Str Biol 5:491
71. Schiffer M, Chang CH, Stevens FJ (1992) Protein Eng 5:213
72. Ippolito JA, Alexander RS, Christianson DW (1990) J Mol Biol 215:457
73. Fauchere JL (1985) Trends Biochem Sci 10:268
74. Fauchere JL, Pliska V (1983) Eur J Med Chem 18:369
75. Radzicka A, Wolfenden R (1988) Biochemistry 27:1664
76. Wimley WC, Whitc SH (1992) Biochemistry 31:12813
77. Wimley WC, White SH (1996) Nature Struct Biol 3:842
78. Burley SK, Petsko GA (1988) Adv Protein Chem 39:125
79. Fonseca V, Daumas P, Ranjalahy-Rasoloarijao L, Heitz F, Lazaro R, Trudelle Y, Andersen OS (1992) Biochemistry 31:5340
80. Becker MD, Greathouse DV, Koeppe RE, Andersen OS (1991) Biochemistry 30:8830
81. Shinitzky M, Dianoux AC, Gitler C, Weber G (1971) Biochemistry 10:2106
82. Kalyanasundaram K, Thomas JK (1977) J Phys Chem 81:2176
83. Mukerjee P, Cardinal JR (1978) J Phys Chem 82:1620
84. Leung R, Shah DO (1986) J Colloid Interface Sci 113:484
85. Nery H, Soderman O, Canet D, Walderhaug H, Lindman B (1986) J Phys Chem 90:5802
86. Maiti NC, Mazumdar S, Periasamy N (1995) J Phys Chem 99:10708
87. Saroja G, Samanta A (1995) Chem Phys Lett 246:506
88. Gruen DWR (1985) J Phys Chem 89:153
89. Ganesh KN, Mitra P, Balasubramanian D (1982) J Phys Chem 86:4291
90. Shobha J, Balasubramanian D. (1986) J Phys Chem 90:2800

91. Shobha J, Srinivas V, Balasubramanian D (1989) J Phys.Chem 93:17
92. Missel PJ, Mazer NA, Carey MC, Benedek GB (1982) Thermodynamics of the sphere-to-rod transition in alkyl sulfate micelles. In: Mittal KL, Fendler EJ (eds) Solution behavior of surfactants: theoretical and applied aspects, vol 1. Plenum Press, New York, p 373
93. Crowe JH, Crowe LM (1984) Biol Membr 5:57
94. Rand RP, Parsegian VA (1989) Biochim Biophys Acta 988:351
95. Ho C, Stubbs CD (1992) Biophys J 63:897
96. Ho C, Kelly MB, Stubbs CD (1994) Biochim Biophys Acta 1193:307
97. Fischer WB, Sonar S, Marti T, Khorana HG, Rothschild KJ (1994) Biochemistry 33:12757
98. Kandori H, Yamazaki Y, Sasaki J, Needleman R, Lanyi JK, Maeda A (1995) J Am Chem Soc 117:2118
99. Sankararamakrishnan R, Sansom MSP (1995) FEBS Lett 377:377

Fluorescence Approaches for the Characterization of the Peripheral Membrane Binding of Proteins Applied for the Blood Coagulation Protein Prothrombin

R. HUTTERER AND M. HOF

Different fluorescence spectroscopic approaches for the characterization of the peripheral membrane binding of proteins are summarized and compared for the blood coagulation protein prothrombin. Basically, the problem of membrane binding can be approached from both the protein side, using either intrinsic tryptophan fluorescence or dye-labeled protein, as typically employed in fluorescence recovery after photobleaching experiments, or from the membrane side, studying solvent relaxation, fluorescence anisotropy or excimer formation of suitable membrane probes. For the case of prothrombin and its fragment 1 the combined use of these fluorescence methods shows a slightly tighter binding of the fragment 1 portion in the case of the entire prothrombin molecule compared to the isolated fragment 1 while no evidence could be obtained for hydrophobic membrane binding sites in the "nonfragment 1" part of prothrombin. Tryptophan fluorescence of fragment 1 is able to detect differences in binding in response to the kind of procoagulant membrane lipid.

13.1
Introduction

The blood coagulation protein prothrombin, the zymogene of thrombin, is a single polypeptide chain glycoprotein which requires a high calcium stoichometry for membrane binding. It serves as the substrate of the prothrombinase complex consisting of the serine protease factor Xa that associates with cofactor Va on membrane surfaces containing negatively charged phospholipids in the presence of Ca^{2+} [1]. The catalytic activity of the prothrombinase complex strongly depends on the chemical structure of the phospholipid headgroup. Although several anionic phospholipids accelerate the prothrombin activation, the *L*-serine headgroup has been shown to be by far the most efficient one, promoting phosphatidyl-*L*-serine (PS) to the outstanding prothrombin-activating lipid [2–5]. The reason for this high lipid specificity is not known yet. However, the elucidation of the molecular mechanism of the interaction between the three proteins and the membrane surface will be a crucial step towards a more thorough understanding.

Fluorescence spectroscopy has been an important tool in the study of protein–membrane interactions due to its versatility and high sensitivity. Principally, the problem of protein–membrane interactions can be approached from two different sides. The first possibility is to take advantage of the intrinsic tryptophan (Trp) fluorescence or of fluorescent labels attached to the protein. In tryptophan studies the fluorescence properties of the protein in free and membrane-bound state are compared [6]. Fluorescence changes in the intrinsic fluorescence may be interpreted in terms of a specific conformational change occurring during membrane binding. The use of bright covalently attached fluorescent labels allows for the application of techniques like fluorescence resonance energy transfer (FRET) [7, 8] or fluorescence recovery after photobleaching (FRAP) [9–11]. The second possibility is to focus on the phospholipid organization of the membrane, which might change due to protein binding. The characterization of pyrene excimer formation [12, 13] as well as the determination of fluorescence anisotropy of embedded membrane probes like 1,6-diphenylhexatriene (DPH) [14] have been widely used in studies of biological and model membranes. While both methods are well suited to monitor the dynamics of the interior part of a lipid bilayer [14], they turned out much less useful for the study of the organization of the lipid headgroup region [14–16]. A fluorescence method which overcomes this limitation is the solvent relaxation method. Originally designed for the study of very fast solvation processes in nonviscous solvents [17], the method has been successfully adapted to the study of both the hydrophilic headgroup region and the hydrophobic interior of phospholipid bilayers [18–23]. This review summarizes several contributions describing the use of fluorescence techniques in the characterization of the calcium-mediated membrane binding of prothrombin and its *N*-terminal peptide fragment 1 (F1).

13.2
Prothrombin Binding to Negatively Charged Membrane Surface Characterized by Protein Fluorescence

The main advantages of exploiting the intrinsic (Trp) fluorescence in protein studies are the high sensitivity of the Trp fluorescence to changes in its microenvironment and the fact that the protein has not to be modified and, thus, the spectroscopic measurements can be performed using the native protein. Since the photophysics of even a single Trp residue in proteins appears complex in most cases, the applicability of tryptophan studies is limited to proteins with a low number of Trp residues. Moreover, the low quantum yield and the need for ultraviolet excitation (ideal excitation wavelength for Trp $\lambda_{ex} = 295$ nm) of Trp prevent the combination of Trp fluorescence with techniques employing fluorescence microscopy at the present time. Application of multiphoton excitation might be one way to overcome this limitation in the future [24]. Since prothrombin contains a rather high number of Trp residues (12 and 14 for human and bovine prothrombin, respectively [25]), tryptophan studies have been limited to its fragment 1 (F1), which contains only three Trp residues [6, 26].

The application of bright covalently attached fluorescent labels in proteins studies, on the other hand, is not limited by the size of the protein or protein fragment. Moreover, this approach allows for the application of various fluorescence techniques and the use of microscopes. The main disadvantages, however, are that the covalent labeling might be rather unspecific regarding the site of attachment and that the label might change structure and function of the protein.

13.2.1
Intrinsic Protein Fluorescence: Picosecond Tryptophan Fluorescence of Membrane-bound Prothrombin Fragment 1 (F1)

The 1-156 *N*-terminal polypeptide F1 is believed to be the region predominantly responsible for the metal ion and membrane binding properties of prothrombin. Besides small but significant differences in the desorption rate, it displays basically very similar membrane binding characteristics as the entire protein [10]. The structure of F1 is commonly divided into the *N*-terminal "Gla domain", characterized by 10 γ-carboxyglutamic acid residues (Gla) and a region of disulfide linkages known as the "kringle region". Calcium ions bind (almost) exclusively to the Gla domain [27] and form the native conformation required for membrane binding [28]. Fig. 13.1 depicts a sketch of the X-ray structure of Ca-bovine (B)F1 and shows the location of its three tryptophan residues and the seven calcium ions bound to the Gla domain. Investigating possible molecular differences in the binding of the calcium–prothrombin complex to differently composed, negatively charged membrane surfaces, one might speculate whether such differences can be found in the conformation of membrane-bound Gla domains.

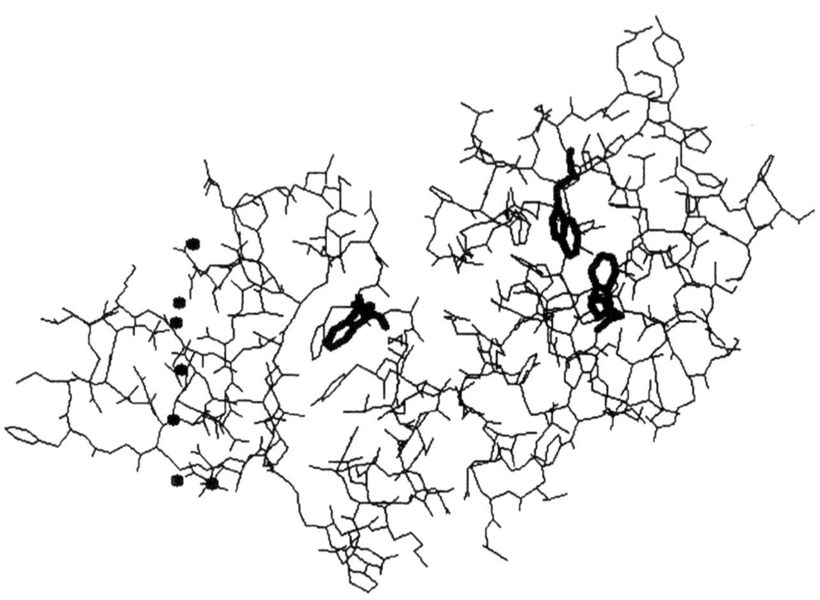

Fig. 13.1. A depiction of the X-ray structure of Ca-BF1. The right part of the protein is the kringle-domain, where Trp90 and Trp126 are located. The Gla-domain is the left part of the protein, containing Trp42 and seven calcium ions (dots). The coordinates where taken from the Brookhaven Protein Data Bank entries [29] and displayed via Rasmol V2.6 software (Glaxo Welcome Research Development, Sterenage, United Kingdom)

Using picosecond fluorescence time-correlated single photon counting, changes in the microenvironment of the three individual tryptophan residues can be separated from each other without either cleaving BF1 into the isolated Gla (containing Trp42) and kringle domains (containing Trp90 and Trp126) or modifying the protein by site-directed mutagenesis. The sensitivity of this approach to conformational changes has been demonstrated by a time-resolved study of the calcium-induced conformational change in BF1 [26]. The cited work [26] comprises a detailed analysis of the wavelength-dependent fluorescence decays of apo-BF1 as well as of Ca-BF1. Fluorescence lifetime distribution (see Fig. 13.2) and conventional multiexponential analysis, as well as acrylamide quenching studies led to the identification of six distinguishable tryptophan excited states for the apo-form as well as for the Ca-form of BF1. Accessibility to the quencher and the known structure have been used to associate the fluorescence decay of the tryptophan present in the Gla domain (Trp42) with two red-shifted components (2.3 and 4.9 ns for apo-BF1). The two kringle domain tryptophans (Trp90 and Trp126) exhibit three decay times (0.24, 0.68 and 2.3 ns for apo-BF1) which are blue shifted. The 0.06 ns component remained unassigned due to the limited time-resolution of the experiment. It should be noted that the mentioned lifetime values result from a global analysis of the fluorescence decays detected at 305 to 425 nm.

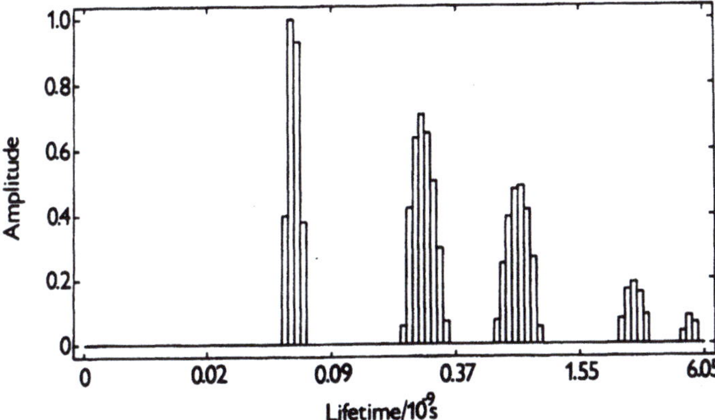

Fig. 13.2. Amplitude profile of the fluorescence lifetimes distribution of apo-BF1 fluorescence decay (λ_{ex} = 295 nm; λ_{em} = 330 nm; 4 µM in Tris-buffer) resulting from Edinburgh Analytical Instruments distribution analysis illustrating the existence of five separate classes of fluorescence lifetimes [26]. Note that the 2.3 ns component is due to the fluorescence of Gla and kringle tryptophan residues

Addition of calcium ions did not change fluorescence lifetimes and intensities of those components which have been assigned exclusively to the two kringle tryptophans Trp90 and Trp126 (0.24 ns and 0.68 ns) – a result that argues against a calcium binding site in the kringle domain [27]. On the other hand, we found that the overall fluorescence quenching is due to a "static-like" quenching of the Gla-Trp (Trp42) components D and E (2.3 ns and 5.1 ns for Ca-BF1, respectively) as a consequence of a ground state interaction between Trp42 and the Cys18-Cys23 disulfide bridge. For illustration see Fig. 13.3, comparing the decay-associated spectra (DAS) of the two Gla-components in absence and presence of calcium. It is important to note that an observed decrease in the fluorescence intensity in the 5.1 ns component of 85% is by far larger than the calcium-induced changes in the parameters determined by circular dichroism [30], by antibody binding experiments [31], differential scanning calometric studies [32, 33], and fouricr trans form infrared spectroscopy [34].

The high sensitivity of the time-resolved Trp42 fluorescence to conformational changes in the Gla domain was one motivation to reexamine the hypothesis of possible lipid-induced conformational changes in the Gla-domain of prothrombin by picosecond tryptophan fluorescence spectroscopy of BF1 [6]. Therefore, the wavelength-dependent tryptophan fluorescence decays of Ca-BF1 in presence of pure phosphatidylcholine (PC) small unilamellar vesicles (SUV) and PC-SUV containing either 25% phosphatidylserine (PS) or 40% phosphatidylglycerol (PG) were characterized. Based on the determined apparent membrane dissociation constant K_d (The K_d values are 0.9 ±0.1 µM and 0.8 ±0.1 µM for 25 mol % PS, and 40 mol % PG, respectively), Ca-BF1 (4 µM) should have been >90% bound to the membrane surface at a lipid concentration of 1.3 mM in both investigated lipid systems. In both cases the lifetime analysis identified the existence of five wavelength-independent lifetimes. Specific binding to PS-containing membranes

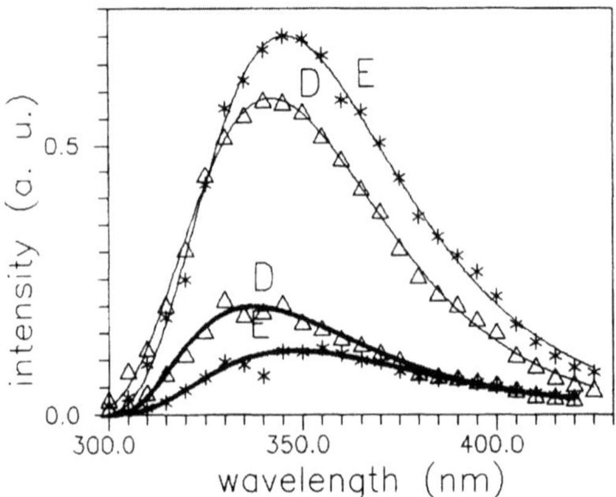

Fig. 13.3. Decay-associated spectra of components D (Δ) and E (∗) for apo-BF1 (2.3 ±0.2 ns (D) and 4.9 ±0.3 ns (E)) and for Ca-BF1 (2.3 ±0.3 ns (D) and 5.1 ±0.4 ns (E)). Shown are the experimental points and the log-normal fits to the data. Thick lines represent the decay-associated spectra of Ca-BF1. The emission maxima are 341.9 nm (D) and 345.5 nm (E) for apo-BF1 and 337.4 nm (D) and 347.6 nm (E) for Ca-BF1

did neither change the fluorescence lifetimes nor the corresponding wavelength-dependent amplitudes. Apparently, the membrane binding part of calcium-prothrombin remains in its native conformation when bound to the highly procoagulant PS-containing membrane surface. In contrast to the PS results, the tryptophan studies of the PG-bound BF1 yielded an interesting new result, i.e., a lipid-induced conformational change in the Gla-domain, observed by a significant prolongation of lifetime E. At protein/lipid concentration ratios ensuring that the majority of the protein is bound to the 40% PG/60% PC surface, the lifetime of component E shifts from 5.1 to 7.5 ns, when compared with Ca-BF1 in solution. The prolongation of the Trp42 fluorescence lifetime can be observed as well by an apparent shift of the component D from 2.2 ns to 2.8 ns. On the other hand, as in the case of the binding to PS containing surfaces, the kringle components B and C remain unchanged. Since component D is due to the fluorescence of Gla and kringle tryptophan residues, the constant kringle tryptophan fluorescence portion in D might mask the entire magnitude of the lifetime shift in the Gla-portion of component D.

For the *N*-terminal membrane binding part the comparison of the PS results with the PG results leads to the conclusion that Ca-BF1 exhibits already the "perfect" conformation for binding and proteolysis and thus retains its conformation when bound to PS surfaces. The PG-induced conformational disruption of the Gla domain possibly might affect the protein conformation in the nonfragment 1 part of the protein and/or the lateral diffusion on the membrane surface. Both scenarios could yield a possible explanation for the lower procoagulant activity of PG when compared to PS.

13.2.2
Overview on Investigations Applying Fluorescently Labeled Prothrombin

In the majority of the studies on fluorescently labeled prothrombin a combination of substrate-supported planar model membranes and the use of evanescent illumination with laser-based, quantitative fluorescence microscopy has been employed [9–11, 35, 36]. Measurement of the steady-state, surface-associated fluorescence can be used to examine the thermodynamic properties of prothrombin at membranes [35]. Combined with fluorescence recovery after photobleaching (FRAP), this technique provides information about membrane-binding kinetics. On the other hand, combined with fluorescence pattern photobleaching recovery (FPPR) [9, 10], measurement of the translational diffusion coefficients of prothrombin bound to membranes is possible [11, 36].

In the first paper of this series, fluorescein-labeled bovine prothrombin and its amino and carboxy-terminal peptides, prothrombin fragment 1 and prothrombin 1, were added at various concentrations in the presence or absence of Ca^{2+} to substrate-supported planar membranes composed of 1-palmitoyl-2-oleoyl-3-sn-phosphatidylcholine (POPC), POPC/PS (70:30 mol/mol), or POPC/1,2-dioleoyl-3-sn-phosphatidylglycerol (DOPG) (70:30 mol/mol). Measurement of the steady-state surface-associated fluorescence was employed for the determination of equilibrium dissociation constants (k_D). Although the calcium-independent binding constant (k_D) of prothrombin has been shown to be only slightly higher than the k_D of unspecific binding to neutral membranes, the authors postulated a calcium-independent binding site of prothrombin in its nonfragment 1 portion, when bound to PS-containing membranes [35]. In the subsequent two studies of this group evanescent field excitation has been combined with FRAP (see Fig. 13.4) and used to compare the membrane binding characteristics of fluorescein-labeled bovine prothrombin and fluorescein-labeled bovine prothrombin fragment 1 to supported planar membranes composed of mixtures of PS (2–10 mol %) and DOPC in presence of Ca^{2+}. Equilibrium binding measurements showed that the k_D-values increased with decreasing molar fractions of PS and that the dissociation constants were somewhat lower for intact prothrombin. Kinetic measurements, using FRAP, showed that the measured dissociation rates were approximately equivalent for prothrombin and fragment 1 and did not change with the protein solution concentration or the molar fraction of phosphatidylserine. The kinetic data also implied that the surface binding mechanism for both proteins is more complex than a simple reversible reaction between monovalent proteins and monovalent surface sites.

In the last work of this series [11], fluorescence pattern photobleaching recovery (FPPR) with evanescent interference patterns has been used to measure the translational diffusion coefficients of membrane-bound prothrombin fragment 1. In this method, two internally reflected laser beams are collided to create a periodic evanescent intensity pattern that illuminates a region of a solid/liquid interface [11]. The results show that the translational diffusion coefficient on fluid-like PS/PC planar membranes is about 5×10^{-9} cm^2/s and is reduced when the fragment 1 surface density is increased. In addition, no translational diffusion was de-

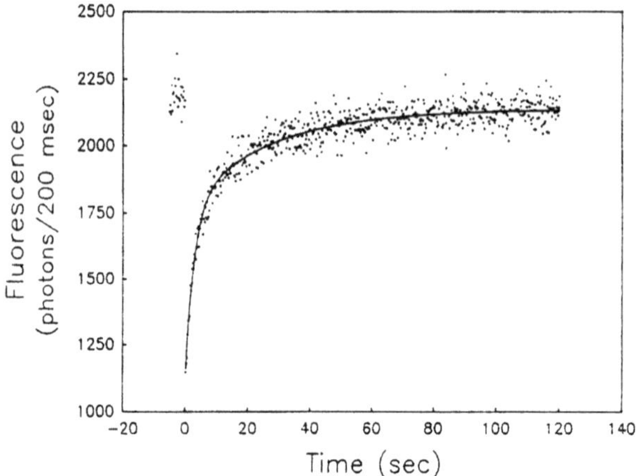

Fig. 13.4. Representative FRAP intensity versus time profile for the binding of prothrombin (1 µM) to planar PC-bilayers containing 6 mol % PS in the presence of 10 mM Ca^{2+} using evanescent field excitation. Experimental data points are given together with the best fit using a biexponential model [10]

tected for fragment 1 on solid-like (gel phase) membranes. FPPR combined with evanescent field excitation was also used to measure the diffusion coefficient of a fluorescent lipid in PS/PC planar membranes. These measurements yielded a diffusion coefficient of approximately 10^{-8} cm^2/s which is consistent with that measured by conventional FRAP [37] or fluorescence correlation spectroscopy (FCS) [38]. The authors conclude that the translational diffusion of fragment 1 requires and mimics membrane fluidity suggesting that fragment 1 does not skim over the membrane surface [11].

An important issue in the study of membrane binding of prothrombin and other coagulation proteins are possible conformational changes of the protein during Ca^{2+}-mediated association with negatively charged lipids, especially PS. Fluorescence resonance energy transfer (FRET) from a donor-labeled protein to an acceptor-labeled bilayer can be used to study protein–membrane interactions and changes in protein conformation. Chen and Lentz [8] applied this method to test the hypothesis of a membrane-induced conformational change in prothrombin compared to meizothrombin, an active intermediate in prothrombin activation formed by initial cleavage at Arg^{323}-Ile [39]. Based on a R_0-value for the given donor-acceptor pair of 52 Å the authors calculated the distance from the selectively fluorescein-labeled C terminus of prothrombin to the rhodamine-labeled phosphatidylethanolamin in the PC/PS bilayer to be 94 ±3 Å versus 114 ±2 Å for meizothrombin. As the overall length of prothrombin in solution was estimated to be about 113 Å these data suggest that binding with PS-containing membranes induces the prothrombin molecule to "fold up" internally to achieve the shorter fluorescein-to-rhodamine distance. It was concluded that these conformational changes might help to align the bond Arg^{323}-Ile in prothrombin with the active site of membrane-bound factor Xa which is known to be 61 Å above the membrane surface in the absence of another cofactor, factor Va [7]. It should be stated, however, that the employed method requires several assumptions, e.g., concerning the value of κ^2 used in the calculation of R_0, and data corrections; thus, the accuracy of the donor–acceptor distances given by Chen and Lentz [8] appears quite optimistic.

13.3
Prothrombin-induced Changes in the Organization of Phospholipid Bilayers

13.3.1
Solvent Relaxation (SR)

13.3.1.1
Solvent Relaxation Probed by the Headgroup Labels Prodan and Patman

During the last years solvent relaxation monitored by time-resolved fluorescence measurements has become an extremely useful method in membrane research [18–22, 40, 41]. It has been shown that suitable fluorescent dyes allow for direct observation of viscosity and polarity changes in the vicinity of the probe molecule which can be intentionally located in the hydrophobic backbone or in the hydrophilic headgroup region of the phospholipid bilayer. Such dyes described in recent publications were based on 2-amino-substituted naphthalene [19, 20] and anthracene-9-carboxylic acid [21]. Representatives for 2-amino-substituted naphthalenes are the dyes Prodan (6-propionyl-2-(dimethylamino)-naphthalene) and its long alkyl chain derivative Patman (6-palmitoyl-2-[[trimethyl-ammoniumethyl]methylamino]-naphthalene chloride) whose chromophores are known to be located within the hydrophilic headgroup region of the membrane. It might seem obvious considering that Ca^{2+}-assisted membrane binding of peripheral proteins should first of all influence the organization and dynamics of the lipid headgroup region. Thus, Prodan and Patman were considered as ideal probes for the study of the Ca^{2+}-induced prothrombin–membrane interaction.

Since the solvent relaxation method cannot be considered as a standard fluorescence technique up to know, its principles will be shortly outlined. The electronic excitation of a chromophore causes an ultrafast change of the probe's charge distribution, but does not affect the position or orientation of the surrounding solvent molecules. The solvent molecules are thus forced to adapt to the new situation and start to reorient themselves in order to find an energetically favored position with respect to the excited dye. The dynamic process starting from the originally created nonequilibrium Franck-Condon state and gradually establishing a new equilibrium in the excited state is called solvent relaxation (SR). This relaxation red shifts the probe's emission spectrum continuously from the emission maximum frequency of the Franck-Condon state ($v(0)$ for $t = 0$) to the emission maximum of the fully relaxed state ($\nu(\infty)$ for $t = \infty$). The time-resolved emission spectra (TRES) are usually determined by "spectral reconstruction" [17–21]. The TRES are fitted by an empirical "log-normal function" [42]. From the fitted spectra the emission maximum frequencies $\nu(t)$ and the full width at half maximum (FHWM) of the TRES are usually derived. The maximum of the time-zero ($t = 0$) spectrum $\nu(0)$ can be estimated quite accurately when both the absorption and fluorescence spectra in a non-polar reference solvent and the absorption spectrum in the system of interest are known [17, 41, 43]. Since $\nu(t)$ contains both information about polarity ($\Delta\nu$) and viscosity of the reported environment, the spectral shift $\nu(t)$ is normalised to the total shift $\Delta\nu$. The resulting "correlation functions" $C(t)$ (Eq. 13.1) describe the time course of the solvent response.

$$C(t) = \frac{\nu(0) - \nu(\infty)}{\Delta\nu} \qquad (13.1)$$

In order to characterize the overall time scale of the solvation response, we use an (integral) average relaxation time (Eq. 13.2):

$$\langle \tau_r \rangle = \int_0^\infty C(t)\,dt \qquad (13.2)$$

The time-dependent Stokes-shift $\Delta\nu$ ($\Delta\nu = \nu(0) - \nu(\infty)$) depends both on the solvent polarity and on the change in the solute's dipole moment. Since Prodan and Patman contain practically identical chromophores, detected differences in $\Delta\nu$ (e.g., 3750 cm^{-1} and 3000 cm^{-1} for Prodan and Patman in PC-SUV at room temperature, respectively [41]) directly reflect microenvironments of different polarity of the chromophores. It has been shown that the chromophore of Prodan is located in the headgroup region of phospholipid bilayers close to the lipid/water interface [18–20]. Smaller $\Delta\nu$ values for Patman indicate that its chromophore is embedded deeper in the bilayer [18–20, 41]. A comparison of $\nu(0)$ values obtained by the time-zero spectrum estimation with those obtained by TRES reconstruction shows that about 90% of the SR in PC-SUV in the liquid crystalline state probed by both dyes occurs on the subnanosecond to nanosecond timescale [41]. This conclusion is confirmed by the time evolution of the FWHM. In SUV composed of PC as well as of PC/PS mixtures the FWHM increase at early times and reach their maxima followed by a decrease of the spectral width. The observed profiles prove that during the lifetime of the excited state SR completes and almost the whole relaxation process is captured by an equipment providing subnanosecond time-resolution. The resulting characteristic SR time for Prodan (e.g., 1.1 ns in PC-SUV [20] are significantly smaller than obtained for Patman (e.g., 2.1 ns in PC-SUV [20]) in all investigated phospholipid bilayer systems so far [18–20, 22, 23, 40].

13.3.1.2
Influence of Prothrombin and its Fragment 1 on the Phospolipid Headgroup Organization

For the application of the solvent relaxation method described above to the protein–Ca^{2+}–PS interaction we had to assure that the used fluorophores do not bind significantly to the proteins in the presence of lipid. This was done comparing steady-state spectra of Prodan and Patman in Tris-buffer, in the presence of 40 μM BF1 or 30 μM prothrombin, respectively, and after addition of different amounts of vesicle suspension. While the addition of the proteins alone did neither change the emission spectra nor the decay times, addition of vesicles in the presence of the protein led to both a new blue-shifted emission band caused by the binding of Prodan and Patman, respectively, to the membrane and considerably increased decay times [20].

In order to establish the correlation between the solvent relaxation kinetics of Prodan and Patman with protein coverage of the membrane surface, small unilamellar vesicles (SUV) composed of PC/PS = 80/20 (mol:mol) in the presence of 5 mM Ca^{2+} were titrated with prothrombin. An increase in $<\tau_r>$ for Patman from 2.3 ns in the absence of prothrombin to about 3.0 ns in the presence of saturating concentrations of prothrombin (>12 μM) was observed with the largest increase for the first addition of 2 μM prothrombin. This behavior is in qualitative agreement with binding isotherms established by other methods [44].

To compare the effect of binding of prothrombin and its fragment 1 on the headgroup organization of PC/PS-SUV the SR-kinetics of Prodan and Patman were determined at saturating protein concentrations (16 μM). In Fig. 13.5 some exemplary correlation functions in absence and presence of proteins for Prodan and Patman in SUV composed of PC/PS = 60/40 (mol/mol) are shown. As shown by the data in Table 13.1, protein binding significantly reduces the mobility of the dye microenvironment in all investigated systems.

For Patman both proteins at saturating concentrations yield nearly identical effects on the mean relaxation times, i.e., an increase of 30% (Table 13.1). In contrast to Patman, which obviously could not differentiate between both proteins, a clear difference was detected using Prodan as the fluorophore. The binding of the complete prothrombin induces a higher rigidity than binding of the N-terminal fragment alone. Thus, Prodan which has been shown to be localised closer to the lipid/water interface [19] reacts considerably more sensitive to the binding of different proteins than does Patman.

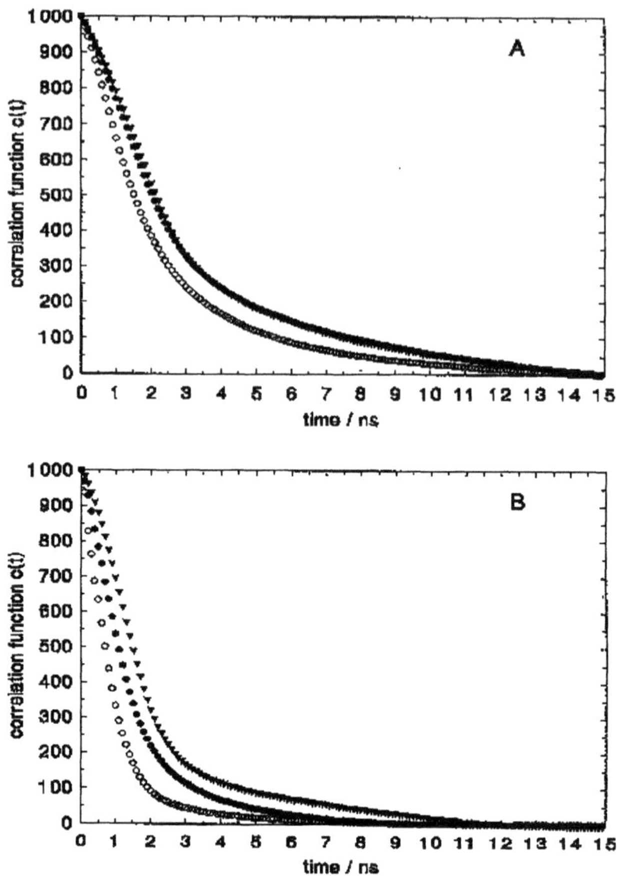

Fig. 13.5. Correlation functions $C(t)$ for Patman (**A**) and Prodan (**B**) in PC/PS 60:40 vesicles in the presence of 5 mM Ca^{2+} without protein (*open circles*), in the presence of 16 μM BF 1 and prothrombin (*triangles*). The temperature was 25°C

Table 13.1. Percentual increases of $\langle \tau_r \rangle$ in the presence of protein relative to the relaxation times in the absence of proteins

PC/PS	Fluorophore	Prothrombin	BF 1
80:20	Patman	30%	30%
60:40	Patman	35%	30%
80:20	Prodan	100%	50%
60:40	Prodan	92%	58%

Final values of $\langle \tau_r \rangle$ are obtained by averaging the $\langle \tau_r \rangle$ values obtained at apparent saturation protein concentrations (8, 12, 16 μM for PC/PS 60:40 and 12, 16, 20 μM for PC/PS 80:20).

The observation that the increase in solvent relaxation times monitored by Patman is identical for both prothrombin and its fragment 1 indicates similarities in the binding mechanism of both proteins. Located closer to the lipid/water interface, Prodan, however, is more affected by prothrombin binding. A possible explanation for the more rigid outermost region of the negatively charged phospholipid headgroup due to binding of the entire protein could be a tighter binding of the fragment 1 portion when associated with the nonfragment 1 part of prothrombin.

13.3.2
Fluorescence Anisotropy: Influence of Prothrombin and its Fragment 1 on PC/PS Membrane Order

It is instructive to compare the effects of protein binding using the solvent relaxation approach with those obtained by steady-state fluorescence anisotropy measurements [15, 20]. In the PC/PS-system used by us adding both proteins to a concentration leading to nearly surface coverage slightly increases the membrane order as monitored by the DPH anisotropy in the temperature range 10–35°C. In the former investigation [15] the authors did not detect any effect of BF 1 binding on the membrane order of fluid phase DMPC/DMPS-SUV. On the other hand they found a substantial increase due to prothrombin binding. Our experiments using PC/PS-SUV showed similar trends, with the major difference, however, that we observed a detectable, though very small effect of BF 1 ($\approx$3%) and a smaller prothrombin induced increase of membrane order ($\approx$5%) than detected by Tendian and Lentz [15]. Summarizing both these investigations, it can be stated that binding of prothrombin and its BF1 leads, if at all, only to a very small increase in packing density of the hydrocarbon region of the bilayer.

13.3.3
Pyrene Fluorescence: Influence of Fragment 1 on Membrane Order Monitored by the Excimer/Monomer Ratio

The excited pyrene molecule can exist as either a monomer or an excited dimer (excimer), both forms exhibiting different fluorescence properties, i.e., different emission maxima. The ratio of excimer to monomer emission (E/M) increases with increasing pyrene concentration and with increasing rates of diffusion [45]. Thus, changes in the E/M-ratio have been taken as a qualitative reflection of changes in the probe's microenvironment, i.e., changes in the bilayer microviscosity. For example, the presence of cholesterol in DMPC vesicles results in a lower E/M-ratio, reflecting slower pyrene diffusion [46]. Pyrene-labeled phospholipids have been used to detect temperature-induced lateral phase separations and changes in lipid phase transitions associated with the interaction of membranes with Ca^{2+} [47] and extrinsic membrane proteins [48].

Jones and Lentz [16] reported the use of pyrene-labeled phospholipids with either anionic or neutral headgroups to probe the effect of prothrombin fragment 1 binding on the lipid organization within phosphatidylglycerol/phosphatidylcholine (PG/PC) bilayers. Similar to PS, PG supports the binding of prothrombin and its fragment 1 in the presence of Ca^{2+}. Saturating amounts of fragment 1 and 5 mM Ca^{2+} have been added to pure dioleoylphosphatidylglycerol (DOPG) vesicles containing pyrene-PG. No change in the E/M-ratio could been observed, indicating that the diffusion of pyrene-PG was not significantly decreased by the presence of fragment 1. These data are in agreement with the results obtained by DPH anisotropy measurements, supporting the view that the membrane fluidity in the inner part of the bilayer is not substantially affected by the binding of prothrombin or its fragment 1.

On the other hand, the phospholipid headgroup region becomes considerably more rigid, as demonstrated by the solvent relaxation data. The comparison of all three membrane-focussed methods to characterize protein–membrane interactions supports our view that the solvent relaxation method is superior to both steady-state anisotropy and pyrene excimer formation for the study of headgroup organization and dynamics.

13.4
Conclusions

The described investigations employing fluorescence spectroscopy clearly demonstrate that calcium ions as well as negatively charged lipid surfaces are essential for the membrane binding of prothrombin. Kinetic data obtained by fluorescence recovery after photobleaching imply that the surface binding mechanism is more complex than a simple reversible reaction between monovalent proteins and monovalent surface sites. The *N*-terminal peptide fragment 1 displays basically very similar membrane binding characteristics as the entire prothrombin. However, fluorescence recovery after photobleaching experiments as well as solvent relaxation experiments using the membrane labels Prodan and Patman indicate a slightly tighter binding of the fragment 1 portion in the case of the entire prothrombin molecule compared to the isolated fragment 1. Investigations based on the fluorescence of membrane labels show that the negatively charged headgroup region of phosphatidylserine-containing membranes is predominantly responsible for the calcium-mediated prothrombin binding; there are no hydrophobic membrane binding sites in the nonfragment 1 part of prothrombin. The lateral diffusion coefficient of fragment 1 bound to phosphatidylserine-containing membranes appears to be within the same range as that determined for lateral lipid diffusion within fluid bilayer systems.

The tryptophan fluorescence characteristics of fragment 1 changes when bound to differently composed membrane surfaces. Binding to a phospholipid surface with high prothrombin binding affinity but low procoagulant activity disrupts the

Gla-domain conformation. The membrane binding part of the calcium–prothrombin, on the other hand, remains in its native conformation when bound to highly procoagulant (phosphatidylserine)-containing surfaces.

Acknowledgement. We thank the Ministry of Education, Youth and Sports of the Czech Republic (via LN 00A032) for financial support.

References

1. Mann KG, Nesheim ME, Church WR, Haley P, Krishnaswamy S (1990) Blood 76:1
2. Jones ME, Lentz BR, Dombrose FA, Sandberg H (1985) Thrombos Res 39:711
3. Rosing J, Tans G (1988) Thromb Haemost 60:355
4. Rosing J, Tans G (1988) In: Zwaal RFA (ed) Coagulation and lipids. CRC Press, Boca Raton, FL, p 4224
5. Comfurius P, Smeets EF, Willems GM, Bevers EM, Zwaal RFA (1994) Biochemistry 33:10311
6. Hof M (1998), Biochim Biophys Acta 1388:143
7. Husten EJ, Esmon CT, Johnson AE (1987) J Biol Chem 262:1295
8. Chen Q, Lentz BR (1997) Biochemistry 36:4701
9. Pearce KH, Hiskey RG, Thompson NL (1992) Biochemistry 31:5983
10. Pearce KH, Hof M, Lentz BR, Thompson NL (1993) J Biol Chem 268:22984
11. Huang Z, Pearce KH, Thompson NL (1994) Biophys J 67:1754
12. Duportail G, Lianos P (1996) In: Rosoff M (ed) Vesicles. Marcell Dekker, New York, p 296
13. Hutterer R, Haefner A, Schneider FW, Hof M (1997) In: Slavik J (ed) Fluorescence microscopy and fluorescence probes, vol. 2. Plenum Press, New York, p 93
14. Stubbs CD, Williams BW (1992) In: Lakowicz JR (ed) Topics in fluorescence spectroscopy: biochemical applications. Plenum Press, New York, p 231
15. Tendian SW, Lentz BR (1990) Biochemistry 29:6720
16. Jones, ME, Lentz, BR (1986) Biochemistry 25:567
17. Horng ML, Gardecki JA, Papazyan A, Maroncelli M (1995) J Phys Chem 99:17311
18. Hof M (1999) In: Rettig W et al. (Eds) Applied fluorescence in chemistry, biology and medicine. Springer, Berlin, p 439
19. Hutterer R, Schneider FW, Sprinz H, Hof, M (1996) Biophys Chem 61:151
20. Hutterer R, Schneider FW, Hermens WT, Wagenvoord R, Hof M (1998) Biochim Biophys Acta 1414:155
21. Hutterer R, Schneider FW, Lanig H, Hof M (1997) Biochim Biophys Acta 1323:195
22. Hutterer R, Schneider FW, Hof M (1997) J Fluorescence 7:27
23. Hutterer R, Parusel A, Hof M (1998) J Fluorescence 8:389
24. Shear JB, Xu C, Webb WW (1997) Photochem and Photobiol 65:931
25. Mann KG, Elion J, Butkowski RJ, Downing M, Nesheim ME (1981) In: Methods of enzymology, vol 80. Academic Press, New York, p 286
26. Hof M, Fleming GR, Fidler V (1996) Proteins Struct Funct Genet 24:485
27. Berkowitz P, Huh NW, Brostrom KE, Panek MG, Weber DJ, Tulinsky A, Pedersen LG, Hiskey RG (1992) J Biol Chem 267:4570
28. Nelsestuen GL (1976) J Mol Biol 251:5648

29. Soriano-Garcia M, Park CH, Tulinsky A, Ravichandran KG, Skrzypczak-Jankun E (1989) Biochemistry 28:6805
30. Bloom JW, Mann KG (1978) Biochemistry 17:4430
31. Borowski M, Furie BC, Bauminger S, Furie B (1986) J Biol Chem 261:14969
32. Lentz BR, Wu JR, Sorrentiono AM, Charleton JN (1991) Biophys J 60:942
33. Plopis VA, Strickland DK, Castellino FJ (1981) Biochemistry 20:15
34. Wu JR, Lentz BR (1991) Biophys J 60:70
35. Tendian SW, Lentz BR, Thompson NL (1991) Biochemistry 30:10991
36. Thompson NL, Drake AW, Chen L, Van den Broek W (1997) Photochem Photobiol 65:39
37. Huang Z, Pearce KH, Thompson NL (1992) Biochim Biophys Acta 1112:259
38. Benes M, Billy D, Hermens W, Hof M (2001) Biol Chem. (in press)
39. Krishnaswamy S, Mann KG, Nesheim ME (1986) J Biol Chem 261:8977
40. Hutterer R, Hof M. (2001) Z Phys Chem (in press)
41. Sykora J, Kapusta P, Fidler V, Hof M (2002) Langmuir (submitted)
42. Siano DB, Metzler DF (1969) J Chem Phys 51:1856
43. Fee RS, Maroncelli M (1994) Chem Phys 183:235
44. Cutsforth GA, Whitaker RN, Hermans J, Lentz BR (1989) Biochemistry 28:7453
45. Lakowicz JR (1999) Principles of fluorescence spectroscopy. Kluwer Academic / Plenum Publishers, New York.
46. Soutar AK, Pownall HJ, He AS, Smith LC (1974) Biochemistry 13:2828
47. Galla HJ, Sackmann E (1975) Biochim Biophys Acta 559:103
48. Wiener JR, Wagner RR, Freire E (1983) Biochemistry 22:6117

Assessment of Membrane Fluidity in Individual Yeast Cells by Laurdan Generalised Polarisation and Multi-photon Scanning Fluorescence Microscopy

R. P. LEARMONTH AND E. GRATTON

Here we describe techniques that we developed for monitoring membrane fluidity of individual yeast cells during environmental adaptation and physiological changes. Multi-photon scanning fluorescence microscopy using laurdan as a membrane probe enables determination whether fluidity changes seen by spectroscopy reflect universal responses or changes only of subpopulations.

Yeast membranes are a primary site of environmental response and adaptation. Using fluorescence spectroscopy with DPH polarization and laurdan Generalized Polarization (GP), we previously found rapid "average" membrane fluidity modulation in yeast populations during growth and in response to nutrients or environmental stresses. To determine whether such responses reflect all cells we conducted the first multi-photon scanning fluorescence microscopy study of yeasts, measuring laurdan GP. We assessed membrane fluidity responses of individual yeasts related to growth phase, heat stress and ethanol stress.

Average fluidity decreased as cultures aged, however the decreased fluidity was due in some cases to an increasing proportion of uniformly low fluidity (high GP) cells, which were shown by vital dye to be dead. When yeasts were heat stressed, the mean laurdan GP increased in all cells, thus the entire population evidenced damage (viz. decreased membrane fluidity) to the same degree. On the other hand, with ethanol stress fluidity increased (GP decreased) on exposure of cells. All cells were affected although not to the same degree, and with variable recovery. The recovery assessed from GP microscopy was highly variable, and greater by that seen by spectroscopy.

14.1
Introduction

In order to provide background to these studies, we need to introduce three major concepts: the rationale for studying yeast membranes and their fluidity; the use of multi-photon scanning fluorescence microscopy; and the analysis of laurdan Generalized Polarization (GP).

In brief, using fluorescence spectroscopy we previously studied membrane fluidity modulation in baker's yeast, brewer's yeast, and in *Saccharomyces cerevisiae* S288c for which the genome sequence is known. We determined membrane fluidity by fluorimetry, measuring polarization of DPH (1,6-diphenyl-1,3,5-hexatriene) fluorescence. More recently, to reduce problems with cell density-dependent scattering of the polarized light, we utilized the environmentally sensitive spectra of laurdan (6-lauroyl-2-dimethylamino naphthalene). We found rapid membrane fluidity modulation in yeasts in response to environmental stresses (heat and ethanol), during growth in batch culture and as a physiological response to glucose availability.

14.1.1
Yeast Membrane Fluidity

The plasma membrane provides the semi-permeable barrier that allows all cells to exist. When yeasts encounter changes in the environment such as nutrient depletion, metabolite accumulation or temperature variation, the plasma membrane must adapt prior to internal structures. Thus the membrane is the primary site of response to environmental change.

Fluidity of membranes is a key factor in their function, affecting cell permeability and important activities such as nutrient transport and pH maintenance. Membrane function, which is largely dependent on membrane enzymes and transport proteins such as the yeast plasma membrane H^+-transporting ATPase, is also affected by changes in membrane fluidity [1].

Elucidation of molecular mechanisms of adaptation of yeasts to stress is highly relevant to commercial production of yeast products, with temperature and ethanol stresses of most economic significance. Furthermore, knowledge of yeast adaptation processes at the molecular level is applicable to studies of adaptive responses of higher organisms. Stressful environmental change may directly affect membrane fluidity, either transiently or permanently [2–4]. Cells that survive and adapt must therefore modulate their membrane fluidity to compensate.

In heat stress the plasma membrane is considered a primary site of heat damage, although the mechanism is controversial. The "fluidization" hypothesis maintains that heat increases fluidity to unstable levels, resulting in breakdown of structure and function [5]. On the other hand, we believe that rather than fluidization, heat damage to membrane proteins impacts on membrane function. The latter theory is supported by data from a number of studies [6–8]. It is thought that heat responses are triggered by damaged cytosolic proteins, although membrane proteins may be more vulnerable since heat damage may involve oxygen-derived free radicals, which partition into membranes [9]. In support of the former theory we found yeast membrane fluidity relates inversely to heat tolerance [1, 10–12]. However, more strikingly supporting the latter theory, we found that during heat stress membrane fluidity progressively and irreversibly decreased [1].

When considering ethanol stress, survival and adaptation to ethanol (from fermentation of sugars) are important criteria in beer and wine production. Like heat, ethanol causes accumulation of damaged proteins and induces similar responses [13], although it definitely increases membrane fluidity due to its solvent action. In preliminary experiments, we found that yeasts exposed to high concentrations of ethanol showed an immediate increase in fluidity. In the case of cells early in the growth phase the fluidity increase was followed by a marked, sustained decrease that may reflect denaturation of membrane proteins [2, 3, 14].

However, while these studies provided useful information on rapid membrane responses of yeasts, a disadvantage is that average responses of whole populations were measured. Furthermore, information of subcellular location of responses was not available. Recently, the technique of multi-photon scanning fluorescence microscopy has become available. This technique has a number of advantages over confocal microscopy, as described below. Multi-photon fluorescence microscopy enables immediate and direct visualization of events at the individual cell and subcellular levels. Therefore, we conducted the first study of yeasts by multi-photon scanning fluorescence microscopy.

14.1.2
Multi-photon Scanning Fluorescence Microscopy

The technique of multi-photon scanning fluorescence microscopy has been applied to study of laurdan GP in model membranes and mammalian cells [15, 16]. Multi-photon scanning fluorescence microscopy involves using laser beams and optics to focus the incident beam on a small point, rather than illuminating the whole specimen. In addition, using the intense radiation allows the possibility of multiple photons simultaneously exciting a fluorophore. Consequently, for two-photon excitation the wavelength of the incident light may be approximately doubled, reducing the energy of the incident radiation from the high energy UV region to the lower energy red region of the visible spectrum. This technique has a number of advantages over conventional confocal fluorescence microcopy [15]. Photobleaching is dramatically reduced by point illumination of specimen (fluorescence excitation is localized to the focus region), and the use of lower energy red to near-infrared irradiation (700–1050 nm). This also allows for study of UV-excitable dyes without the expense of UV-laser systems. In addition, the penetration of longer wavelength radiation into biofilms of microorganisms or animal or plant tissues is much greater, allowing analysis of events at much greater depth.

14.1.3
Determination of Membrane Fluidity Using Laurdan Generalized Polarization

Laurdan [17] in membranes is known to be sensitive to the polarity of the environment, exhibiting a 50 nm red shift in emission spectrum over the gel to liquid-crystalline phase transition [15, 18]. The spectroscopic property Generalized Polarization (GP), derived from fluorescence intensities at critical wavelengths, can be considered as an index of membrane fluidity [19]. The GP is calculated in an analogous way to the Polarization parameter, exchanging values at critical wavelengths for the polarization orientations. Thus the GP is calculated from relative fluorescence intensities at wavelengths at the red and blue edges of the spectrum, representing gel (approx. 440 nm) and liquid crystalline (approx 490 nm) phases of bilayer systems. The GP is calculated as follows [19]:

$$ GP = \frac{I_{gel} - I_{lc}}{I_{gel} + I_{lc}} \tag{14.1}$$

The GP may theoretically range from -1 to $+1$. This GP parameter is inversely related to membrane fluidity; high GP values are found in gel phase, and low in liquid-crystalline phase [15]. Laurdan has been used to analyze membrane structure and organization in model phospholipid systems [19–21] and mammalian cell membranes [22–25] by cuvette fluorimetry and two-photon microscopy [15, 16]. As noted above, we reported the first use of laurdan to detect rapid membrane fluidity modulation in microorganisms by fluorimetry – specifically in yeast populations during growth and under stress. [2, 3, 26]. In addition we pioneered the use of multi-photon scanning fluorescence microscopy of microorganisms [26, 27].

14.2
Materials and Methods

These studies included three *Saccharomyces cerevisiae* strains: a wild type baker's yeast [28], a wild type brewer's yeast [27], and strain S288c for which the genome sequence is known [29]. Cells from a slope were inoculated into YNBG broth containing 1% glucose and 0.67% Yeast Nitrogen Base (Difco), and incubated overnight on an orbital shaker (30 $\pm1°C$, 180 rpm). Growth media was filter sterilized as autoclaving resulted in increased background fluorescence. Culture grown overnight (starter culture) was inoculated in fresh YNBG broth at 0.1 OD_{600nm} and continued on the shaker (experimental culture). This provided a starter culture population in relatively similar growth phase for all experimentation. It is important to briefly define the phases of aerobic growth in diauxic yeast cultures. We previously defined these stages [30] as initial lag (adaptation), respiro-fermentative growth on glucose (about 12 h under our conditions), diauxic lag phase (adaptation), respiratory growth on the accumulated ethanol (up to 80 or more h, note this would not occur under anaerobic conditions), stationary phase (maintenance phase – nutrients exhausted), followed eventually by death phase (duration of stationary phase depends upon cell conditions). Experimental cultures were analyzed at various stages of growth, as indicated below.

The fluorescent membrane probe laurdan was obtained from Molecular Probes, Eugene, OR, (USA). Cells at appropriate culture phases were labeled by mixing with laurdan stock in ethanol (1 μL per mL cells at $OD_{600nm} = 0.4$, final laurdan concentration 5μM) and incubating for 1 h in the dark, under the same conditions as the main culture. Where required, cells were diluted in filtered culture supernatant. The negligible addition of ethanol was previously shown not to affect the yeast cells.

Two-photon fluorescence scanning microscopy and GP analysis was performed as described previously [15] with excitation at 770 nm (equivalent to 1-photon excitation at 385 nm), and emission analyzed using two 46-nm bandpass filters centred at 446 and 499 nm. The two filters were exchanged after each full scan. To compensate for photobleaching, three scans in the sequence red-blue-red were taken, and the red scans averaged. Laurdan-labeled cells were analyzed against unlabeled cells to subtract background autofluorescence.

As GP image was derived from 3 independent images, immobilization of cells was important, as movement would result in nonoverlapping images. In our experiments we devised and partially optimized two methods for fixing live yeast cells, to give comparable results. In the first method, pelleted cells were mixed with low temperature melting agar (to avoid unintended heat shock of cells) and set to consistent depth under cover slips on microscope slides. The cover slips were then sealed to prevent specimens drying out during microscopy. This method provided good immobilization of cells, although it precluded addition of reagents during analysis. The second method involved attaching cells to poly-D-lysine coated cover slips in 8-chambered format (Fisher Scientific). This allowed up to 500 μL of culture supernatant or other solutions to be added to the fixed cells, and replacement if required with fresh solutions. However the immobilization of cells by the latter technique was variable and worked well for some, but not all yeast strains. The adhesion of cells was growth phase dependent, reduced over time on the cover slip, and was affected by ethanol addition. In some experiments, while cells remained adherent to the cover slip, they tended to wobble, thus leading to poor image quality. However, when conditions were optimal the cover slip/well method provided useful data. Further studies are progressing to optimize live cell fixation, utilizing techniques such as a poly-lysine based flow cell biofilm preparation method [31], a mucin-based biofilm preparation method [32], and a biotin-streptavidin method [33]. Since this study was conducted, the microscopes in the laboratory of E.G. have been modified to enable simultaneous 2-channel measurements. R.P.L. has also obtained a 2-channel system.

14.3
Results and Discussion

Using the fluorescent probe laurdan, we viewed optical sections of cells to assess membrane fluidity, represented as GP. We assessed membrane fluidity of the three yeast strains as a function of growth phase in batch culture, heat stress and ethanol stress. Laurdan-labeled cells were analyzed against unlabeled cells to subtract background autofluorescence. Fig. 14.1 displays GP images of three yeast strains at various stages of the growth curve. The 6 h images represent respiro-fermentative growth on glucose, 24 h respiratory growth on ethanol, 3–4 d late respiratory growth to stationary phase, and 8 d stationary phase. At 6 h and less at 24 h, one can see the typical morphology of growing yeasts, which devide by budding – a larger parent cell with a smaller bud growing from it. One can also see the large cell vacuole in some cell sections, viewed as a large black area (i.e., no fluorescence above background detected). The microscopy study confirmed our previous fluorimetry studies, in that average fluidity decreased (GP increased) as cultures progressed.

Fig. 14.2 compares the GP values for the yeast strain A9, obtained by microscopy and spectroscopy. It can be seen that the standard deviations of the former are much greater, as would be expected since a limited number of cells are analyzed per scan, where values are averaged from thousands of cells in the cuvette.

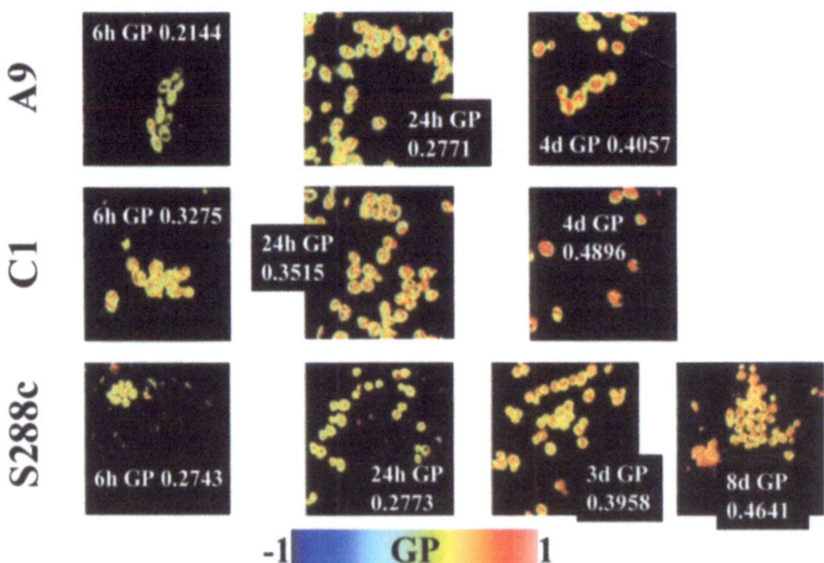

Fig. 14.1. Generalized Polarization images of three yeast strains related to culture time. The scanned intensity data were obtained using a Zeiss 40X 1.25 NA oil emulsion objective and corrected for photobleaching and unlabeled cell autofluorescence. The 256×256 pixel scans represent 23×23 µm, the depth scanned was 1 µm

For technical reasons including time of cell preparation for scanning, the data from microscopy are not as sensitive to change around the time of exhaustion of glucose, however the trend of both curves is of increased GP as cultures age. It should be noted that the absolute values of GP vary between different instruments, due e.g., to differing optics, differing wavelength-associated transmission characteristics of monochromators and/or differing wavelength-dependent efficiencies of photomultiplier tubes in detectors.

Unfortunately, in organisms as small as yeast, the resolution was not adequate to precisely define subcellular membranes. However, there seems to be clustering of like GP values (particularly high GP values). While we were unable to assign this clustering to particular subcellular structures, it may reflect membrane lipid rafts (clustering of particular species of membrane lipids into defined domains). Further studies will attempt to establish whether the clustering of GP values represents lipid rafts, and whether membrane proteins are specifically associated with these.

Notwithstanding the lack of ability to delineate subcellular membranes, valuable information was gained on individual cell responses in a population. Examination of microscope GP images revealed in some (but not all) cases the decreased average fluidity may be due to an increasing proportion of uniformly low fluidity (high GP) cells with the remaining cells unchanged (with variable fluidity across the cell). This is not obvious in Fig. 14.1, but can be seen well in Fig. 14.4, comparing the control cells at 6 and 24 h. Although double labeling with a vital dye (methylene blue) caused quenching of fluorescence, it indicated that cells with uniform low fluidity were dead. Thus cells seemed to be entirely healthy or dead, with "frozen" membranes, with no intermediate form detectable.

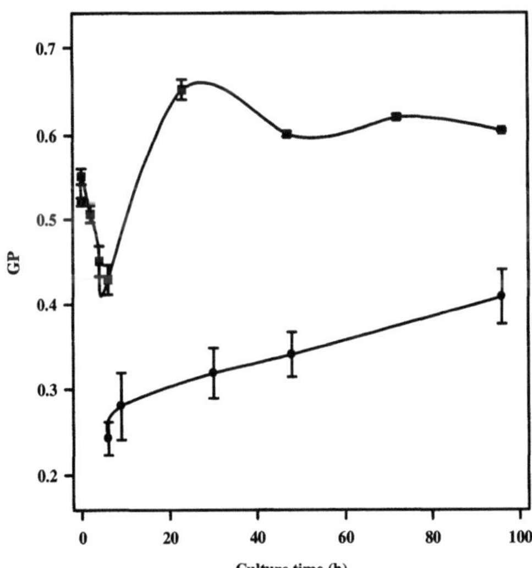

Fig. 14.2. Mean GP of cells from yeast strain A9 related to culture time. ● by microscopy, ■ by spectroscopy (using an ISS PC2 spectrofluorimeter). Error bars represent standard deviation. Respiro-fermentative growth on glucose to 12 h, respiratory growth on ethanol to 72 h

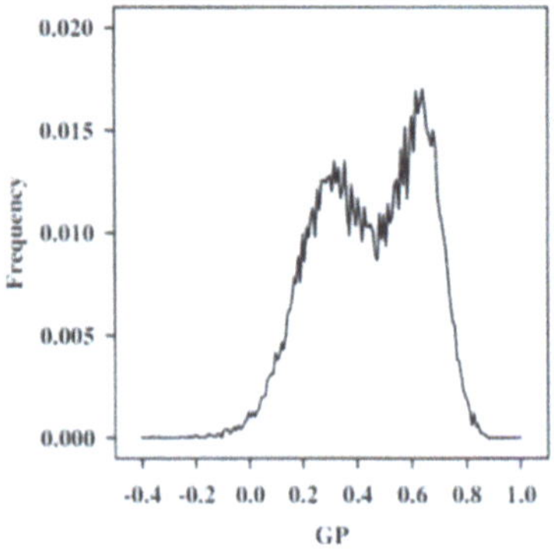
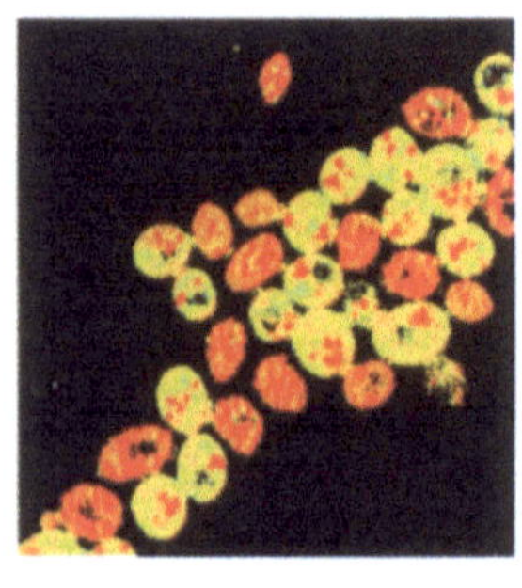

Fig. 14.3. GP frequency histogram and image of A9 cells in stationary phase at 4 d

Numerical analysis of histograms of GP images (Fig. 14.3) showed that there were only two classes of cells: live cells with mean low GP, and dead cells with mean high GP. The histogram in Fig. 14.3 shows a peak centred at GP 0.3176 representing a total of 57% of the data, with a second peak centred at GP 0.6392 representing 43% of the values. This correlates well with the fractions obtained by simply counting low average GP (live) cells and high average GP (dead) cells, giving 56% and 44%, respectively.

It should be noted that this phenomenon might not occur, for example it is not indicated in 6–24 h images of Fig. 14.1. Nevertheless, cuvette-based studies must be interpreted with caution and viability closely monitored, as increased GP may be due to death of a proportion of cells rather than an "average" cell response. It should be possible for investigators to determine a "correction factor" to determine the effect of skewing GP by the proportion of nonviable cells. However, this would need to be done for each instrument, due to the individual instrument factors discussed above.

In relation to heat stress, our previous DPH polarization data indicated that membrane fluidity decreased progressively in cells exposed to heat, and that the decrease was greater in respiro-fermentative cells. When cells were heat stressed (52°C, 5 min) and assessed by GP microscopy (Fig. 14.4), the mean GP of respiro-fermentative cells increased from 0.199 ±0.040 to 0.360 ±0.028. The increase was seen consistently across all cells, rather than a subpopulation change, thus the entire population was damaged to the same degree. Respiratory cells had higher mean GP (0.379 ±0.029) and were more resistant to heat stress, as the 5 min heat exposure did not markedly increase the mean GP of the latter cells (0.392 ±0.012). In this case the microscopy data confirmed the previous fluorimetry data.

When ethanol stress was studied, our previous DPH polarization and laurdan GP data indicated that fluidity was increased within seconds by exposure of respiro-fermentative and respiratory cells to ethanol, followed by only slight recovery. By GP microscopy (Fig. 14.5) however, the results were inconsistent, but indicated potentially greater recovery of membrane fluidity, and generalized effect across all cells.

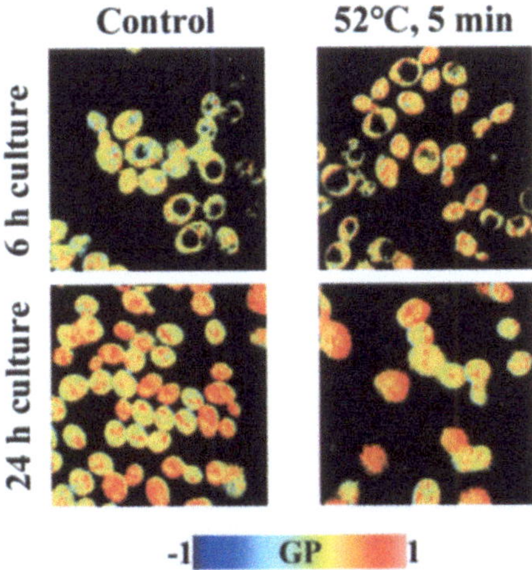

Fig. 14.4. Generalized Polarization images of yeast strain A9 growing on glucose (respiro-fermentative, 6h) and ethanol (respiratory, 24h) before and after heat stress (52°C, 5 min). The data were obtained using a Zeiss 63X 1.25 NA oil immersion objective. The 256×256 pixel scans represent 36.2×36.2 μm

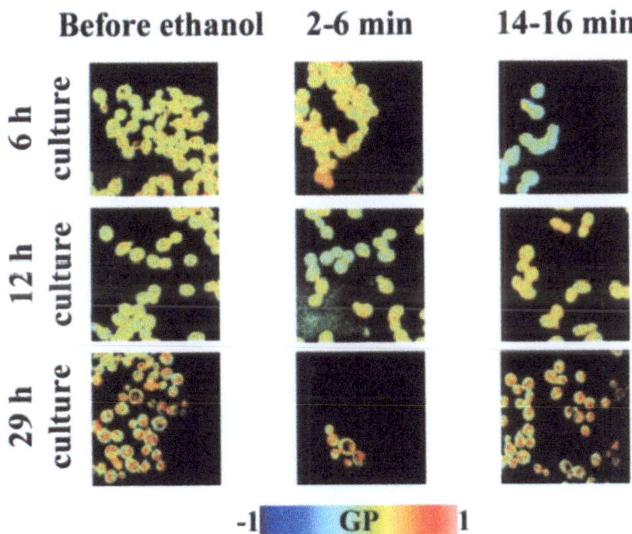

Fig. 14.5. Generalized Polarization images of yeast strain C1 during and at the end of growth on glucose (respiro-fermentative, 6 h, 12 h), and growth on ethanol (respiratory, 29 h) before and after addition of ethanol. The scanned intensity data were obtained using a Zeiss 40X 1.25 NA oil immersion objective as in Fig. 14.1. The 256×256 pixel scans represent 23×23 μm

The inconsistency may reflect suboptimal immobilization of cells in presence of ethanol, and further studies aim to improve immobilization of cells in the cover slip/well system. Application of simultaneous 2-channel recording should also improve this situation.

14.4
Conclusions

In conclusion, we developed techniques for monitoring membrane fluidity in individual yeast cells during environmental adaptation and physiological change. The microscopy studies allow us to determine whether changes reflect ubiquitous responses, or changes only in subpopulations of cells. The GP images show evidence of subpopulation changes in relation to age of culture and to some extent for ethanol damage, although heat damage is more universal.

Multi-photon scanning fluorescence microscopy adds a further dimension to our analyses, providing information on responses of individual cells within a population. The observation in some situations that a variable proportion of dead cells skews mean GP values demonstrates that cuvette-based studies must be interpreted with caution. Increased GP may be due to death of a proportion of cells rather than an "average" cell response, necessitating simultaneous assessment of cell viability during experiments.

The ability to monitor responses of individual live cells is of critical importance in establishing adaptation of populations to minor or major environmental change. For example a measured average response may reflect high activity of a subpopulation of cells at a certain stage of their growth cycle. In the case of major stresses (e.g., heat, ethanol, osmotic, pH) it is not uncommon for a small proportion (0.001 to 10 %) of cells to survive. Whether this is simply a stochastic process, or related to the precise nature of the surviving cells, may be elucidated by following individual cell responses. Here we concentrated on membrane fluidity responses, although the technique could be used to assess a range of responses for which fluorescent probes are available.

Acknowledgements. The Laboratory for Fluorescence Dynamics is a National Research Resource, supported by the National Institutes of Health and the University of Illinois. E.G. acknowledges support from grant PHS P41 RR03155. R.P.L. acknowledges support from the Australian Research Council grants C29600212, IP97050 and R00107932.

References

1. Learmonth RP, Carlin SM, Shah DN, Attfield PV (2002) Effects of growth phase and heat stress on membrane fluidity in Saccharomyces cerevisiae. (in preparation)
2. Shah DN, Learmonth RP (1998) Use of laurdan fluorescence to detect rapid membrane fluidity changes in Saccharomyces cerevisiae. Proc Aust Soc Biochem Molec Biol 30: POS-TUE-155 Abstract
3. Shah DN, Butcher B, Learmonth RP (2002) Use of laurdan fluorescence to detect rapid membrane fluidity changes in Saccharomyces cerevisiae. (in preparation)
4. Learmonth RP, Carlin SM (1997) Yeast membrane fluidity, potassium fluxes and extracellular pH in relation to glucose availability. FASEB J 11:A1101 Abstract
5. Van Uden N (1984) Adv Microbial Physiol 25:195
6. Konings AWT, Ruifrok ACC (1985) Radiat Res 102:86
7. Lepock JR (1982) Radiat Res 92:433
8. Gille G, Sigler K, Hofer M (1993) J Gen Micro 139:1627
9. Steels EL, Learmonth RP, Watson K (1994) Microbiol 140:569
10. Carlin SM, Learmonth RP (1994) Fluidity of yeast membranes. Proc Aust Soc Biochem Molec Biol 26: POS1-27 Abstract
11. Learmonth RP, Carlin SM (1995) Membrane fluidity and stress tolerance are related to mode of metabolism in yeast. Proc Aust Soc Biochem Molec Biol 27: COL9-4 Abstract
12. Learmonth RP, Carlin SM (1996) Effects of growth phase and heat stress on membrane fluidity in S. cerevisiae. Proc 9th Int Symp Yeasts, p 44, Abstract
13. Piper PW (1995) FEMS Microbiol Lett 134:121
14. Learmonth RP, Carlin SM (1994) Commercial Research Report, pp 1–13
15. Yu W, So PTC, French T, Gratton, E (1996) Biophys J 70:626
16. Parasassi T, Gratton E, Yu WM, Wilson P, Levi M (1997) Biophys J 72:2413
17. Weber G, Farris FJ (1979) Biochemistry 18:3075
18. Parasassi T, Gratton E (1995) J Fluorescence 5:59
19. Parasassi T, De Stasio G, d'Ubaldo A, Gratton E (1990) Biophys J 57:1179
20. Parasassi T, Di Stasio M, Ravagnan G, Rusch RM, Gratton E (1991) Biophys J 60:179
21. Parasassi T, Di Stefano M, Loiero M, Ravagnan G, Gratton E (1994) Biophys J 66:120
22. Fiorini R, Curatola G, Kantar A, Giorgi PL, Gratton E (1993) Photochem Photobiol 57:438
23. Levi M, Wilson PV, Cooper OJ, Gratton E (1993) Photochem Photobiol 57:420
24. Parasassi T, Di Stefano M, Ravagnan G, Sapora O, Gratton E (1992) Exp Cell Res 202:432
25. Parasassi T, Loiero M, Raimondi M, Ravagnan G, Gratton E (1993) Biochim Biophys Acta 1153:143
26. Learmonth RP, Yu W, Shah DN, Gratton E (1999) Study of membrane fluidity modulation in microbial populations using laurdan, by cuvette fluorometry and 2-photon scanning fluorescence microscopy. In Proc 4th International Weber Symposium on Innovative Fluorescence Methodologies in Biochemistry and Medicine, 1999, Hawaii
27. Learmonth RP, Gratton E (2000) Membrane fluidity modulation during environmental adaptation of populations and individual yeast cells by 2-photon scanning fluorescence microscopy. Proc 10th Internat Symp Yeasts: The Rising Power of Yeasts in Science and Industry, p 66–67 Abstract

28. Lewis JG, Learmonth RP, Attfield PV, Watson K (1997) J Indust Microbiol Biotech 18:30
29. Mewes HW, Albermann K, Bähr M, Frishman D, Gleissner A, Hani J, Heumann K, Kleine K, Maierl A, Oliver SG, Pfeiffer F, Zollner A (1997) Nature 387:7
30. Lewis JG, Northcott CJ, Learmonth RP, Attfield PV, Watson K (1993) J Gen Micro 139:835
31. Cowan SE, Gilbert E, Khlebnikov A, Keasling JD (2000) Appl Env Micro 66:413
32. Vroom JM, De Grauw K J, Gerritsen HC, Bradshaw DC, Marsh PD, Watson GK, Birmingham JJ (1999) Appl Env Micro 65:3502
33. Sinclair DA, Guarente L (1997) Cell 91:1033

Formation of Higher Order Signal Transduction Complexes as Seen by Fluorescence Spectroscopy

L. DOWAL AND S. SCARLATA

Fluorescence spectroscopy is the primary method to view the interactions between membrane-bound proteins in real time. We have been using fluorescence homo- and heterotransfer methods to follow the associations and oligomerization of membrane associated proteins. One system we have focused on is the G-protein-phospholipase Cβ signaling system. This pathway is responsible for transducing extracellular signals that bind to heptahelical surface receptors such as ions, hormones and neurotransmitters. Binding of these agents activate heterotrimeric G proteins which activates phospholipase Cβ (PLCβ), which in turn causes an increase in intracellular calcium and an activation of protein kinase C. Using a combination of fluorescence, biochemical and molecular biology techniques, we have measured the interaction energies of these proteins taking into account their dependence on the surface area of the lipid membrane. We find that activation of PLCβ is mediated by the pleckstrin homology domain, a structural module found in a wide variety of signal transduction proteins. Activation of the catalytic domain by the PH domain plays an analogous role in proteins that are not activated by G proteins. Using energy transfer, we also find that both PLCβ and G proteins have secondary binding sites for each other and for a protein that regulates G protein signaling (RGS). Our data suggest that in cells these proteins form stable signaling complexes capable of mediating high output, rapid and localized signals. We are currently investigating the role that lipid domain or rafts have in stabilizing these protein interactions.

15.1
Introduction

The plasma membrane of a cell serves as a barrier which mediates the contact cells have with their external environment. One mechanism through which cells receive extracellular signals is through binding of an agonist to an extracellular site of receptor protein. One of the largest families of extracellular receptors is the seven transmembrane G protein coupled receptors. These receptors include ones for light, ions and neurotransmitters (for review, see [1]).

Ligand binding to the receptor changes their interaction with the heterotrimeric (GTP binding) proteins on the membrane surface. G proteins consist of three subunits, α, β and γ. The α subunit contains the GTP/GDP binding site. The β and γ subunits form a tight dimer than can only be dissociated by denaturation. In the unactivated state (i.e., nothing is binding to the receptor), GDP is bound to the α subunit which allows it to interact strongly with the $\beta\gamma$ subunit. Binding of ligand to the receptor allows the receptor to catalyze the exchange of GDP for GTP on the α subunit. GTP-bound α subunit has a much-fold weaker affinity for $\beta\gamma$. These two species can then change the catalytic activity of a number of cellular effectors, which ultimately results in overall changes in the cell such as proliferation and differentiation. The α subunit has an intrinsic GTPase and as GTP is hydrolyzed to GDP, activation is turned-off. Also, there has recently been discovered a new class of proteins that are called GAPs (GTPase activating proteins). These proteins increase the GTP hydrolyzing activity of α thereby reducing the time of the activated state [2].

We have been studying a G protein system for the past several years using fluorescence spectroscopy. Fluorescence is one of the few techniques that allows one to monitor the interactions between membrane proteins in real time. Fluorescence measurements can also determine whether a particular protein has multiple binding partners. We have focused on the G protein activation of the effector phospholipase C-β (PLCβ). PLCβ can be activated by either the q family of α subunits or by G$\beta\gamma$ subunits [3]. PLCβ isoenzymes are a family of four known species, PLCβ1-4, that differ in their ability to be activated by the q family of α subunits or by G$\beta\gamma$ subunits ([3]). PLCβs are soluble but bind tightly to membranes where they catalyze the hydrolysis of a minor lipid component in membranes, phosphatidylinositol-4,5-bisphosphate. This reaction releases two second messengers in the cell that result in mitogenic and proliferative changes.

Adding to the list of players in the inositol-signal transduction is a newly discovered class of proteins called regulators of G protein signaling (RGS) proteins [4]. RGS proteins have no known effector activities but they bind to the α subunits of G proteins and increase the rate of GTP hydrolysis by stabilizing the transition state of the complex as seen in the crystal structure of the complex [5]. This stabilization allows RGSs to turn-off the effector signal without diminishing the strength of the signal. There are several members of the RGS family that have varying selectivity for different types of α subunits. Of interest here is RGS4,

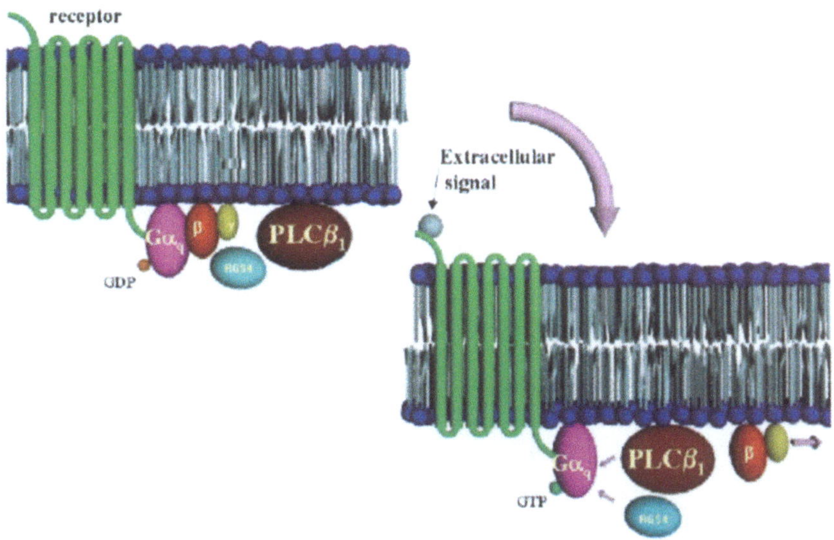

Fig. 15.1. The classic picture of signaling. (Based on [7])

which increases the GTPase activity of $G\alpha_q$ 1000 fold [6]. Family members of RGS all have a core "RGS box" responsible for accelerating GTP hydrolysis of α(GTP) subunits with varying flanking regions that are distinctive for each of the RGS family members.

The classic picture of signaling is shown in cartoon in Fig. 15.1 based on the one in [7]. In this figure the proteins are freely diffusing along the membrane surface and the cytosol. For signals to occur, the rate of diffusion of each interacting species must be considered along with the area in which they can diffuse during the time of activation. Thus, each signal will involve a certain time constant and affect a certain area of the cell.

In recent years, several lines of evidence have convinced us that this picture of freely diffusing peripheral proteins is not appropriate. Although once thought to be homogeneous, natural membranes appear to contain specialized domains that are just beginning to be discovered. The most notable of these are the domains that remain after disruption of cell membranes by detergents (for review, see [8–10]). The detergent-insoluble domains have been found to be aggregates of glycosphingolipids, sphingomyelein (SM) and cholesterol (CH). These domains appear to exist in the liquid-ordered phase (L_o) rather than in the fluid, liquid-disordered phase (L_d) that is characteristic of most cell membranes [11]. The L_o phase is characterized by tight chain packing, reduced fluidity and extended lipid chains, although the lipid mobility is still high. This phase can be formed in model membranes in vitro by mixing cholesterol with sphingomyelein, or with lipids that contain at least one saturated acyl chain. Phase diagrams of these domains have been characterized [12, 13].

Proteins that contain an acyl modification will localize to membrane domains. A characteristic feature of domain-bound proteins is an acyl modification such as palmitoylation. Localization by palmitoylation is though to be due to an increase in retention time because of the gel-like nature of the microdomains [12]. Also, a subset of detergent-insoluble domains are caveolae, which are flask-shaped invaginations on the plasma membrane visualized by electron microscopy [8]. Caveolae are enriched in members of the caveolin family of the transmembrane proteins. There is evidence that PIP_2, G protein subunits and other proteins involved in signal transduction are localized in caveolae [14, 15]. However, because these studies involve the disruption of the plasma membrane and subsequent fractionization, it is not clear whether G proteins are actually localized in these domains.

There are several lines of evidence suggesting that the proteins involved in the inositol-signaling pathway exist in signaling complexes and may be localized in domains. Association between seven transmembrane receptors and $G\alpha_q$ has been observed in cells [16]. Notably, in *Drosophila* a signaling complex involving receptor, PLCβ and protein kinase C and a scaffold protein have been identified [17]. In model systems, we have found that Gα(GDP) will form a complex with PLCβ$_2$-Gβγ which rapidly shuts down the signal without physical dissociation (see below), showing that G protein heterotrimers can complex with their effectors [18]. In studies of PLCβ activation, Ross and coworkers use kinetic data to argue that $G\alpha_q$ must remain bound to receptor [19]. More evidence for signaling complexes comes from patch clamping studies of Wilkie and coworkers in which they observe RGS4-dependent Ca^{2+} oscillations in cells using a PLCβ agonist [20]. These authors suggest that the G protein coupled receptor, the G protein heterotrimer and PLC are localized in a signaling complex.

15.2
Experimental

Our long term goal has been to determine the factors that govern PLCβ activation by G protein subunits and we have expressed these proteins in Sf9 cells to study their interactions in model systems by fluorescence spectroscopy and biochemical methods. We have shown that PLCβ enzymes bind strongly to model membranes where they laterally associate to G protein subunits [21]. Binding to membranes concentrates and localizes PLCβ on the same surface as the G proteins, thus promoting activation.

15.2.1
Membrane Binding

We initially measured the membrane binding properties of PLCβ to rafted and nonrafted lipids by the decrease in intrinsic fluorescence of the enzyme when it binds to the membrane surface. While the underlying reason for this decrease is unclear, we speculate that it is caused by quenching of interfacial Tyr or Trp residues by the ionic groups on the membrane surface. In previous studies, we have characterized the binding of PLCβ enzymes to various membrane surfaces using intrinsic fluorescence, by changes in the emission properties in a fluorescence lipid probe doped into the membrane, and by the changes in fluorescence properties of a probe covalently labeled to the protein [21]. We find that PLCβ binds strongly and nonspecifically (i.e., $K(p)$ ~5–50 µM) to lipid bilayers.

15.2.2
Interactions of PLCβ and G Protein Subunits on Membrane Surface

We measured the membrane binding of PLCβ to fluid purified lipids extruded through a 0.1 µm filter. Binding of PLCβ to PC:PS (2:1) and membranes with lipid rafts (PS:SM:Ch 1:1:1) was measured to fluid phase bilayers of various composition. Binding to the rafted lipid was reduced but within range of the affinity previously obtained for the uniformed PC:PS bilayers (Fig. 15.2).

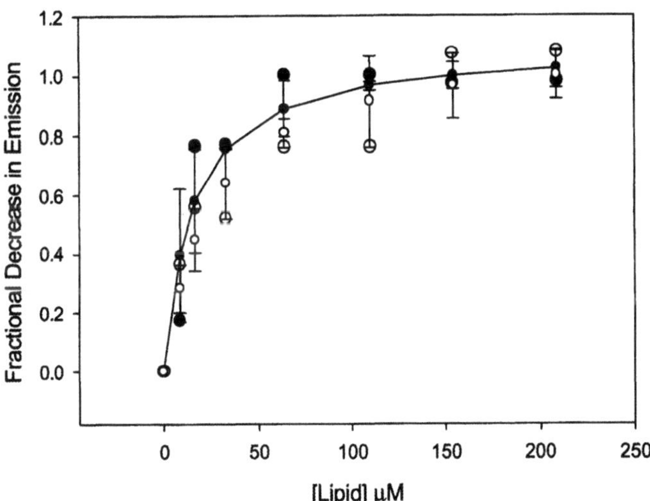

Fig. 15.2. Binding of PLCβ to membranes composed of PC:SM:Ch (*filled circles*) or SM:Ch (*open circles*) as determined by the 10% decrease in intrinsic fluorescence measured by exciting at 280 nm and scanning from 300–420 nm

15.2.3
PLCβ₂-Gβγ Associations on Membrane Surfaces

We directly measured the binding of $PLC\beta_2$ to $G\beta\gamma$ on membrane surfaces using fluorescence resonance energy transfer [18]. Fluorescence energy transfer is the only direct technique to view protein–protein associations on membrane surfaces in real time. For these studies we labeled $G\beta\gamma$ at a 1:1 molar ratio with an amine-reactive coumarin. The high quantum yield of coumarin allows us to detect nanomolar quantities of protein. We used DABCYL probes as nonfluorescent energy transfer acceptors. The absorption band of DABCYL has good overlap with the emission of coumarin and the R_o for this pair is ~34 Å (see [22]). DABCYL has the advantage of being nonfluorescent, thereby omitting the need for corrections due to overlap with donor fluorescence.

Experimentally, we used FRET to determine the apparent affinity between these proteins and the kinetics. Affinities were determined by labeling of $G\beta\gamma$, reconstituting it into bilayers by simple addition and titrating DABCYL-PLCβ into the solution. We have found that these proteins associate strongly on the membrane surface with a $K_{d(app)}$ that depends on the lipid concentration (i.e., the higher the lipid concentration, the weaker the $K_{d(app)}$ due to dilution of the proteins on the membrane surface; for a full discussion, see [23]). An example of these affinity measurements for protein association on lipid rafts (see below) is shown in Fig. 15.3.

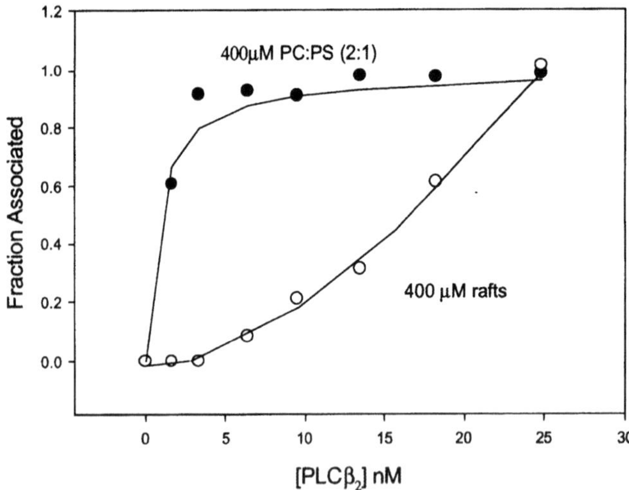

Fig. 15.3. Fraction of PLCβ₂–Gβγ formed on uniform surfaces versus rafted surfaces. Fraction associated was determined by the 10% loss of donor (coumarin-Gβγ) fluorescence, due to the addition of DABCYL- PLCβ₂. Fluorescence was determined by the integrated area of the peak scanned from 380–500 nm exciting at 340 nm

We have also used FRET to determine the on and off rates of the $PLC\beta_2$–$G\beta\gamma$ interactions to gauge the lifetime of the signal in cells [18]. The on rate was measured by directly monitoring the time dependence of the protein associations whereas the off rate was measured by forming the energy transfer complex, adding a large excess of unlabeled protein and following the loss in transfer as a function of time. We found that k_{on} rate of $PLC\beta_2$–$G\beta\gamma$ is slower than diffusion controlled, indicating that conformational changes occur upon protein docking. However, the time that the complex remains together is very long (2 minutes) meaning that this signal would be sustained without some deactivating factor. We tested whether $G\alpha_i$–GDP may induce deactivation by displacing $G\beta\gamma$ from the $PLC\beta_2$ ($G\alpha_i$ subunits do not activate $PLC\beta$s). We instead found that $G\alpha_i$–GDP causes very rapid (>20 s) enzymatic deactivation $PLC\beta_2$ but does not physically dissociate $PLC\beta_2$–$G\beta\gamma$, implying that $G\alpha_i$–GDP–$PLC\beta_2$–$G\beta\gamma$ complexes form. We interpret this complex as being the result of multiple interaction sites on the $G\beta\gamma$ propellor that may change when $G\alpha$(GDP) binds and switch off the activating $PLC\beta_2$ interaction for a weaker $PLC\beta$ secondary interaction site. Thus, corraling the proteins may allow for rapid deactivation of PLC-β signals by this pathway.

We were interested to determine whether RGS4, which increases the rate of α_q deactivation, will interact with other proteins in the $G\alpha_q\beta\gamma$–$PLC\beta_1$ signaling system. Since these proteins are membrane-bound, we first determined whether RSG4 will significantly bind to model membranes. We found that RGS4 prefers binding to negatively charged lipid, but that its association is very weak (K_p ~500 μM where K_p is the apparent partition coefficient and refers to the lipid concentration where 50% of the protein is bound) [7].

Not surprisingly, RSG4 binds strongly to $G\alpha_q$ subunits. However, we found that RGS4 strongly associates to $G\beta\gamma$ and also to $PLC\beta_1$ indicating that RGS4 may also bind to these secondary weaker sites when activated $G\alpha_q$ is unavailable (see Figure 15.1). Thus, during signaling RGS4 may be displaced from $G\beta\gamma$ and $PLC\beta_1$ but remains in close proximity to $G\alpha_q$. These data support the notion that RGS4 is part of a signaling complex and we have determined the local concentrations at which these proteins will aggregate into a higher order complex [7].

15.2.4
Effects of Lipid Rafts on Protein Association

We measured the effect that lipid rafts may have on the lateral association of $G\beta\gamma$ and $PLC\beta_2$. It has been proposed that proteins with acyl chain modifications, such as $G\beta\gamma$, become diffusionally trapped in lipid domains that have reduced fluidity [11]. This trapping in turn serves as a nucleus for protein domain formation. Thus, it is possible that lipid domains promote the formation of protein signaling complexes independent of scaffolding proteins.

The presence of lipid domains may or may not affect the location of the proteins. Proteins may preferentially partition either inside, outside, or on the borders of the lipid domains. Preferential partitioning of the proteins would have the effect of increasing their apparent affinity due to the reduced area for the proteins to diffuse freely. Alternately, the individual proteins may partition differently in the domains and the presence of domain would then decrease the apparent affinity.

To determine the role of lipid rafts, we compared the affinity between PLCβ_2–G$\beta\gamma$ on uniform PC:PS (2:1) and (SM:Ch), and "rafted" lipids (PS:SM:Ch = 1:1:1) under conditions where all of the proteins should be membrane-bound (see above). We find that association on the rafted lipids is greatly reduced as compared to the other lipids. These data indicate that the two proteins partition differently in rafted and uniform lipids. We are in the process of better characterizing this behavior.

Acknowledgements. The authors are grateful for support from the National Institutes of Health GM53132 and the American Heart Association.

References

1. Neer EJ (1995) Heterotrimeric G proteins: organizers of transmembrane signals. Cell 80(2):249–257
2. Ross EM, Wilke TM (2000) GTPase-activiating proteins for heterotrimeric G proteins: Regulators of G protein signaling (RGS) and RGS-like proteins. Annu Rev Biochem: 69:1–33
3. Rebecchi M, Pentylana M (2000) Structure, function and control of phosphoinositide-specific phospholipase, C Physiol Reviews (in press)
4. Berman D, Gilman A (1998) Mammalian RGS proteins: barbarians at the gate. J Biol Chem: 273:1269–1272
5. Tesmer J, Berman D, Gilman A, Sprang S (1997) Structure of RGS4 bound to alumminun flouride-activated G(ialpha1): stabilization of the transition state of GTP hydrolysis. Cell 89:251–261
6. Chidiac P, Ross EM (1999) Phospholipase C-beta1 directly accelerates GTP hydrolysis to G(alpha)q and acceleration is inhibited by G(beta-gamma) subunits. J Biol Chem 274:19639-19643
7. Dowal L, Elliott J, Popov S, Wilkie T, Scarlata S (2001) Determination of the contact energies between a regulator of G protein signaling and G protein subunits and phospholipase C-beta1. Biochemistry (in press)
8. Okamoto T, Schlegal A, Scherer P, Lisanti M (1998) Caveolins, a family of scaffolding proteins for organizing "preassembled signaling complexes" at the plasma membrane. J Biol Chem 273:5419–5422
9. Schlegel A, Volante D, Engelman J, Galbiata F, Mehta P, Zang XL, Scherer P, Lisanti M (1998) Crowded little caves: structure and function of caveolae. Cell Signal 10:457–463
10. Anderson RG (1998) The caveolae membrane system. Annu Rev Biochem 67:199–225
11. Schoeder R, London E, Brown D (1994) Interactions between saturated acyl chains confer detergent resistance on lipid and glycosylphosphatidylinositiol-anchored proteins. Proc Natl Acad Sci USA 91:12130–12134
12. Ahmed SN, Brown DA, London E (1997) On the origin of sphingolipid/cholesterol-rich detergent-insoluble cell membranes: physiological concentrations of cholesterol

and sphingolipid induce formation of a detergent-insoluble, liquid-ordered lipid phase in model membranes. Biochemistry 36:10944–10953

13. Silvius J, del Giudice D, LaFleur M (1996) Cholesterol at different bilayer concentrations can promote or antagonize lateral segregation of phospholipids of differing acyl chain length. Biochemistry 35:15198–15208

14. Hannan L, Lisanti M, Rodriguez-Boulan E, Edinin M (1993) Correctly sorted molecules of a GPI-anchored protein are clustered and immobile when they arrive at the apical surface of MDCK cells. J Cell Biol 120:353–358

15. Pike LJ, Casey L (1996) Localization and turnover of phosphatidylinositol-4,5-bisphosphate in caveolin-enriched membrane domains. J Biol Chem 271:26453–26456

16. De Weerd W, Leeb-Lundberg L (1997) Bradykinin sequesters B2 bradykinin receptors and the receptor coupled G alpha sunuits G alpha(q) and G alpha (i) in caveolae in DDT1 MF-2 smooth muscle cells. J Biol Chem 272:17858–17866

17. Tsunoda S, Sierralta J, Sun Y, Bodner R, Suzuki E, Becker A, Socolich M, Zucker C (1997) A multivalent PDZ-domain protein assembles signalling complexes in a G-protein-coupled cascade. Nature 388:243–249

18. Runnels LW, Scarlata SF (1998) Regulation of the rate and extent of phospholipase C--beta2 effector activation of the beta-gamma subunits of heterotrimeric G proteins. Biochemistry 37:15563–15574

19. Biddlecome GH, Berstein G, Ross EM (1996) Regulation of phospholipase C-β1 by G_q and m1 muscarinic cholinergic receptor: steady-state balance of receptor-mediated activation and GTPase-activating protein-promoted deactivation. J Biol Chem 271(14): 7999–8007

20. Zeng W, Xu X, Popov S, Mukhopadhyay S, Chidic P, Swistok J, Danho W, Yagalooff K, Fisher S, Ross EM, Muallem S, Wilkie T (1998) The N-terminal domain of RGS4 confers receptor-selective inhibition of G protein signaling. J Biol Chem 273:34687–34690

21. Runnels LW, Jenco J, Morris A, Scarlata S (1996) Membrane binding of phospholipases C-beta 1 and C-beta 2 is independent of phosphatidylinositol-4,5-bisphosphate and the alpha and beta gamma subunits of G proteins. Biochemistry 35:16824–16832

22. Van der Meer W, Coker G and Chen SS-Y (1994) Resonance energy transfer, theory and data. VCH, New York

23. Runnels LW (1997) Regulation of phospholipase C-beta isozymes by heterotrimeric GTP-binding proteins. Doctoral thesis. State University of New York at Stony Brook

Mechanisms of the Modulation of Membrane Interfacial Enzyme Catalysis by Non-lamellar Forming Lipids: Comparison with the Behavior of a Fluorescent Probe in Membranes

R. M. EPAND, R. CORNELL, S. M.A. DAVIES, AND R. KRAAYENHOF

The activities of several membrane proteins are modulated by the presence of non-lamellar forming lipids. Although several proteins are activated by the presence of these lipids, the molecular mechanism by which this activation occurs is different for different proteins. We have studied two enzymes that are activated by non-lamellar forming lipids in different ways; protein kinase C (PKC) and CTP:phosphocholine cytidylyltransferase (CT). CT is the enzyme that regulates the rate of synthesis of phosphatidylcholine. For many lipid systems, there is a quantitative correlation between the calculated curvature strain of the membrane and the activation of this enzyme. For many lipid systems, including liposomes containing a series of homologous di-18:1 phosphatidylethanolamines, there is a quantitative correlation between the extent of activation of CT and the curvature strain in the membrane. In contrast, the order in which this series of di-18:1 phosphatidylethanolamines enhances the activity of protein kinase C does not correlate with membrane curvature strain. However, the order in which these lipids affect the quenching of the fluorescent probe 4-[(n-dodecylthio)-methyl]-7-(N,N-dimethylamino)-coumarin by doxyl groups positioned in the acyl chain region of the membrane is well correlated with the extent of activation of protein kinase C. This fluorescent probe is monitoring a property of the membrane that is affected by the presence of non-lamellar forming lipids. Curiously, changes in the fluorescence properties of the probe do not correlate exactly with membrane curvature strain, but they do correlate with changes in the extent of activation of protein kinase C. The physical properties that may be affecting both the behavior of the probe, as well as the activation of protein kinase C, appear to be dependent on the nature of the membrane-water interface and may include the lateral pressure at the membrane interface, the extent of hydration, the presence of defects and others. The studies demonstrate that there is more than a single mechanism by which the activity of membrane-bound enzymes are affected by non-lamellar forming lipids.

16.1
Introduction

16.1.1
Specific vs. Non-specific Modulation of Protein Activity

Many proteins express their biological activity when attached to membranes. These proteins have diverse functions including being receptors, serving as specific ion channels or acting as enzyme catalysts. In general, the activities of these proteins will be affected by the nature of the membrane lipids in their direct vicinity. There are two general mechanisms by which lipids can modulate the activity of proteins. One of these is by the lipid binding to a specific site on the protein and acting as a cofactor. The other way in which the lipid can act is by altering the physical properties of the membrane and thereby affecting protein function by changing the general physical environment of the protein. The two mechanisms are not mutually exclusive and may represent two extremes of a range of specificity. A cofactor that binds to a site on the protein will, to some degree at least, alter the physical environment of the protein. Conversely, a substance that affects the activity of the protein through non-specific changes in the properties of the membrane will be dependent on the degree to which it colocalizes in the membrane with the protein.

One can assess the degree of specificity by which the activity of the protein is modulated by determining the sensitivity to the chemical structure of the modulator. For example, changes in the stereochemistry of a lipid modulator would generally be expected to affect protein activity only if the interactions are specific, since different stereoisomers should have the same effects on the physical properties of a membrane. This, however, is not always true and different stereoisomers may interact differently with the surrounding chiral phospholipid of the bilayer and therefore not produce identical changes in the physical properties of the membrane [1]. Nevertheless, in many cases the activity of the protein will not be sensitive to minor variations in the chemical structure of a number of modulators, indicating that the mechanism is non-specific.

In cases for which the mechanism is suggested to be non-specific there is a question as to what physical property the protein is responding to. Some of the properties that may affect protein activity include membrane fluidity, membrane surface charge, membrane curvature strain, membrane hydration, ability of the protein to penetrate into the membrane, bilayer thickness, etc. It is not a simple matter to assess the relative importance of these factors. It is likely that in many cases several of these physical features will modulate the activity of a particular protein. In addition, any comparisons made to assess the role of a single physical property on protein activity must also take into consideration that it is not possible to alter only one physical feature without affecting others. Two kinds of evidence, however, can provide strong support for the importance of a particular physical property in modulating the activity of a protein. One is to show that there are many substances of diverse chemical structure that modulate the activity of the protein in a manner that correlates with a change in a particular physical property of the membrane. Another approach is to measure the effect on protein activity of two or more substances that are structurally related but differ in the effects that they have on a particular membrane property.

16.1.2
Non-lamellar Forming Lipids

Non-bilayer forming lipids have been suggested to play an important role in biological membranes by establishing an environment with an optimal balance between stability and lack of rigidity [2]. In particular, studies with bacteria have demonstrated that changing the growth temperature of the organism will lead to an alteration in the lipid composition of the membrane, both in *E. coli* [3] as well as in *Acholeplasma laidlawii* [4], so as to maintain a constant curvature instability. For the latter organism, it has been suggested that this sensitivity to curvature strain is a consequence of a specific enzyme, diglucosyldiacylglycerol synthase, whose activity is sensitive to the balance between lamellar and non-lamellar lipids [5].

The activity of several membrane and amphitropic proteins has been shown to be sensitive to membrane curvature strain. These proteins include rhodopsin [6] as well as the G-protein α-subunits [7], among others. We will focus on the properties of two amphitropic enzymes whose activity is modulated by membrane curvature, but through apparently different mechanisms. These enzymes are protein kinase C (PKC) and CTP:phosphocholine cytidylyltransferase (CT). The activity of both PKC [8] and CT [9, 10] are enhanced by increasing the negative curvature strain of the membrane. However, each of these two enzymes has a very different mode of association with the membrane. While both proteins bind interfacially, their mechanisms of interaction with membranes are quite different. Classical PKCs bind membranes via the cooperation of the C1 and C2 domains. The C1 domain of PKC binds diacyl glycerol, whereas the C2 domain binds Ca^{2+}. Both domains also interact with anionic lipids. CT, on the other hand, has no pocket for binding a lipid monomer; instead, it contains a ~50 residue amphipathic α-helix which partitions into the membrane interfacial region with the helix axis parallel to the surface. A comparison of the membrane binding mechanisms and modules of PKC and CT is shown in Fig. 16.1. PKC interacts with lipid, in part, by the insertion of a protein domain into the membrane, while CT associates primarily by binding of an amphipathic helix segment.

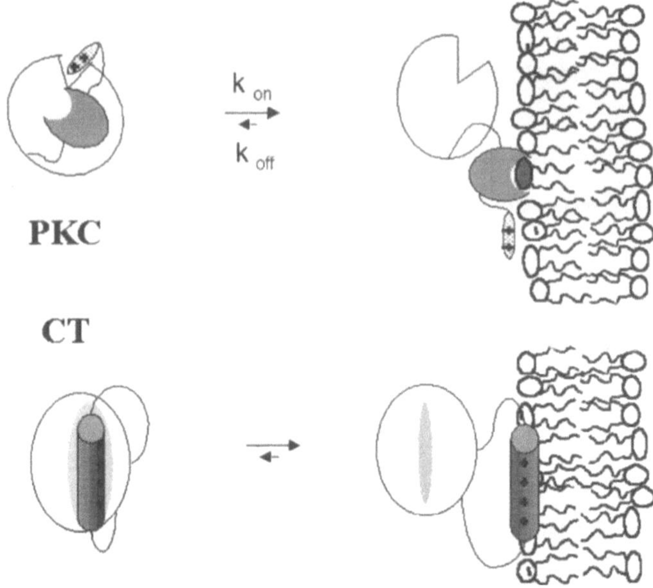

Fig. 16.1. Model for the binding of PKC and CT with a membrane. Adapted from [38]

16.2
Specific Protein Examples

16.2.1
Protein Kinase C

Protein kinase C represents a family of enzymes that are related by similarities in their amino acid sequence. These enzymes play important roles in a number of signal transduction pathways [11]. One of the physical properties that can influence the binding and activity of PKC is the intrinsic curvature of a membrane. Negative curvature strain exists when each monolayer of a bilayer can more readily form an inverted phase such as the inverted cubic or hexagonal (H_{II}) phase. Phosphatidylethanolamines (PE) with unsaturated acyl chains readily form inverted phases. The lamellar phase of PE will be converted into a hexagonal phase upon heating to a particular temperature. This phase transition temperature (T_H) is very sensitive to the presence of certain additives in the membrane [12]. Substances that promote the formation of inverted phases will lower T_H. This property has been associated with the activation of PKC, independently of the chemical nature of the curvature-modulating substance [13]. In addition, two phosphatidylcholines that have very similar chemical structures but which have opposite effects on membrane curvature, modulate the activity of PKC in a manner that correlates with their effects on T_H, i.e., the analog which lowers T_H activates PKC, while the one that raises T_H is an inhibitor [14]. Even relatively small changes in membrane structure, such as the change from dioleoyl (di 18:1) to 1-palmitoyl-2-oleoyl chains (1-16:0-2-18:1) elicit differences in the behavior of PKC [15]. The dioleoyl forms, which have greater tendency to form negatively curved structures, both bind PKC at a lower mol fraction of PS and also exhibit a higher maximal activity compared with that of the palmitoyl, oleoyl forms.

There is a maximal activation of PKC by phosphatidylethanolamine and at higher mole fractions of this lipid, the activity of the enzyme declines [16]. This property is exhibited by several individual isoforms of PKC, including the calcium dependent c-PKC. There is evidence that the sensitivity to membrane curvature is a result of interaction of the C1 domain of PKC with the membrane [17]. The decline in activity at higher concentrations of PE is not because of a loss of enzyme binding to the lipid, but it may be a consequence of the enzyme's inability to penetrate as deeply into the membrane. Alternatively, there may be an optimal curvature required to promote PKC activity and membranes with either too high or too low of an intrinsic membrane curvature do not optimally activate PKC. Although the change in concentration of PE does not alter the fraction of PKC bound to the membrane, it does appear to affect the nature of the interaction between the enzyme and lipid so as to produce a biphasic response of the fluorescence anisotropy of dansyl-PS to the concentration of PE [17].

16.2.2
Phosphocholine Cytidylyltransferase

Phosphocholine cytidylyltransferase catalyses the rate-limiting step in the *de novo* synthesis of PC, the most abundant lipid in mammalian cell membranes [18]. Like PKC, CT is also an amphitropic enzyme, present in both soluble and membrane-bound forms [19]. Also like PKC, CT is activated by binding to membranes [20, 21]. The association of CT with a membrane is greatly influenced by the lipid composition [18]. The membrane-binding domain (domain M) of CT is a single, long, uninterrupted amphipathic α-helix [22]. Electrostatic interactions between the negatively charged lipid headgroups and the positively charged amino acids in domain M of CT are thought to promote CT's membrane binding [23].

The binding of CT to membranes is also promoted by the presence of non-lamellar forming lipids [9, 10]. In addition, the lipid lysoPC that promotes positive membrane curvature counters the activating effects of the negative curvature lipid, diacylglycerol [9, 10, 24].

16.3
Molecular Mechanisms of Enzyme Activation by Non-lamellar Forming Lipids

16.3.1
Curvature Strain

It is useful to consider the curvature properties of a lipid monolayer. A bilayer in a large unilamellar vesicle (LUV) or a multilamellar vesicle (MLV) will have no physical curvature on a molecular length scale because locally the bilayer will be flat. However, the constituent monolayers of the bilayer may have an intrinsic curvature that is not flat if the monolayer is composed of non-lamellar forming lipid. If each monolayer could bend to its preferred shape, without a change in the polarity of the environment of any of its groups, it would then achieve its intrinsic curvature. This cannot happen in a symmetrical bilayer since each monolayer would bend in opposite directions, leaving a void between the ends of the acyl chains. However, the arrangement of lipid in the H_{II} phase can approximate that of a monolayer that has attained its intrinsic curvature. The extent of this curvature can readily be calculated from the lattice spacing of the H_{II} phase, measured by diffraction. To be accurate, this measured intrinsic curvature requires a correction resulting from the fact that the H_{II} phase does not fill all of the space if each unit is a perfectly round cylinder. There will be voids between the cylinders that have to be filled by gauche to trans isomerization of the acyl chains that point toward these voids. This will require energy and as a consequence the H_{II} cylinder diameter will decrease to lower the extent of this hydrocarbon packing problem. There are experimental ways to correct for this factor and an accurate value of the intrinsic curvature can be obtained. This curvature is usually expressed as an intrinsic radius of curvature, R_0, defined as the distance from the center of the cylinder to the pivitol plane; the pivitol plane being the position in the lipid structure whose cross-sectional area does not change when the monolayer bends.

A monolayer organized in a structure in which its physical curvature is equal to its intrinsic curvature will not possess any curvature strain. Lipids that spontaneously form highly curved structures, such as the non-lamellar phases, will have instability due to curvature strain when they form a planar bilayer. The curvature energy associated with each monolayer of the bilayer will depend on two factors, the intrinsic radius of curvature, R_0, and the elastic bending modulus, K_c. The curvature energy per unit area of interface is given by:

$$\frac{0.5 K_c}{R_0^2} \tag{16.1}$$

The curvature strain of a monolayer will be equal to the energy required to unbend it from the form in which it has achieved its intrinsic curvature to the flat structure of the bilayer. The elastic bending modulus is a measure of the stiffness of the monolayer. The easier it is to bend the monolayer, the less curvature strain will be acquired by forming a structure whose curvature is different from the intrinsic curvature.

16.3.2
Lateral Pressure Profile

An alternative formulation of curvature strain has been recently proposed by Cantor [25, 26]. This approach focuses on the variation of the lateral pressure as a function of the position in the bilayer. In general, lateral pressure profiles and curvature strain are alternative ways of expressing the same phenomenon. If there is a higher lateral pressure at the methyl terminus of the acyl chains than in other regions of the bilayer, this is equivalent to the bilayer having negative curvature strain. Each of the two formulations has its own advantage. Curvature strain is a simpler idea and is more amenable to direct experimental measurement. However, the concept of lateral pressure profile provides a more detailed molecular description and can be more useful to correlate with the location of substances within the membrane [27].

16.3.3
Other Mechanisms

There may be other mechanisms by which non-lamellar lipids affect the interactions with proteins. These include the extent of hydration of the membrane. Bonding among lipid headgroups reduces the cross-sectional area at the water-membrane interface and concomitantly reduces the degree of hydration as well as promoting the formation of inverted phases. The reduced hydration will alter the way proteins interact with the membrane. In addition, the tendency to form non-bilayer structures may result in the bilayer having defects into which a protein may more easily penetrate. Hence, observing effects on protein function that correlate with the presence of non-lamellar forming lipids, may not necessarily be a direct consequence of negative curvature strain.

16.4
Criteria for the Role of Membrane Curvature Strain

16.4.1
Cubic Phases

The bicontinuous cubic phase has less curvature strain than the lamellar phase, because the bilayers in the bicontinuous cubic phase bend to partly relieve curvature strain, but they have more curvature strain than the hexagonal phase since the structure is further from the intrinsic radius of curvature [28]. One can compare the activity of proteins in the cubic phase with that in the lamellar and hexagonal phases. For lipids with a negative intrinsic radius of curvature, the lamellar phase will have the largest curvature strain, followed by the cubic phase and the hexagonal phase having the least curvature strain (i.e., the lipid monolayer having a curved morphology close to that of the intrinsic curvature). If non-lamellar lipids activate the protein through a mechanism coupled with membrane curvature strain, then the activation should be greatest in the lamellar phase and least in the hexagonal phase.

Unfortunately, it is not that straight-forward to make this comparison. There are two fundamental problems. One is that under a particular set of physical conditions, such as temperature and pH, and for a particular lipid, the phase content of the sample will be fixed. Thus, in order to compare systems in different phases, something in addition to the phase, such as temperature or lipid composition, has to also change and one must be able to distinguish changes caused by the difference in phase from those caused by the change in the variable required to induce a change in phase. The other complication is that the aqueous compartments are internal to the inverted phases. There has to be a path by which the largest molecules can enter and be sterically accommodated in the internal aqueous compartments.

To our knowledge, this test of the curvature strain mechanism has been done only with PKC [29]. A well studied lipid system that forms a bicontinuous cubic phase in the presence of excess water is that of monoolein. PKC does not bind to pure monoolein but requires the presence of phosphatidylserine (PS). Above 13 mol % PS, mixtures of monoolein and PS form lamellar phases, but below 10 mol % PS, the mixture remains in the cubic phase. A relatively small increase of PS concentration in the range of 10 to 13 mol % causes a change in phase from cubic to lamellar. This phase change is accompanied by a large increase in membrane binding of PKC but a decrease in the catalytic activity of the enzyme. There is on the order of a 5 to 10-fold decrease in the specific activity of the membrane-bound form of PKC when the cubic phase is converted to a lamellar phase. Since the lamellar phase has more negative curvature strain than does the cubic phase, these results indicate that although non-lamellar forming lipids activate PKC it is not by a mechanism that is directly coupled with curvature strain [29]. To test this relationship in a different lipid system we took advantage of the fact that it was shown that the addition of small amounts of alamethicin to dielaidoylphosphatidylethanolamine (DEPE) facilitates the formation of a bicontinuous cubic phase [30]. Although in lamellar systems alamethicin slightly inhibits PKC, this peptide activates the enzyme at concentrations at which it converts the lamellar phase of DEPE to the cubic phase. Thus, it has been demonstrated in two different systems that the activity of PKC with lipid in a bicontinuous cubic phase is greater than with lamellar phase lipid of comparable chemical composition. Therefore, the activation of PKC by non-lamellar lipids is not attributable to increased negative curvature strain.

16.4.2
Homologous Lipids

A second method for testing the role of curvature strain in the modulation of enzyme activity is to compare the properties of membranes having a series of lipids with very similar chemical structure but altered physical properties. Initially we did this in a qualitative manner, comparing the shift of T_H caused by a series of homologous phosphatidylcholines having branched acyl chains with their effect on the activity of PKC [14]. More recently, we have determined the curvature properties of three PEs, all with C18:1 acyl chains, but with the double bond in different positions of the acyl chain. The polymorphic behavior of these three PEs, as well as their curvature and elasticity properties have been determined [31]. As with other amphiphiles, there was a relationship between the T_H of the lipid and the activation of PKC, with the PE having the highest T_H being the one that supported the lowest activation of PKC [32]. However, there was no correlation between a fundamental parameter of curvature, such as the curvature strain energy, the intrinsic radius of curvature, or the elastic bending modulus [33]. This series of three PEs was also used to test the relationship between membrane curvature and the activity of CT. Unlike the case for PKC, there is quantitative agreement between the calculated curvature strain of a membrane containing one of these three PE lipids and the activity of CT [10]. This indicates that at least by this criterion, there is a direct relationship between membrane curvature strain caused by the presence of non-lamellar forming lipids and the activation of CT.

16.5
Relationship of the Properties of a Fluorescent Interfacial Membrane Probe

The fluorescent probe 4-(n-dodecylthiomethyl)-7-(dimethylamino)coumarin (DTMAC) partitions into membranes to a depth at which the fluorophore encounters an environment of a particular polarity. This is indicated by the observation that, although the fluorescence emission spectrum of this probe exhibits a marked Stokes shift when dissolved in different solvents, its emission is insensitive to the type of lipid bilayer into which it is inserted [10, 32, 34, 35]. Nevertheless, the quenching of the fluorescence of DTMAC by the nitroxide-labeled lipid, 1-palmitoyl-2-(5-doxyl)stearoylphosphatidyl-choline (5-doxyl PC) varies among these lipids, indicating that the extent of penetration of this fluorophore into the membrane interfacial region is dependent on the lipid composition of the bilayer (Fig. 16.2). In bilayers of PE, a lipid with a more hydrophobic surface and greater interaction among headgroups, there is less fluorescence quenching of DTMAC by 5-doxyl PC, compared with DTMAC in a bilayer of phosphatidylcholine that has the same acyl chain composition [34]. In the case of the three di-18:1-PEs, referred to above, the order of quenching by 5-doxyl PC is $\Delta 9 > \Delta 11 > \Delta 6$ [32]. It was found that this relationship also held when the nitroxide was linked at either

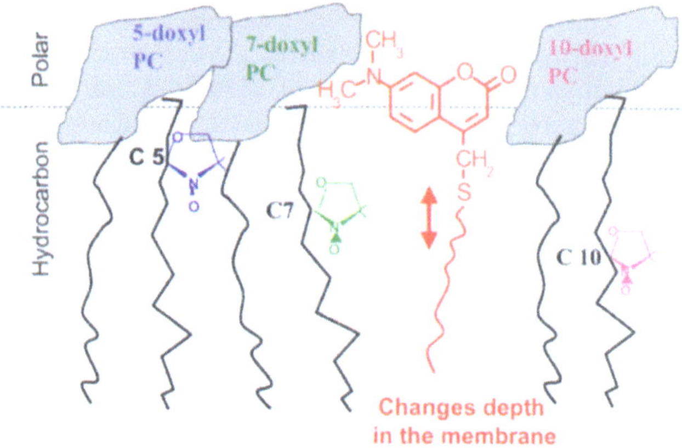

Fig. 16.2. Model of the insertion of DTMAC into a bilayer, illustrating the relationship between the depth of insertion of the probe and the extent to which it is quenched by nitroxide-labeled phospholipids. Adapted from [10]

the 7 or the 10 position of the acyl chain [10]. This order of quenching efficiency is different from the order of curvature stain energy, which is $\Delta 9 > \Delta 6 > \Delta 11$. Interestingly, the order of DTMAC quenching with these three PEs parallels their ability to increase the activity of PKC [32]. This indicates that whatever property is required for the insertion of DTMAC into the membrane, also affects the activity of PKC. This likely involves a property of the membrane-water interface such as polarity or lateral tension of the membrane surface. It should be noted that the weakest activator, $\Delta 6$, does not function by reducing the binding of PKC to the membrane, but in fact augments the membrane affinity of PKC. A different situation was found with CT, in which the order of activation of this enzyme by the three PEs parallels their curvature strain energies and not the insertion depth of the DTMAC fluorescent probe [10].

16.6
Comparison of the Mechanisms of Activation of PKC and CT by Non-lamellar Forming Lipids

It is interesting to compare the lipid modulation of the activities of the enzymes PKC and CT. These two enzymes have many similarities in their properties. Both enzymes are amphitropic, i.e., they can exist in both soluble and membrane-bound forms. Both enzymes are activated when bound to a membrane and this activation is greater in the presence of non-lamellar forming lipids. However, in the case of

CT there appears to be a direct relationship between negative membrane curvature strain and activation of the enzyme, while this is not the case with PKC. An initial indication of this is the finding that the activity of PKC can be assayed in Triton micelles [36] and membrane additives that promote negative curvature still activate PKC [13], despite the fact that micelles have positive membrane curvature. This is not the case, however, for CT whose activity is not enhanced by negative curvature agents when the enzyme is assayed in the presence of Triton micelles [37]. Further indication that the activity of PKC is not modulated directly by curvature strain is the observation that its activity is higher in the presence of lipids arranged in cubic phases than it is with lamellar phase lipid. Since the cubic phase is formed at the expense of the lamellar phase in order to reduce curvature strain, the fact that PKC activity is higher in the cubic phase shows that the enhanced activity is not a consequence of increased curvature strain.

There is an interesting comparison between the properties of the fluorescence probe DTMAC, curvature strain and the activities of PKC and CT. We have maintained the components of the lipid systems similar so as to be able to compare the consequences of altered physical properties. This was done by varying the position of the only double bond in the PE component of the lipid mixture. There was no relationship between curvature strain and the properties of this fluorescent probe. However, there was a correlation between the properties of the probe and the ability to activate PKC, again indicating that the activity of PKC is not directly modulated by curvature strain. The situation is different with CT, for which the activity correlates better with curvature strain than with the properties of this fluorescent probe.

16.7
Conclusions

Non-lamellar forming lipids activate a number of membrane-bound proteins. In some cases, such as CT, this phenomenon can be explained by negative curvature stress of the membrane being directly coupled to the activity of the enzyme. However, in other cases, such as PKC, for which there is a correlation between the presence of non-lamellar forming lipids and activation of the enzyme, there is not a direct relationship with curvature strain.

For both of these enzymes, even for CT for which curvature strain appears to be the explanation for activation of the enzyme by non-lamellar forming lipids, there are also other physical properties of the membrane that modulate activity. Both CT and PKC are activated by the presence of negatively charged lipids and acyl chain order also appears to play a role in CT activation since the effects of cholesterol cannot be explained simply by curvature strain [10].

The fluorescent probe, DTMAC, appears to be sensitive to membrane properties in a manner that is mimicked by PKC. This provides a tool for trying to identify the mechanism by which the presence of non-lamellar forming lipids is coupled with the activation of PKC.

274 R. M. Epand et al.

Acknowledgement. This work was supported by the Canadian Institutes of Health Research and The Wellcome Trust. SMAD is an International Wellcome Fellow.

References

1. Epand RM, Stevenson C, Bruins R, Schram V, Glaser M (1998) The chirality of phosphatidylserine and the activation of protein kinase C. Biochemistry 37:12068–12073
2. Garab G, Lohner K, Laggner P, Farkas T (2000) Self-regulation of the lipid content of membranes by non-bilayer lipids: a hypothesis. Trends Plant Sci 5:489–494
3. Morein S, Andersson A, Rilfors L, Lindblom G (1996) Wild-type Escherichia coli cells regulate the membrane lipid composition in a "window" between gel and non-lamellar structures. J Biol Chem 271:6801–6809
4. Osterberg F, Rilfors L, Wieslander A, Lindblom G, Gruner SM (1995) Lipid extracts from membranes of Acholeplasma laidlawii A grown with different fatty acids have a nearly constant spontaneous curvature. Biochim Biophys Acta 1257:18–24
5. Vikström S, Li L, Karlsson OP, Wieslander A (1999) Key role of the diglucosyldiacylglycerol synthase for the nonbilayer–bilayer lipid balance of Acholeplasma laidlawii membranes. Biochemistry 38:5511–5520
6. Brown MF (1994) Modulation of rhodopsin function by properties of the membrane bilayer. Chem Phys Lipids 73:159–180
7. Escriba PV, Ozaita A, Ribas C, Miralles A, Fodor E, Farkas T, Garcia-Sevilla JA (1997) Role of lipid polymorphism in G protein-membrane interactions: nonlamellar-prone phospholipids and peripheral protein binding to membranes. Proc Natl Acad Sci USA 94:11375–11380
8. Epand RM, Lester DS (1990) The role of membrane biophysical properties in the regulation of protein kinase C activity. Trends Pharmacol Sci 11:317–320
9. Attard GS, Templer RH, Smith WS, A.N.Hunt, Jackowski S (2000) Modulation of CTP:phosphocholine cytidylyltransferase by membrane curvature elastic stress. Proc Natl Acad Sci USA 97:9032–9036
10. Davies SMA, Epand RM, Kraayenhof R, Cornell RB (2001) Regulation of CTP:phosphocholine cytidylyltransferase activity by the physical properties of lipid membranes: an important role for stored curvature strain energy. Biochemistry 40: 10522–10531
11. Toker A (1998) Signaling through protein kinase C. Front Biosci 3:D1134–D1147.
12. Epand RM (1985) Diacylglycerols, lysolecithin, or hydrocarbons markedly alter the bilayer to hexagonal phase transition temperature of phosphatidylethanolamines. Biochemistry 24:7092–7095
13. Epand RM (1987) The relationship between the effects of drugs on bilayer stability and on protein kinase C activity. Chem Biol Interact 63:239–247
14. Epand,R.M., R.F.Epand, B.T.Leon, F.M.Menger, and J.F.Kuo. 1991. Evidence for the regulation of the activity of protein kinase C through changes in membrane properties. Biosci Rep 11:59–64
15. Mosior M, Golini ES, Epand RM (1996) Chemical specificity and physical properties of the lipid bilayer in the regulation of protein kinase C by anionic phospholipids: evidence for the lack of a specific binding site for phosphatidylserine. Proc Natl Acad Sci USA 93:1907–1912

16. Slater SJ, Kelly MB, Taddeo FJ, Ho C, Rubin E, and Stubbs CD (1994) The modulation of protein kinase C activity by membrane lipid bilayer structure. J Biol Chem 269: 4866–4871
17. Ho C, Slater SJ, Stagliano B, Stubbs CD (2001) The c1 domain of protein kinase c as a lipid bilayer surface sensing module. Biochemistry 40:10334–10341
18. Kent C (1997) CTP:phosphocholine cytidylyltransferase. Biochim Biophys Acta 1348: 79–90
19. Cornell RB. Northwood IC (2000) Regulation of CTP:phosphocholine cytidylyltransferase by amphitropism and relocalization. Trends Biochem Sci 25:441–447
20. Friesen JA, Campbell HA, Kent C (1999) Enzymatic and cellular characterization of a catalytic fragment of CTP:phosphocholine cytidylyltransferase alpha. J Biol Chem 274:13384–13389
21. Yang W, Boggs KP, Jackowski S (1995) The association of lipid activators with the amphipathic helical domain of CTP:phosphocholine cytidylyltransferase accelerates catalysis by increasing the affinity of the enzyme for CTP. J Biol Chem 270:23951–23957
22. Dunne SJ, Cornell RB, Johnson JE, Glover NR, and Tracey AS (1996) Structure of the membrane binding domain of CTP:phosphocholine cytidylyltransferase. Biochemistry 35:11975–11984
23. Arnold RS, Cornell RB (1996) Lipid regulation of CTP: phosphocholine cytidylyltransferase: electrostatic, hydrophobic, and synergistic interactions of anionic phospholipids and diacylglycerol. Biochemistry 35:9917–9924
24. Jamil H, Hatch GM, Vance DE (1993) Evidence that binding of CTP:phosphocholine cytidylyltransferase to membranes in rat hepatocytes is modulated by the ratio of bil. Biochem J 291 (Pt 2):419–427
25. Cantor RS (1999a) The influence of membrane lateral pressures on simple geometric models of protein conformational equilibria. Chem Phys Lipids 101:45–56
26. Cantor RS (1999b) Lipid composition and the lateral pressure profile in bilayers. Biophys J 76:2625–2639
27. Cantor RS (1997) The lateral pressure profile in membranes: a physical mechanism of general anesthesia. Biochemistry 36:2339–2344
28. Anderson DM, Gruner SM, Leibler S (1988) Geometrical aspects of the frustration in the cubic phases of lyotropic liquid crystals. Proc Natl Acad Sci USA 85:5364–5368
29. Giorgione JR, Huang Z, Epand RM (1998a) Increased activation of protein kinase C with cubic phase lipid compared with liposomes. Biochemistry 37:2384–2392
30. Keller SL, Gruner SM, Gawrisch K (1996) Small concentrations of alamethicin induce a cubic phase in bulk phosphatidylethanolamine mixtures. Biochim Biophys Acta 1278:241–246
31. Epand RM, Fuller N, Rand RP (1996a) Role of the position of unsaturation on the phase behavior and intrinsic curvature of phosphatidylethanolamines. Biophys J 71: 1806–1810
32. Giorgione JR, Kraayenhof R, and Epand RM (1998b) Interfacial membrane properties modulate protein kinase C activation: role of the position of acyl chain unsaturation. Biochemistry 37:10956–10960
33. Mosior M, Epand RM (1999) Role of the membrane in the modulation of the activity of protein kinase C. J Liposome Res. 9:21–42
34. Epand RF, Kraayenhof R, Sterk GJ, Wong Fong Sang HW, Epand RM (1996b) Fluorescent probes of membrane surface properties. Biochim Biophys Acta 1284:191–195

35. Sterk GJ, Thijsse PA, Epand RF, Wong Fong Sang HW, Kraayenhof R, Epand RM (1997) New fluorescent probes for polarity estimations at different distances from the membrane interface. J Fluorescence 7:115S–118S
36. Hannun YA, Loomis CR, Bell RM (1985) Activation of protein kinase C by Triton X-100 mixed micelles containing diacylglycerol and phosphatidylserine. J Biol Chem 260:10039–10043
37. Cornell RB (1991) Regulation of CTP:phosphocholine cytidylyltransferase by lipids. 1. Negative surface charge dependence for activation. Biochemistry 30:5873–5880
38. Johnson JE, Cornell RB (1999) Mol Membr Biol 16:217–235

Emission Spectroscopy of Complex Formation between *Escherichia coli* Purine Nucleoside Phosphorylase (PNP) and Identified Tautomeric Species of Formycin Inhibitors Resolves Ambiguities Found in Crystallographic Studies

B. KIERDASZUK

The ubiquitous purine nucleoside phosphorylases (PNPs) play a key role in the purine salvage pathway, and are widely considered as major targets in the chemotherapy of immune system diseases, as well as in potentiation of the antitumor and antiviral activities of therapeutically active nucleoside analogues, by preventing them from phosphorolytic cleavage. Therefore, wide attention is devoted to development of more potent and specific inhibitors of the enzyme from various sources, and to studies of the mechanism of enzyme action. This review recalls the results of studies by steady-state and time-resolved emission (fluorescence and phosphorescence) spectroscopy, X-ray crystallography and enzyme kinetics on the interaction of highly purified bacterial (*E. coli*) purine nucleoside phosphorylase with a specific formycin A inhibitor (antibiotic) and its *N*-methylated analogues. The red shift of absorption and emission spectra of the ligands versus the enzyme permits selective excitation of ligand in the enzyme-ligand complex, as well as selective detection of enzyme or ligand emission.

Complex formation between *E. coli* PNP and formycin A (FA) and *N*(6)-methylformycin A (m6FA) by emission spectroscopy demonstrated preferential binding of identified tautomeric species, and resolved ambiguities found in crystallographic studies, where tautomeric forms were indistuinguishable. Analysis of the emission, excitation and absorption spectra of enzyme-ligand mixtures pointed to fluorescence resonance energy transfer (FRET) from protein tyrosine residue(s) to FA and m6FA base moieties, as a major mechanism of protein fluorescence quenching. Förster radius (R0 = 7 Å) was determined for FRET in the complex with FA, and Tyr160 is most probably an energy donor. Effect of enzyme-inhibitor interactions on nucleoside excitation and emission spectra of fluorescence and phosphorescence revealed shifts in tautomeric equilibria of the bound ligands. With FA, the proton N(1)-H is shifted to N(2), independently of the presence of orthophosphate (Pi, substrate). Complex formation with m6FA led to a shift of the amino-imino tautomeric equilibrium in favour of the imino species in the absence of Pi, and in favour of the amino species in the presence of Pi. This provides an excellent example of the complementation of solution and crystallographic studies, as well as further insight into physicochemical factors responsible for selective binding of inhibitors with potential therapeutic applications.

17.1
Introduction

The ubiquitous purine nucleoside phosphorylase (PNP, purine nucleoside:orthophosphate ribosyl transferase, EC 2.4.2.1), a key enzyme of the anabolic and catabolic pathways of purine nucleosides, catalyzes the reversible cleavage of the glycosidic bond of ribo- and 2'-deoxyribonucleosides of guanine and hypoxanthine in higher organisms, as well as of adenine in some prokaryotes, e.g., *Escherichia coli* and *Salmonella typhimurium*, in the presence of inorganic orthophosphate (P_i), as illustrated by the scheme in Fig. 17.1.[1]

The equilibrium of the in vitro reaction is shifted in favor of nucleoside synthesis (Fig. 17.1). However, in vivo phosphorolysis is highly favored over synthesis, due to its coupling with two additional enzymatic reactions, viz. oxidation and phosphoribosylation of the purine bases by xanthine oxidase and hypoxanthine-guanine phosphoribosyltransferase, respectively.

PNP functions intracellularly in the salvage pathway of purine nucleosides, including even those of protozoan parasites such as *Plasmodium*, *Leishmania* and *Trypanosoma*, thus enabling the cells to utilize purine bases recovered from metabolized purine ribo- and deoxyribonucleosides to synthesize purine nucleosides. Since these parasites are incapable of *de novo* purine biosynthesis, selective inhibitors of these enzymes should therefore be good antiparasitic agents. The PNPs

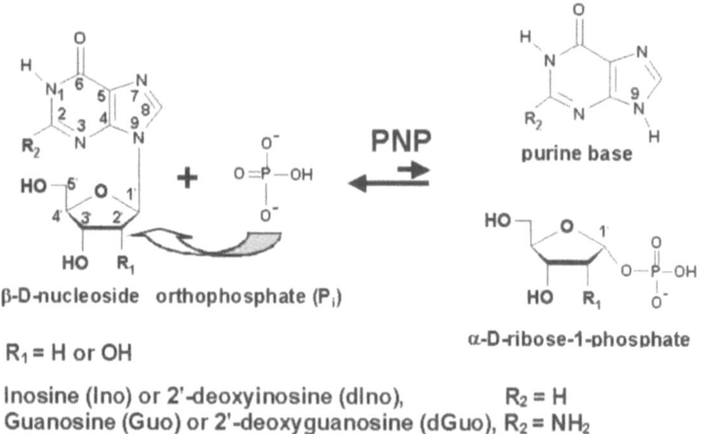

Fig. 17.1. PNP catalyzed phosphorolysis, and reverse synthetic reaction, of β-*D*-purine nucleosides

---

[1] *Abbreviations:* PNP, purine nucleoside phosphorylase; FA, formycin A, 3-(β-*D*-ribofuranosyl)-7-aminopyrazolo[4,3-d]pyrimidine; m¹FA, N(1)-methylformycin A; m²FA, N(2)-methylformycin A; m⁴FA, N(4)-methylformycin A; m⁶FA, N(6)-methylformycin A; FB, formycin B; Ino, inosine; Guo, guanosine; Hepes, N-2-hydroxyethylpiperazine-N'-2-ethanesulfonic acid; P_i, orthophosphate; DMSO, dimethylsulfoxide; FRET, fluorescence resonance energy transfer.

from various sources are members of a broader class of *N*-ribohydrolases and transferases, the transition states for which share ribosyl oxocarbenium-like character with cleavage of the C–N ribosyl bond by an S_N1-like mechanism ([1], and references cited therein).

Major metabolic effect of PNP deficiency, or PNP inhibition, is an accumulation of dGTP, a potent feedback inhibitor of ribonucleotide reductase and, consequently, DNA synthesis, which was long ago considered to account for selective T-cell immune deficiency. Symptoms of PNP-deficiency in humans suggest possible chemotherapeutic applications of potent inhibitors of this enzyme as selective immunosuppressive agents [2, 3] to suppress the host versus graft response in organ transplantation, treatment of T-cell mediated autoimmune diseases such as erythematosus or rheumatoid arthritis ([4], and references cited therein). PNP inhibitors are also the center of interest for potentiation of the antitumor and antiviral activities of therapeutically active nucleoside analogs. Following administration, they must reach the target cell and undergo phosphorylation by the nucleoside kinases [5] in order to manifest their activities, while PNP inhibitors may prevent them from phosphorolytic cleavage. Furthermore, the much broader substrate specificity of PNP from various microorganisms, e.g., *E. coli* [6], as compared to that from human erythrocytes [7, 8], has beeen employed in tumor-directed gene therapy [9–11], again pointing to studies on the mechanism of catalysis and specificity of ligand binding.

Enzyme kinetics and crystal structure data have been extensively profited from delineating the mechanism of action of the enzymes from various sources, and to design more potent and specific inhibitors, some with K_i values in the nanomolar range [12–14]. However, most successful in current efforts has been an approach based on the transition-state structure of the enzyme-ligand complex, determined from kinetic isotope effect investigations [15–17], leading to the design of nonsubstrate transition-state nucleoside analog inhibitors (immunicillins) of the calf spleen enzyme with K_i values in the picomolar range [18, 19]. The crystal structure of the enzyme complexed with immucillin-H significantly contributed to further clarification of the mechanism of action of the enzyme, and the mode of binding of inhibitors [19].

However electron densities around hydrogen atoms are too low to be unequivocally determined by X-ray crystallography. For example, hydrogen bonds are usually detected by an appropriate distance between hydrogen donor and acceptor, because the proton is not detectable. For the same reason, the neutral, cationic, anionic, or tautomeric forms of ligands in the enzyme-active site are not distinguishable by X-ray crystallography, although determination of the structure preferred by the enzyme is essential for understanding the binding and catalysis. To solve such ambiguities found in the X-ray crystallographic structure of formycin inhibitor bound by *E. coli* PNP (Sect. 17.3), emission spectroscopy of the enzyme, ligand and enzyme–ligand complexes (Sect. 17.4) was employed, with results that are complementary to X-ray crystallography.

17.2
Formycin A and its *N*-methyl Analogs, Specific Inhibitors of *E. coli* PNP

Particularly interesting inhibitors, discovered so far, are C-nucleosides, with a noncleavable C–C glycosidic bond. These include 8-aza-9-deaza analogues, a class of pyrazolo[4,3-*d*]pyrimidine nucleosides, i.e., naturally occurring antibiotic formycin A (FA), formycin B (FB) and $N(6)$-methylformycin A (m^{6}FA), close structural analogs of adenosine, inosine and $N(1)$-methyladenosine (Fig. 17.2), respectively.

They all have the same modifications on the aglycone, embracing observations that 9-deaza [20] and/or proton donor at C(8), usually furnishing by C(8)-amino [21], lead to analogs with higher affinity than the parent nucleoside substrate. FA and FB are good competitive inhibitors of the *E. coli* enzyme [22], with K_i values ~5 µM versus Ino (Fig. 17.2), whereas FA was totally inactive versus the enzymes from calf spleen and human erythrocytes. Unexpectedly, FB also exhibited poor affinity (K_i ~100 µM), although this is the 6-keto analogue, usually accepted by the mammalian enzymes. Furthermore, the neutral (but not the protonated) form of m^{6}FA (Fig. 17.2), a structural analog of the neutral form of m^{1}Ado (which, unlike the protonated form, pK_a ~8.9, is a substrate) was a much better inhibitor (K_i = 0.3 µM) of the *E. coli* enzyme, while m^{1}FA is a moderate inhibitor (K_i = 27 µM), and m^{2}FA and m^{4}FA were both inactive.

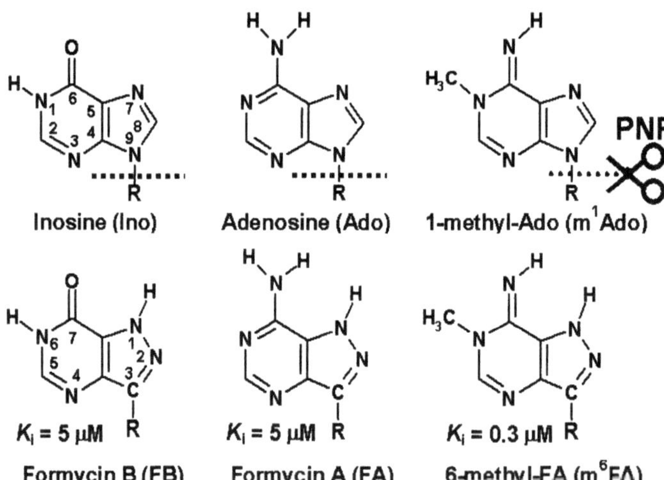

Fig. 17.2. The structurally similar, neutral forms of formycin B (FB) and inosine, formycin A (FA) and adenosine, and 6-methylformycin A (m^{6}FA) and 1-methyladenosine (m^{1}Ado) (R = ribose). Each of the formycins is a competitive inhibitor of *E. coli* PNP, with K_i values versus Inosine, as indicated. Note the different numbering systems for the purine nucleosides and formycins, according to IUPAC

17.3
Ambiguities Found in the Crystallographic Structure of Enzyme–Ligand Complex

The crystal structure of the ternary complex of FB and sulfate (phosphate) bound by *E. coli* PNP is the only structure of bacterial enzyme with nucleoside inhibitor obtained so far by X-ray crystallography [23]. The authors are not sure, if phosphate-binding site is occupied by phosphate, which was present during enzyme purification, or sulfate, added as a precipitating agent during crystallization. This is not important for the main problem addressed by this review, because, as was shown by Kierdaszuk et al. [24] in aqueous solution, sulphate and phosphate interact competitively with the same binding site in *E. coli* PNP, and maybe in the mammalian PNPs, since sulphate was detected in the phosphate-binding site of their solid state structures, as long as phosphate was not added during purification and crystallisation ([4], and references cited therein).

Although, bacterial enzyme revealed a conserved topology with the human erythrocyte PNP [25], their active sites are significantly different, especially in the base-binding site, which accounts for differences in specificity versus FB, FA, and *N*-methylated analogs of the latter [22]. The base-binding site of the bacterial enzyme (Fig. 17.3) is formed mainly with Ser90 and Asp243, responsible for the hydrogen-bond contacts with N(2), and N(1) and the exocyclic O^7, respectively (see Fig. 17.2, for numbering of the formycin aglycone). Additionally, a number of hydrophobic residues (Val78, Met180, Leu206) are also found in the base-binding site [25] and the base is involved in a π–π stacking interaction with two aromatic amino acids, Phe159 and Tyr160.

Bearing in mind that, in solution, FB consists of an equilibrium mixture of two prototropic tautomers, N(1)-H and N(2)-H (Sect. 17.4.1), with population of 77% and 23% [26], respectively, there are only two possible alternative, sets of hydrogen bonds between the FB aglycone and amino acid residues in the base-binding site (Fig. 17.3). For the N(2)-H tautomer of FB and the neutral form of Asp204 carboxyl, the $O^{\delta 1}H$ of Asp204 may donate a hydrogen to N(1) of the FB aglycone, while the N(2)-H may donate a hydrogen to O^γ of Ser90. The N(2)-H tautomer does not carry a donor hydrogen at N(1), so its contact with the anionic form of Asp204 carboxyl is less favored unless N(1) is protonated. For the N(1)-H tautomer of FB, Asp204 may rotate in such a way that $O^{\delta 1}H$ of Asp204 may donate a hydrogen to the exocyclic O^7, and N(1)H of the FB aglycone may donate a hydrogen to $O^{\delta 2}$ of Asp204. The N(1)-H tautomer does not carry a donor hydrogen at N(2), so its hydrogen bond with Ser90 residue could not be formed unless the latter may donate a hydrogen. The contact of the anionic form of Asp204 carboxyl with the N(1)-H tautomer of FB is less favored, but one hydrogen bond may still be formed with N(1)-H of the latter.

In the case of FA, when exocyclic O^7 is replaced by NH_2, its hydrogen-bond contacts with the neutral form of Asp204 carboxyl may be better for the N(2)-H than N(1)-H tautomers of FA, while the anionic form of Asp204 carboxyl may form only one hydrogen bond with the N(2)-H tautomer, and two hydrogen bonds

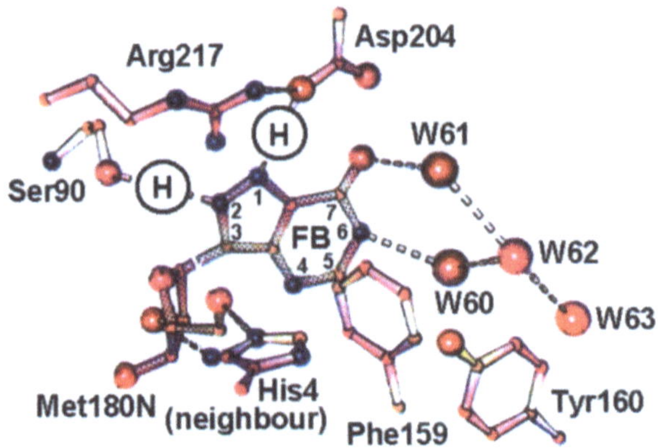

Fig. 17.3. Mode of binding of the inhibitor formycin B (FB) in the active site of *E. coli* PNP [23]. Note the numbering system for the ring of the FB aglycone, which is different than that of purine nucleosides, according to IUPAC. Indicated are two hydrogen atoms at the base N(1) and N(2), which may be alternatively involved in one of two possible hydrogen bonds with O^δ of Asp204 or O^γ of Ser90, which depend on which tautomer N(1)-H or N(2)-H is present in the enzyme active site. Two aromatic amino acids, Phe159 and Tyr160, participate in a π–π stacking interaction with the base. Four water molecules, shown as largest spheres, form a water chain in the region of the active site exposed to solvent. Three of them (W60, W62 and W61) participate in a water bridging between N(6) and exocyclic O^7, and account for the fact that m^6FA is a specific inhibitor, and m^1Ado is a substrate, of *E. coli* PNP but not the mammalian enzymes. Adapted from [23], with permission of the copyright holder

with the N(1)-H tautomer of the ligand, i.e., both N(1)H and C(7)-NH_2 may donate hydrogen atoms to the $O^{\delta 1}$ and $O^{\delta 2}$ of Asp204.

X-ray crystallography is not able to distinguish between two possible tautomeric forms of FB in the solid state complex with *E. coli* PNP [23], because they both may alternatively fit the hydrogen bond contacts with protein (Fig. 17.3). Furthermore, ionic forms of amino acid residues, usually overlooked, also seem to affect ligand binding. Therefore, alternative studies are needed to elucidate ligand structure preferentially bound by the enzyme.

A difference between active site of *E. coli* PNP and the mammalian enzymes is that Asp204, which accepts both 6-oxo and 6-amino purines, is replaced by Asn243, which exclusively prefers 6-oxo analogues. Furthermore, the active site of *E. coli* PNP does not include a Glu residue, which, in the mammalian enzymes (Glu201), plays a key role in the binding and catalysis, i.e., forms two hydrogen bonds with the N(1)-H and C(2)-NH_2 of guanine or one hydrogen bond with the N(1)-H of hypoxanthine, and their nucleosides, and stabilize transition state [4]. Thus the active site of *E. coli* PNP is more "open", and accounts for its broader specificity, e.g. accepts as substrates N(1)-methyl analogues of guanosine and inosine, as well as neutral form of *N*(1)-methyladenosine [22] and its structural analog inhibitor m^6FA (Fig. 17.2, see also above). In the crystal structure of FB bound by *E. coli* PNP (Fig. 17.3), a water bridge (W60-W62-W61) between the base ring

N(6) (N(1) in the purine ring) and exocyclic O^7 (O^6 in the purine base) was observed [23]. The foregoing indicates that the ring N(1) of purines (N(6) of pyrazolopyrimidines) are not the binding sites in the bacterial enzyme, so that N(6)-methyl would not directly affect the binding of m^6FA, as actually observed below (Sect. 17.4.3).

In addition to the foregoing, a nonspecific π–π stacking interaction between the formycin B aglycone, and the aromatic Phe159 and Tyr160 is present (Fig. 17.3) with a centroid distance between aromatic moieties ~5 Å. This also accounts for the broader specificity of the *E. coli* enzyme versus mammalian PNPs, as well as for FRET, as a major mechanism of protein fluorescence quenching (Sect. 17.4.3).

17.4
Solution Structure of Inhibitors Bound by the Enzyme

17.4.1
Tautomeric Equilibria, and Absorption and Emission Spectra of the Tautomeric Species in Solution

The prototropic tautomerism of FA, with an equilibrium between the N(1)-H and N(2)-H species (Fig. 17.4) was extensively investigated with the aid of ^{13}C NMR spectroscopy in DMSO [26], and by UV absorption and emission (fluorescence and phosphorescence) spectroscopy in aqueous medium [27], in both cases with the use of the fixed tautomeric species m^1FA and m^2FA as reference compounds, which exist as single tautomeric forms [27]. The results show that the minor N(2)-H tautomers of FA and FB (Fig. 17.4) exist at approximately 15% and 13% in aqueous medium, respectively. It follows that both tautomeric forms contribute to the fluorescence and phosphorescence of FA [27].

The excitation spectra for fluorescence and phosphorescence of m^1FA and m^2FA in aqueous solution are similar to their absorption spectra, and both absorption and fluorescence spectra of m^2FA are red-shifted relative to that of m^1FA by approximately 5 nm (Fig. 17.5). The emission spectra of these compounds in glassy water-glycerol solution at 180 K (Fig. 17.6) differ sufficiently as to permit identification of the emitting tautomeric species of FA. As is shown in Fig. 17.6, the phosphorescence versus fluorescence is about one range of value less intense for m^2FA (the N(2)-H tautomer) than for m^1FA (the N(1)-H tautomer), and red-shifted ~28 nm versus phosphorescence band of m^1FA, much more than fluorescence band (~6 nm).

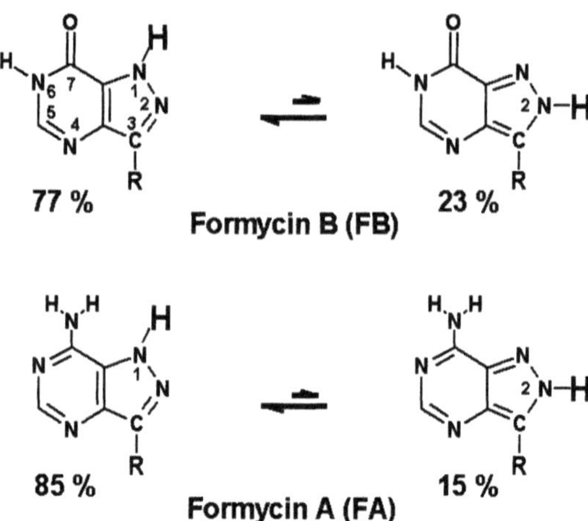

Fig. 17.4. The equilibrium mixture of the N(1)-H and N(2)-H forms of the neutral form of formycin A (FA) and formycin B (FB) in equeous solutions (R = ribose), in the proportions indicated. Note the close structural resemblance of the N(2)-H forms of FA (pK_a 4.3 and 9.7, for protonation and dissociation) and FB (pK_a 0.9 and 8.6) to Ado (pK_a 3.6 and 12.4) and Ino (pK_a =1.0 and 8.8), respectively

Furthermore, as result of the facts mentioned before, fluorescence and phosphorescence spectra of FA depend markedly on the excitation wavelength with regards to both their shape and quantum yield [27]. An increase in excitation wavelength from 260 to 310 nm does not lead to significant changes in the shape and intensity of fluorescence spectrum, while at longer excitation wavelength (315–320 nm), the fluorescence and phosphorescence bands were both red-shifted, and

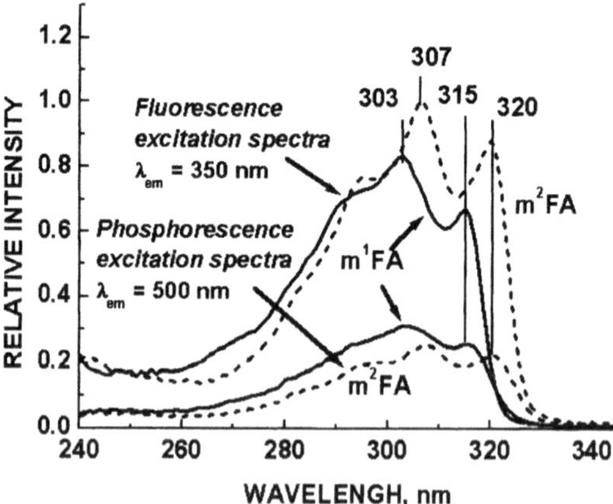

Fig. 17.5. Corrected fluorescence (λ_{em} 350 nm) and phosphorescence (λ_{em} 500 nm) excitation spectra of the fixed tautomeric species m¹FA and m²FA in 50 mM Hepes pH 7 at 298 (fluorescence), and in 50 mM Hepes pH 7 containing 80% of glycerol at 180 K (phosphorescence)

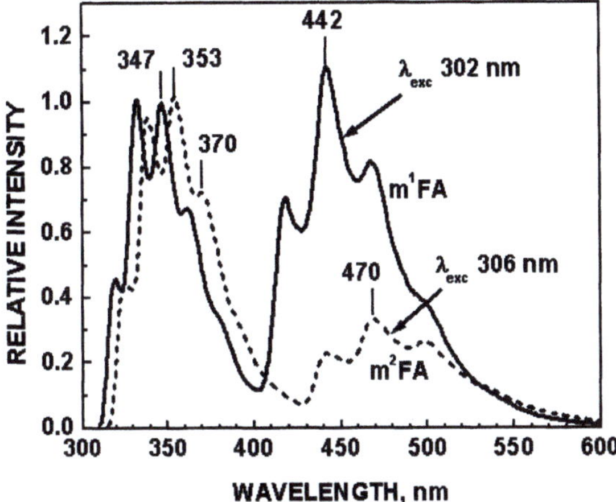

Fig. 17.6. Fluorescence and phosphorescence emission spectra of m^{1}FA (λ_{exc} 302 nm) and m^{2}FA (λ_{exc} 306 nm) in 50 mM Hepes pH 7 containing 80% of glycerol at 180 K

phosphorescence quantum yield decreases about 10-fold relative to excitation at 275 nm, similarly to that observed for m^{2}FA versus m^{1}FA (Fig. 17.6). These findings testify to the existence of two emitting forms, and demonstrate that these tautomeric forms are readily differentiated by their emission properties, both qualitatively and quantitatively, in agreement with data from NMR spectroscopy [31]. This led to identification of the changes in tautomeric equilibrium caused by binding to the enzyme (Sect. 17.4.3).

First observation of the amino-imino tautomerism of N(6)-methylformycin A (m^{6}FA) (Fig. 17.7) was based on the marked solvent-dependency of its UV absorption spectrum [28], which was overlooked in an earlier studies [29], while UV spectra of m^{1}FA and m^{2}FA are virtually solvent independent [27]. The spectrum of m^{6}FA in 90% isopropanol, but not in water, is similar to that of the m^{6}FA aglycone, 3-propyl-6-methyl-7-iminopyrazolo[4,3-*d*]pyrimidine, which spectrally resembles the model of the fixed imino form in all solvents [27]. The UV absorption and fluorescence spectra showed that m^{6}FA is predominantly imino in 90% isopropanol, and addition of water shifts this equilibrium towards amino form, so that it exists in aqueous medium (Fig. 17.7) at 55% [28]. Interpretation of the absorption and emission spectra of m^{6}FA is based on the spectra of its N-methylated analogs, and revealed that both the 340-nm and 440-nm fluorescence bands were observed for the amino form, whereas only one the 340-nm band are visible for the imino species. It follows that the imino form of m^{6}FA may exist in the equilibrium of the N(1)-H and N(2)-H tautomers (Fig. 17.7), with so far unknown proportion.

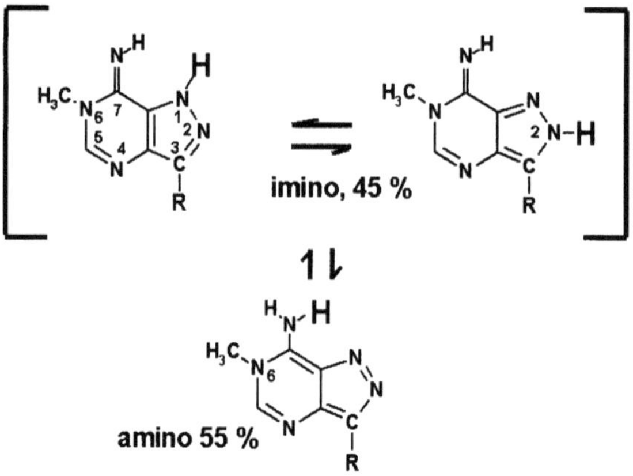

N(6)-methylformycin A (m⁶FA)

Fig. 17.7. The equilibrium mixture of the amino and imino forms of the neutral form of 6-
-methylformycin A (m^{6}FA) (pK_a 6.9 and 10.3 for protonation and dissociation) in aqueous
solutions (R = ribose), in the proportions indicated. Note that the imino form is an equilibrium mixture of two tautomeric species, the N(1)-H and the N(2)-H (proportions not
known). Note also the close structural resemblance of the N(2)-H form of the imino species
to the fixed imino form of m^{1}Ado (pK_a ~8.9), which exists in the protonated (nonsubstrate)
form at neutral pH

17.4.2
Shifts Between Absorption and Emission Spectra of the Enzyme and Ligands

The maxima of the absorption, and emission fluorescence and phosphorescence
spectra of PNP at 277 nm, 304 nm, and 404 nm, respectively (Fig. 17.8), are typical for proteins containing tyrosine, but no tryptophan [29]. This is consistent with
the fact that each of the subunits of the *E. coli* enzyme contains six tyrosine residues and no tryptophan [25, 30]. This is in striking contrast to the enzymes from
mammalian sources (calf spleen and human erythrocytes), the spectra of which are
dominated by the presence of tryptophan [31].

By contrast, the nucleoside ligand FA (Fig. 17.8), as well as FB, m^{1}FA and
m^{6}FA (data not shown), exhibit absorption and emission spectra red-shifted relative to the spectra of PNP [27, 28, 32]. These permit a selective excitation of
ligands in enzyme–ligand mixture at excitation wavelengths above 295 nm (where
tyrosine residues do not absorb). Selective observation of PNP fluorescence is
possible in the range 290–300 nm, and fluorescence of ligands at wavelengths
above 360 nm, where the contribution of protein fluorescence is usually negligible. In the case of phosphorescence emission, selective observation of protein
emission is possible in the range 350–390 nm, and phosphorescence emission of
ligands at wavelength above 450 nm. The foregoing is essential for the strategy of
emission studies based on the selective observation of the enzyme and/or ligand,
when enzyme–ligand complexes are formed.

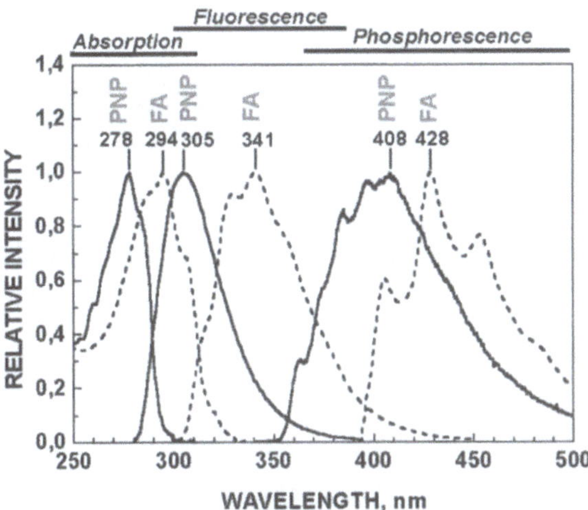

Fig. 17.8. Absorption, and fluorescence and phosphorescence emission spectra of *E. coli* PNP and neutral form of FA in 50 mM Hepes pH 7 (absorption and fluorescence, at 298 K) and 50 mM Hepes pH 7 containing 80% of glycerol (phosphorescence, at 180 K). Maximum intensities of the absorption and emission spectra of PNP (λ_{exc} 270 nm), FA (λ_{exc} 295 nm) are normalized to unity

17.4.3
Effect of Binding of the Identified Tautomeric Species on their Fluorescence and Phosphorescence

The effect of enzyme–ligand interactions on the fluorescence emission spectra of the *E. coli* PNP, and/or ligand was measured by an addition of increasing concentration of ligand (FA, FB, m^{1}FA, m^{2}FA, m^{4}FA and m^{6}FA) to a solution of *E. coli* PNP in 50 mM Hepes buffer (pH 7) at 298 K [32]. With FA (Fig. 17.9) and m^{6}FA (Fig. 17.12), this led to a partial quenching of enzyme tyrosine emission at 304 nm (λ_{exc} < 290 nm) and a concomitant increase in fluorescence of ligands, independently of the presence of P_i. These enable to identify tautomeric species bound by the enzyme, as well as enzyme–ligand FRET (see last paragraph of this section). The existence of an equilibrium between free and bound ligand was reflected in the isoemissive points, at 315 nm (titration with FA), and 330 and 360 nm (titration with m^{6}FA, in the absence and in the presence of P_i), respectively.

The observed changes in fluorescence spectra of FA, upon interactions with the enzyme, mainly account for a shift in the tautomeric equilibrium towards the N(2)-H tautomer, manifested primarily by the red-shift of the emission and excitation spectra (Figs. 17.9 and 17.10), as previously observed for m^{2}FA relative to m^{1}FA analogs (Sect. 17.4.1), which fix the N(2)-H and N(1)-H tautomers, respectively. This is further confirmed by results obtained from time-resolved fluorescence of enzyme, ligand and enzyme–ligand complex. Complex formation resul-

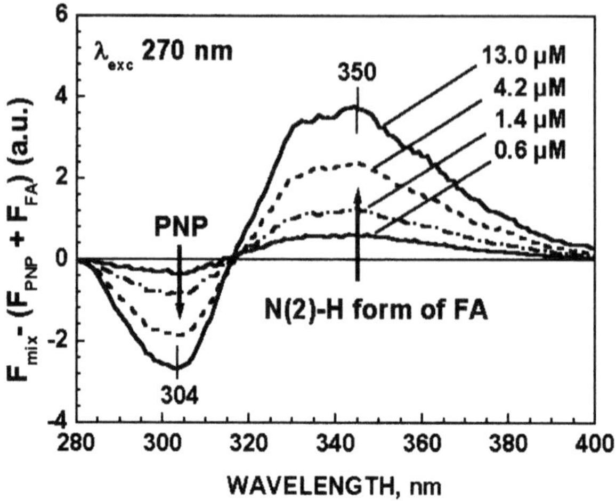

Fig. 17.9. Fluorescence-emission difference spectra (λ_{exc} 270 nm) of 1 µM PNP + increasing concentrations of FA, in 50 mM Hepes buffer pH 7, relative to the arithmetic sum of the two components. Adapted from [32], with permission of the copyright holder

ted in a 2-fold increase in <τ> of FA, and a 2-fold decrease in <τ> of enzyme fluorescence. The amplitude of the long-lifetime component also increased, confirming the shift of the tautomeric equilibrium in favour of the N(2)-H species. The fact that the active center exhibits a preference for the N(2)-H form of FA, which exists predominantly as the N(1)-H tautomer in solution, is in line with the close resemblance of the N(2)-H form to the substrate adenosine, with the N(2)-H and N(1) of FA aglycone corresponding to the C(8)-H and N(7) of the adenine ring (Fig. 17.2).

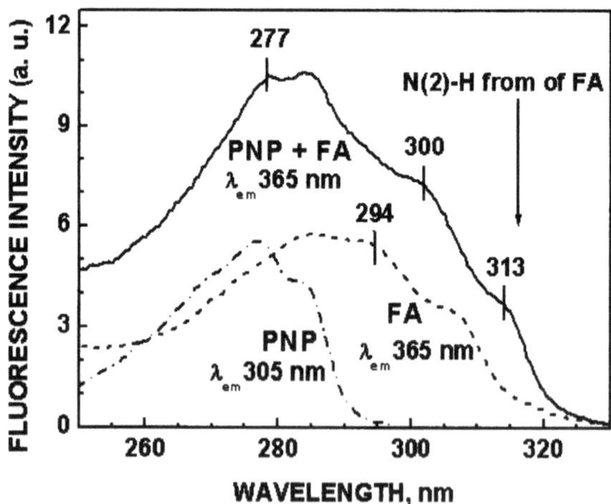

Fig. 17.10. Fluorescence excitation spectra, in 20 mM phosphate pH 7, of 15 µM FA (λ_{em} 365 nm) in the presence (——) and absence (- - -) of 2 µM PNP. Also shown are the fluorescence excitation spectra (λ_{em} 305 nm) of PNP (- · - · - ·)

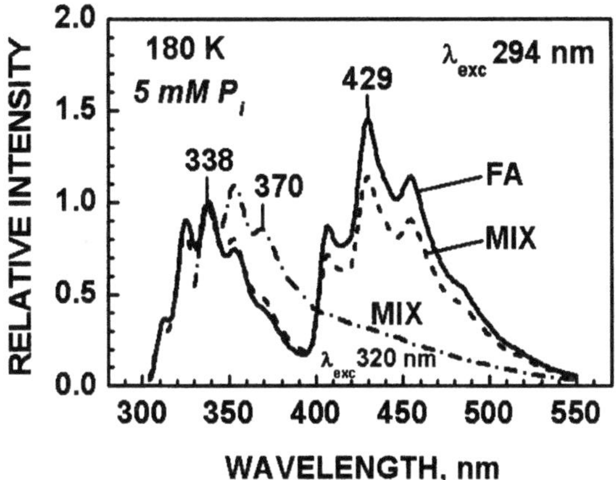

Fig. 17.11. Fluorescence and phosphorescence emission spectra (λ_{exc} 294 and 320 nm, as indicated) of 25 μM FA (——) and their mixture with 1 μM PNP of *E. coli* (- - -) in 50 mM Hepes buffer pH 7 containing 5 mM P_i and 80% glycerol, at 180 K

In order to measure an effect of enzyme-FA interactions on the phosphorescence emission spectra of FA, emission spectra of the free FA were compared with that in the mixture with *E. coli* PNP in 50 mM Hepes buffer (pH 7) containing 80% glycerol at 180 K (Wlodarczyk J, Stoychev G, Kierdaszuk B, in preparation). Enzyme-FA complex formation led to a much lower intensity of FA phosphorescence versus its fluorescence (both at λ_{exc} 294 nm), and to a red-shift of the fluorescence band (Fig. 17.11). When excitation wavelength of the enzyme–ligand mixture increased to 320 nm, where the N(1)-H form of FA do not absorb, resulted emission spectrum revealed a significantly reduced phosphorescence, and red-shifted fluorescence band versus that obtained with λ_{exc} 294 nm. These data further confirm the shift of the tautomeric equilibrium of FA in favor of the less phosphorescent N(2)-H species.

Unexpectedly, UV absorption and emission spectroscopy revealed that both prototropic tautomers, amino and imino of m⁶FA, are acceptable by the enzyme and enzyme–m⁶FA binding markedly depends on binding of the second substrate P_i [32]. In the presence of 20 mM P_i, i.e., at saturating concentration for the P_i binding sites [24], ternary m⁶FA/PNP/P_i complex was formed with binding constant (K_d = 0.5 ±0.1 μM), and fluorescence difference spectra (Fig. 17.12B) showed a pronounced 430-nm band equivalent for the amino form of m⁶FA in the complex. At background (≤1 μM) P_i, i.e., in Hepes with no P_i added, enzyme-ligand binding was much weaker (K_d = 46 ±5 μM), and led to an increased emission at 340 nm (Fig. 17.12A), typical for the m⁶FA imino. Note also that the residual emission at 430 nm (Fig. 17.12A) may conceivably be due to minor ternary complex formation with the amino form of m⁶FA (Sect. 17.4.1) as a result of the presence of background P_i.

It should be emphasized, in this context, that the amino structure is not obligatory for enzyme activity, as illustrated by the good substrate properties of the structurally related neutral, but not protonated, form of the fixed imino nucleoside

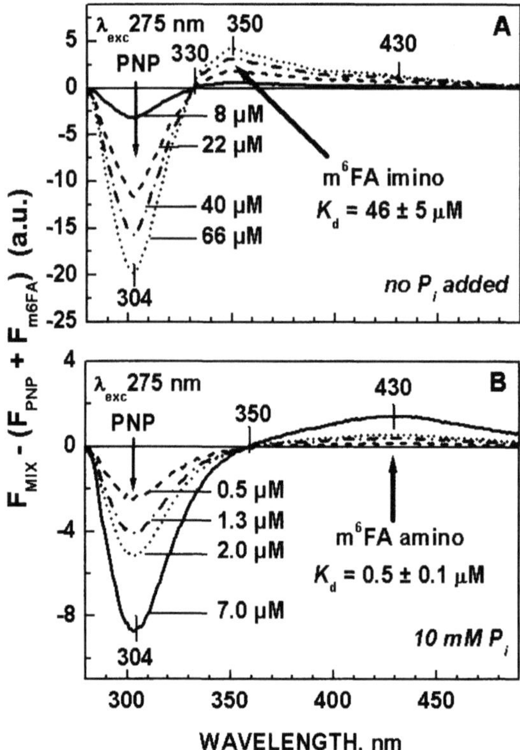

Fig. 17.12. Fluorescence-emission difference spectra (λ_{exc} 275 nm) of 2 μM PNP + increasing concentrations of m⁶FA, relative to the arithmetic sum of the two components, in **A** 50 mM Hepes (phosphate free) pH 8.3, and **B** 20 mM phosphate pH 8.3. Adapted from [32], with permission of the copyright holder

of 1-methyladenosine (Fig. 17.2) [22]. The observed tautomeric shifts may reflect an increase in polarity of the enzyme active center, resulting from the proximity of the negatively charged P_i anion and/or some accompanying structural rearrangements of the binding pocket. In accordance with this is the previously observed marked dependence of the amino–imino tautomeric equilibrium of m⁶FA on solvent polarity [28], and a pronounced shift in the tautomeric equilibrium of the FA aglycone to 50% upon phosphorylation of FA to FA-5′-phosphate, depending whether the phosphate moiety is in the monoanionic or dianionic form [33].

Fluorescence of m¹FA in the complex increased only in the presence of P_i due to the enzyme–ligand FRET, while the weak fluorescence of FB appear unaffected independently of P_i, although the concentration-dependent quenching of enzyme emission was observed [32]. Addition of m²FA to the enzyme was without effect on the fluorescence emission of either the enzyme or the ligand, consistent with the fact that the latter is not an inhibitor of *E. coli* PNP [22]. Similarly, effect of binding of m⁴FA by the enzyme on its fluorescence was too weak to permit determination of the binding constant ($K_d \gg 100$ μM), again consistent with very poor inhibitory properties ($K_i > 500$ μM), of this ligand [22].

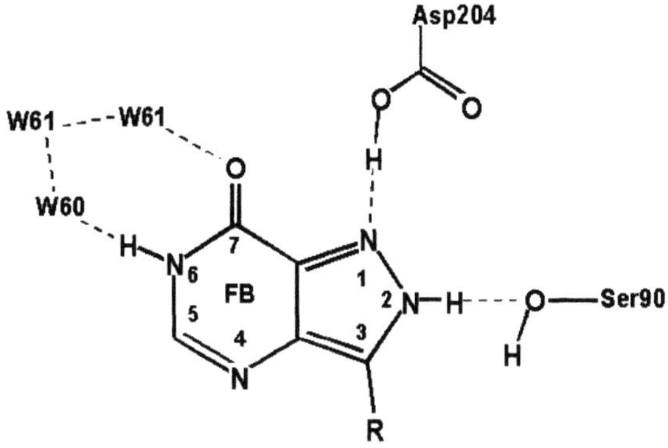

Fig. 17.13. Proposed mode of preferential binding by *E. coli* PNP of the N(2)-H tautomer of the aglycone of FB (R = ribose), based on the results of spectroscopic identification of the N(2)-H tautomeric form of FA in the solution complex with *E. coli* PNP [32], and on the reported crystal structure of FB bound by *E. coli* PNP [23]. See text for further details. Adapted from [32] with permission of the copyright holder

Analysis of the emission, excitation and absorption spectra of enzyme–FA and enzyme–m^{6}FA mixtures pointed to fluorescence resonance energy transfer (FRET) from protein tyrosine residue(s) to FA and m^{6}FA base moieties, as a major mechanism of protein fluorescence quenching. In the complex with FA, a Förster radius (R_o~7Å) was determined (Wlodarczyk J, Stoychev G, Kierdaszuk B, in preparation) and Tyr160 residue is a best candidate to be an energy donor, with centroid separations between the aromatic residues of ~5Å, determined from the X-ray diffraction studies of the FA–enzyme complex [23]. It is worth to identify unequivocally an energy donor with the aid of site-directed mutagenesis, e.g., by replacement of Tyr160 by Ala, unless enzyme activity is changed.

17.5
Conclusions

In conclusion, it is clear that the emission studies resolved ambiguities found in the crystal structure of the active center of *E. coli* PNP in a complex with FB (Fig. 17.3), where FB aglycone donates a hydrogen for one of two possible hydrogen bonds to the protein, i.e., one from the N(1) to O^δ of Asp204, the other from N(2) to O^γ of Ser90. These two are not distinguishable by X-ray crystallography [23], while emission studies revealed the N(2)-H tautomeric form in the enzyme active center, and that the ring N(2)-H donate a hydrogen to O^γ of Ser90, and the N(1) accepts a hydrogen from the O^δH of Asp204 (Fig. 17.13).

Since the exocyclic O^7 is exposed to the solvent, it may be replaced by an NH_2 (to give FA), which may donate hydrogens to both water W61 and O^δ of Asp204, while the ring N(6) can still bond to W60. It happens, that the latter my be removed with no effect on ligand binding, which explains substrate activities of

N(1)-substituted purines (e.g., neutral form of m^1Ado), and inhibitory activity of N(6)-substituted pyrazolopyrimidines (e.g., m^6FA). Emission studies showed that both tautomeric forms of the latter, amino and imino, may be bound in the enzyme active site depending on the binding of second substrate P_i. This is in line with the fact that FA is as good inhibitor of the enzyme as FB, and m^6FA is even much better [22]. Their solid state structures in the complex with *E. coli* PNP await further structural studies by X-ray crystallography.

Acknowledgements. Supported by the State Committee for Scientific Research (KBN, grant no. 6P04A03812), to whom we are also indebted for the purchase of an IBH time-resolved spectrofluorometer (KBN, grant no. 415149101); and by the Foundation for Polish Science (purchase of laser equipment).

References

1. Schramm VL (1999) Methods Enzymol 308:301
2. Kredich NM, Hershfield MS (1989) Immunodeficiency diseases caused by ADA deficiency and PNP deficiency. In: Jeffers JD, Gavret G (eds) The metabolic basis of inherited diseases. McGraw-Hill, New York, p 1045
3. Gilbertsen RB, Sircar JC (1990) Enzyme cascades: purine metabolism and immunosuppression. In: Hansch C, Sammes PG, Taylor JB (eds) Comprehensive medicinal chemistry, vol 2 Pergamon Press, Oxford, p 443
4. Bzowska A, Kulikowska E, Shugar D (2000) Pharmacol Ther 88:349
5. Arner E, Eriksson S (1995) Pharmacol Ther 67:155
6. Doskocil J, Holy A (1977) Collection Czechoslov Chem Commun 42:370
7. Stoeckler JD (1984) Purine nucleoside phosphorylase: a target for chemotherapy. In: Glazer RI (ed) Developments in cancer chemotherapy. CRC Press, Boca Raton, FL, p 35
8. Bzowska A, Kulikowska E, Shugar D (1990) Z Naturforsch Teil C 45:59
9. Hughes BW, Wells AH, Bebok Z, Gadi VK, Garver Jr, Parker WB, Sorscher EJ (1995) Cancer Res 55:3339
10. Sorscher EJ, Peng S, Bebok Z, Allan PW, Bennertt Jr. LL, Parker WB (1994) Gene Therapy 1:233
11. Gadi VK, Alexander SD, Kudlow JE, Allan P, Parker WB, Sorscher EJ (2000) Gene Therapy 7:1738
12. Montgomery JA, Niwas S, Rose JD, Secrist III JA, Babu YS, Bugg CE, Erion MD, Guida WC, Ealick SE (1993) J Med Chem 36:55
13. Kelly JL, Linn JA, McLean EW, Tuttle JV (1993) J Med Chem 36:3455
14. Iwata H, Wada Y, Walsh M, Montgomery JA, Hirose H, Mendez R, Ciccirelli J, Iwaki Y (1998) Transplant Proc 30:983
15. Schramm VL (1998) Annu Rev Biochem 67:693
16. Kline PC, Schramm VL (1993) Biochemistry 32:13212
17. Kline PC, Schramm VL (1995) Biochemistry 34:1153
18. Miles RW, Tyler PC, Furneaux RH, Bagfassarian CK, Schramm VL (1998) Biochemistry 37:8615

19. Fedorov A, Shi W, Kicska G, Fedorov E, Tyler PC, Fruneaux RH, Hanson JC, Gainsford GJ, Larese JZ, Schramm VL, Alamo SC (2001) Biochemistry 40:853
20. Stoeckler JD, Ryden JB, Parks Jr RE, Chu MY, Lim MI, Ren WY, Klein RS (1986) Cancer Res 46:1774
21. Montgomery JA (1993) Med Res Rev 13:209
22. Bzowska A, Kulikowska E, Shugar D (1992) Biochim Biophys Acta 1120:239
23. Koellner G, Luic M, Shugar D, Saenger W, Bzowska A, (1998) J Mol Biol 280:153
24. Kierdaszuk B, Modrak-Wójcik A, Shugar D (1997) Biophys Chem 63:107
25. Mao C, Cook WJ, Zhou M, Koszalka G, Krenitsky TA, Ealick SE (1997) Structure 5: 1373
26. Chenon MT, Panzica RP, Smith JC, Pugmire RJ, Grant DM, Townsend LB (1976) J Am Chem Soc 98:4736
27. Wierzchowski J, Shugar D (1982) Photochem Photobiol 35:445
28. Wierzchowski J, Shugar D (1993) Collect Czech Chem Commun 58:14
29. Ross JBA, Laws WR, Rousslang KW, Wyssbrod HR (1992) In: Lakowicz JR (ed) Topics in fluorescence spectroscopy, biochemical applications, vol 3. Plenum Press, New York, pp. 1–63
30. Hirshfield MS, Chaffe S, Koro-Johnson L, Mary A, Smith AA, Short SA (1991) Proc Natl Acad Sci USA 88: 7185
31. Kierdaszuk B, Gryczynski I, Modrak-Wójcik A, Bzowska A, Shugar D, Lakowicz JR (1995) Photochem Photobiol 61:319
32. Kierdaszuk B, Modrak-Wojcik A, Wierzchowski J, Shugar D (2000) Biochim Biophys Acta 1476:109
33. Wierzchowski J, Lassota P, Shugar D (1984) Biochim Biophys Acta 786:170

**Part 4
Microscopic Imaging Techniques and
their Application for the Study of Living Cells**

Fluorescence Lifetime Imaging Implemented with Resonant Galvanometer Scanners

J. J. BIRMINGHAM

Fluorescence Lifetime Imaging (FLIM) typically utilises specialised image intensifiers to obtain a sequence of images at known times relative to a periodic excitation source. Either time-domain or frequency-domain gating characteristics of such devices have been used to derive fluorescence lifetime images on the nanosecond timescale. However, such devices can be problematical in terms of cost, robustness and complexity. This paper explores an alternative method of obtaining lifetime images by using continuously oscillating scanning elements at defined frequencies. Employing a frequency domain approach, sinusoidally modulated laser excitation at frequencies suitable for nanosecond timescale emission is scanned rapidly and symmetrically over a line by using such resonant scanners. By introducing a sampling frequency on the optical data stream critically related to both the excitation modulation frequency and the scanner frequency, it becomes possible to encode the lifetime-related phase delay and demodulation data as a function of position. The sampling achieves the necessary down shifting of the high frequency data in addition to imposing a continuous instrumental phase shifting function. By combining sampled data for a given pixel across repeated passes of the scanner action, formulas are derived for both the steady-state intensity and lifetime-related phase and demodulation data. The overall method is illustrated by simulations and by experiments on model systems.

18.1
Introduction

The measurement of nanosecond timescale fluorescence lifetimes in an imaging manner has evolved rapidly over recent years [1]. Both time-domain pulse and frequency-domain phase modulation methods [2] have been used successfully to simultaneously measure the fluorescence decay characteristics at a large number of spatial locations. The key step in either method is to introduce imaging devices with fast gating characteristics in order to shift the fast lifetime information onto much slower timescales suitable for low light level camera acquisition, i.e., a strobing process. Image intensifiers with either pulse or RF sinusoidal gain modulation capabilities allow the fast fluorescence decay raw image data to be strobed to either zero (DC, static) or low AC frequencies which are then suitable for integration with imaging devices such as cooled slow-scan CCD cameras. In the time-domain approach, a sequence of such strobed images are acquired using a short gating pulse which is incremented in time between images, thereby scanning the decay curve of the sample [3, 4]. In the frequency-domain approach, the modulated fluorescence emission for the sample is multiplied by an oscillating gain function in the intensifier. The inability of the output phosphor to track high frequencies results in a final image at the difference frequency being presented to the integrating camera. The most usual case is to employ identical frequencies for illumination modulation and for the image intensifier gain modulation, resulting in a zero frequency static image (homodyne approach). The instrumental phase between the illumination and intensifier modulations is then systematically varied giving a sequence of images at known phase increments. This image set is then analysed at each pixel for the steady state and lifetime-related phase and demodulation parameters [5–9]. So, both the time and frequency-domain approaches result in a set of 2D images which have been acquired sequentially but are processed as a single unit to yield lifetime contrast images.

In the frequency-domain case, if the intensifier gain and illumination modulation frequencies are slightly different, then images oscillating at the difference frequency between the two result. Overall the high frequency lifetime information has been downshifted to a low AC frequency with continuous 360-degree phase shifting, occurring every period of this difference frequency. This heterodyne mode of detection, although not widely used in 2D imaging configurations, has however been extensively used in non-imaging frequency-domain fluorimeters as a means of exploiting readily available low frequency signal detection equipment, e.g., lock-in amplifiers. Typically a gain-modulated photomultiplier tube (PMT) is used as the down shifting detector in such "zero-dimensional" (0D) applications rather than a 2D image intensifier.

Rather than employing an all-pixels-parallel imaging approach which then necessitates potentially complex 2D gating devices, imaging tasks can also be performed by scanning approaches [10, 11]. At any one time, the experiment is therefore a 0D type with only devices such as PMTs required for detection. Fluorescence lifetime imaging methods using scanning technologies have been described before, typically the "classical" 0D experimental methods, being time or frequency domain in nature, are performed in one location after another over the sample area required. Typically, a pair of galvanometer mirror scanners are used together to sequentially illuminate each location of an object. The illuminating beam is moved to a particular 2D x,y location, the standard time-resolved measurement performed, the beam then moved on again and so on. This route does not require specialist image intensifier devices or any imaging devices such as cameras, but the measurements are serial rather than parallel and may overall mean longer experimental times. However, compared to the 2D approach, all the available illumination energy is presented to every pixel in turn. Both single-photon and 2-photon laser scanning microscopes have been described which offer lifetime contrast using such an overall strategy [12–15].

This paper instead explores the use of continuously oscillating mirror elements (resonant scanners) in order to perform frequency-domain lifetime imaging. The illumination has one particular characteristic frequency (modulation) which is continuously spread in a 1D fashion over the object at a second characteristic frequency (scanner frequency). These two characteristic frequencies are effectively mixed by a third frequency (data sampling frequency) in a particular way so that the experiment exhibits both down shifting and automatic phase adjustment synchronised to the scanner action. The experiment then has the benefits of the standard heterodyne approach coupled with an easy readout of both the steady-state and lifetime-related parameters at every point along the scan line. The second spatial dimension of the sample is addressed by either sample translation or conventional galvanometer scanning.

18.2
Theory

At any instant of time along the scan line, the instantaneous fluorescence intensity in response to sinusoidally modulated excitation at any frequency ω can be written as

$$s(t) = s_0 + s_1 M \sin(\omega t + \varepsilon - \varphi) \tag{18.1}$$

The fluorescence response is demodulated by a factor M and phase delayed by an amount φ relative to the excitation modulation, ε is an arbitrary initial phase of such excitation. Assuming a single exponential decay model for the fluorescence emission to hold for every pixel along the line, the phase delay and demodulation are related to lifetime τ as

$$\tan(\varphi) = \omega\tau \qquad\qquad M = [1 + \omega^2\tau^2]^{-0.5}$$

Defining a normalised scan line on the interval [0,1] with the scanner oscillating at angular frequency ω_s, we have the following expressions linking x,t

$$x = \frac{1}{2}\left[1 + \sin\left(\omega_s t - \frac{\pi}{2}\right)\right] \qquad t = \frac{1}{\omega_s}\left[\frac{\pi}{2} + \arcsin(2x - 1)\right]$$

In these equations, we assume that at $t = 0$, the scan position is at $x = 0$, then traverses forward in a sinusoidal manner until $x = 1$ is reached at a time t equal to half the period of oscillation. It then reverses back towards $x = 0$ which it reaches at t equal to one period of the scanner action. Resonant galvanometer scanners act in this manner but either electro-optic or acousto-optic deflectors could also be driven in such a fashion. By adjustment of the excitation initial phase ε, any relative alignment of the excitation modulation with respect to the scanner action can be accommodated.

The arcsin function can be expressed in terms of its principal value for any argument z [16] as

$$\arcsin(z) = k\pi + (-1)^k \sin^{-1}(z) \qquad k = 0,1,2,..$$

This then gives the relationship between time t and scanner position x as

$$t = \frac{1}{\omega_s}\left[\frac{\pi}{2} + k\pi + (-1)^k \sin^{-1}(2x - 1)\right] \tag{18.2}$$

This expression for $k = 0$ defines the first forward trace of the scanner, likewise $k = 1$ denotes the scan times that pertain in the first reverse trace of the scanner and so on. This single expression therefore covers all the phases of action of the scanner, one just specifies the value of k in order to isolate which pass through the line of pixels is occurring. So, for a given set of normalised scan positions x, the corresponding scan times t can be calculated for any pass of the scanner simply by specifying the value of k. Inserting Eq. 18.2 into Eq. 18.1, we obtain

$$s(x) = s_0(x) + s_1(x)M(x)\sin\left\{\frac{\omega}{\omega_s}\left[\frac{\pi}{2} + k\pi + (-1)^k \sin^{-1}(2x - 1)\right] + \varepsilon - \varphi(x)\right\} \tag{18.3}$$

Now the steady state intensity, the demodulation factor M and the phase delay φ depend on the scan position x in general, the latter two also dependent as usual on the modulation frequency ω. So, for any given 1D spatial distribution of the lifetime $\tau(x)$, the functions $M(x)$, $\varphi(x)$ can be calculated at modulation frequency ω, and the fluorescence response calculated as a function of scan distance x for any pass of the scanner (k-value) from the above formula. It is implicitly understood that the scanner frequency ω_s is much slower than the excitation modulation frequency ω, so that as the beam oscillates along the line, the fluorescence response can be viewed as instantaneous.

We now consider the effects of sampling. Taking a general time-dependent signal $g(t)$ and assuming ideal equi-spaced sampling at angular frequency ω_0, then the sampled signal, say $S(t)$, in the time domain is just the product of $g(t)$ with a Dirac sampling comb.

$$S(t) = g(t) \sum_{n=-\infty}^{n=\infty} \delta(t - nT_0) \qquad T_0 = \frac{2\pi}{\omega_0}$$

The frequency spectrum of this sampled signal, termed $S(\omega)$, is then the convolution of the frequency spectra of $g(t)$, termed $G(\omega)$ and that of the sampling comb, given by the well known expression [17, 18]:

$$S(\omega) = \sum_{n=-\infty}^{n=\infty} G(\omega - n\omega_0) \qquad |\omega| \leq \frac{\omega_0}{2}$$

This expresses the well-known effect that the bandwidth of the sampled signal is restricted to half the sampling frequency (Nyquist limit). It says that the spectrum of the sampled signal in such a window is a superposition of an infinite number of shifted replicas of the parent spectrum $G(\omega)$. We see that such sampling therefore involves processes of multiplication and filtering, the same two key processes that occur in image intensifier or gain-modulated PMT devices. The $S(\omega)$ expression holds for both the real and imaginary parts of the sampled frequency spectrum and tells us all we need to know in order to calculate properties such as amplitude and phase at any frequency in the Nyquist band.

Now take an incoming sinusoidal signal at angular frequency $(\omega_0 + \Delta\omega)$, which we will sample at frequency ω_0. Taking just the real part of the spectra for illustration purposes, Fig. 18.1 shows a diagrammatic version of the shifting and addition processes in the $S(\omega)$ formula. The parent signal (top panel) has contribu-

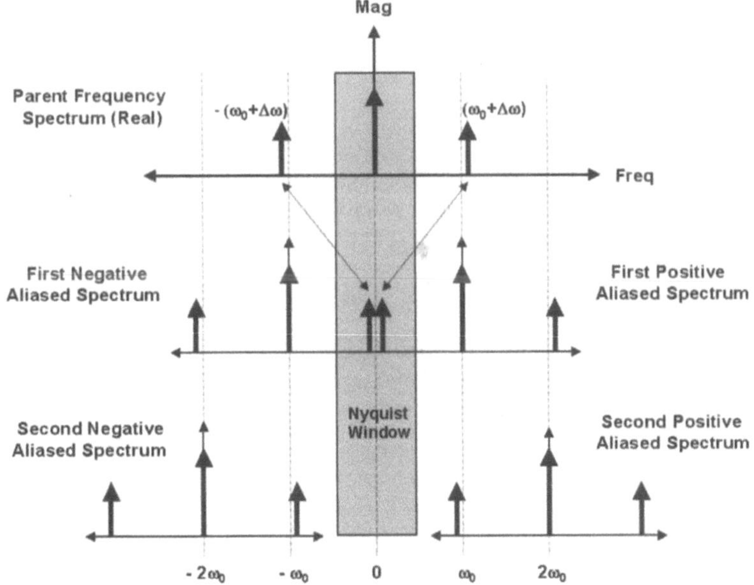

Fig. 18.1. Sampling effects

tions at only the zero frequency (DC) and at $\pm(\omega_0 + \Delta\omega)$, negative frequencies being equally as significant as positive frequencies. To this is added the shifted replicas (lower panels) for all positive and negative n i.e., positioning a copy of the parent at $\pm n\omega_0$ locations and adding to the parent $n = 0$. It is clear that the only AC component that can exist in the Nyquist window after this process is a signal at $\Delta\omega$, describing a pure sinusoidal oscillation at the difference frequency. Thus the incoming signal originally at $(\omega_0 + \Delta\omega)$ now appears down-shifted to the difference frequency $\Delta\omega$ after sampling at frequency ω_0. This aliasing effect in essence transfers information from the original frequency to a new lower frequency and furthermore does so in this case without any reduction in modulation depth. This should be contrasted to the gain-modulation devices where properties that arise solely from the parent high frequency in the $n = 1$ component of the summation, no other terms mix in regardless of how many replicas are added in. The principal restriction is that the difference frequency $\Delta\omega$ has to be chosen to be less than the Nyquist limit ($\omega_0/2$). We also note that the zero frequency DC component of the sampled signal arises directly from the parent DC term. These general arguments also hold for the imaginary part of the spectrum, and together mean that the amplitude and phase parameters (carrying the lifetime information) and originally present at $(\omega_0 + \Delta\omega)$, simply move down without change onto $\Delta\omega$ frequency after sampling.

So if the data stream is sampled in this manner at frequency ω_0 and the original excitation modulation frequency was $(\omega_0 + \Delta\omega)$, we will get a sampled version of Eq. 18.3 describing the fluorescence samples as seen through the scanner action. It is understood that $M(x), \varphi(x)$ now relate to demodulation and phase-delay that pertained at the original high frequency $(\omega_0 + \Delta\omega)$.

$$s(x) = s_0(x) + s_1(x)M(x)\sin\left\{\frac{\Delta\omega}{\omega_s}\left[\frac{\pi}{2} + k\pi + (-1)^k \sin^{-1}(2x-1)\right] + \varepsilon - \varphi(x)\right\} \quad (18.4)$$

Adopting the notation $s(k,x)$ to mean the implementation of Eq. 18.4 for specific values of k i.e., scanner passes, we obtain the sampled fluorescence line profiles for the first three passes of the scanner as

$$s(0, x) = s_0(x) + s_1(x)M(x)\sin\left[\varepsilon - \varphi(x) + \alpha(0, x)\right]$$
$$s(1, x) = s_0(x) + s_1(x)M(x)\sin\left[\varepsilon - \varphi(x) + \alpha(1, x)\right]$$
$$s(2, x) = s_0(x) + s_1(x)M(x)\sin\left[\varepsilon - \varphi(x) + \alpha(2, x)\right] \quad (18.5)$$

$$where \; \alpha(k, x) = \frac{\Delta f}{f_s}\left[\frac{\pi}{2} + k\pi + (-1)^k \sin^{-1}(2x-1)\right] \qquad k = 0,1,2,\ldots$$

and where f denotes the circular frequency of the corresponding ω angular frequency quantities. The set of Eqs. 18.5 is then recognised as the following linear system (18.6):

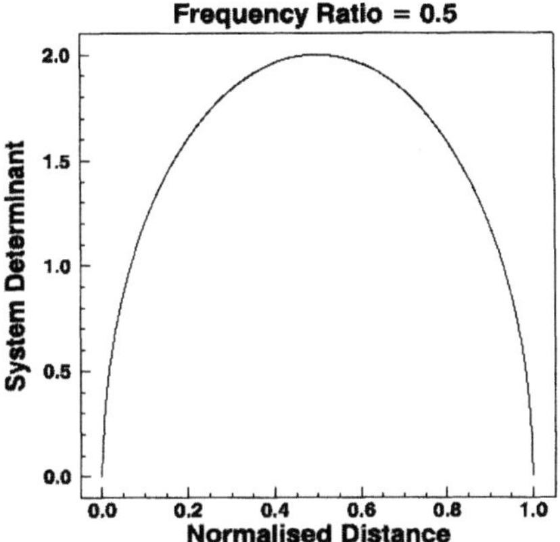

Fig. 18.2. System determinant for frequency ratio 0.5 case

$$
\begin{bmatrix} 1 & cos(\,\alpha(0,x)) & sin(\,\alpha(0,x)) \\ 1 & cos(\,\alpha(1,x)) & sin(\,\alpha(1,x)) \\ 1 & cos(\,\alpha(2,x)) & sin(\,\alpha(2,x)) \end{bmatrix} \times \begin{bmatrix} s_0(x) \\ s_1(x)M(x)sin(\varepsilon - \varphi(x)) \\ s_1(x)M(x)cos(\varepsilon - \varphi(x)) \end{bmatrix} = \begin{bmatrix} s(0,x) \\ s(1,x) \\ s(2,x) \end{bmatrix} \quad (18.6)
$$

The sampled data for each of three passes of the scanner form a vector on the right hand side of this linear system, this being sufficient to solve for the vector of unknowns. There are three basic unknowns per pixel $s_0(x), M(x), \varphi(x)$, so this 3×3 system is sufficient to obtain them via their combinations as written above. We now examine under what conditions this system is solvable, which is equivalent to asking when can one guarantee that the determinant of the system matrix is nonzero.

The determinant of the system 3×3 matrix in Eq. 18.6 is

$$
Det(\,x) = 2sin(\,\pi r)cos\left[2rsin^{-1}(\,2x-1)\right] - sin(\,2\pi r) \quad \text{where } r = \frac{\Delta f}{f_s}
$$

We want this to be nonzero for a general x value and note that it is a function of the frequency ratio r (difference frequency/scanning frequency). We can immediately see that if the frequency difference Δx between excitation and sampling is say any integer multiple of the scanning frequency f_s, then the determinant will vanish for all x values, i.e., no solution possible, the equations are not linearly independent. However if instead we take fractional values for r say $r = 1/2, 1/3, \ldots$ then it will take multiple cycles of the scanner for one period of the difference to occur. The simplest case of this is $r = 0.5$, giving a determinant of

$$
Det(\,x, r = 0.5) = 2cos\left[2rsin^{-1}(\,2x-1)\right] \quad (18.7)
$$

This is nonzero everywhere except at the exact endpoints $x = 0$ and $x = 1$, these are the stationary turning points of the resonant scanner action where the scanner velocity is momentarily zero. So provided we exclude the exact endpoints from consideration, the determinant is nonzero everywhere else for this choice of frequency ratio, and we can solve for the information we require. Fig. 18.2 graphs this determinant function, Eq. 18.7 over the scanline. If instead we chose say $r = 0.25$, then it would take four full cycles of the scanner (a cycle is one forward and one retrace phase) for one period of the difference frequency to occur, then we find that in addition to the endpoints giving concern, that the center of the scan $x = 0.5$ now also gives a singular matrix, again no solution possible there. So it appears that the choice of $r = 0.5$ is a good one which will give us 2 forward and 2 retrace phases of the scanner before the data repeats, we need a minimum of 3 of these "quadrants" of sampled data in order to solve for the sample lifetime data at all pixels along the line, endpoints excluded. Fixing r at 0.5, we proceed to solve Eqs. 18.6 by standard methods [19] and after some algebra we will obtain the final Eqs. 18.8:

$$s_0(x) = \frac{1}{2}\left[s(0, x) + s(2, x)\right]$$

$$\tan\left(\varepsilon - \varphi(x)\right) = \left[\tan\left(\frac{\pi}{4} + \frac{\theta(x)}{2}\right)\right] \cdot \frac{D_1(x)}{D_2(x)}$$

$$s_1(x)M(x) = \frac{\sec\theta(x)}{\sqrt{2}}\sqrt{\left(1 + \sin\theta(x)\right)\left[D_1(x)\right]^2 + \left(1 - \sin\theta(x)\right)\left[D_2(x)\right]^2} \qquad (18.8)$$

$$\textit{where } D_1(x) = s(0, x) - s(1, x) \qquad D_2(x) = s(1, x) - s(2, x)$$

$$\theta(x) = \sin^{-1}(2x - 1) \qquad \textit{and } 0 < x < 1$$

The steady state line profile is just the average of the first and third quadrants ($k = 0, 2$). If we had included the $k = 3$ case, we would equivalently have found that the steady state values are also given by the average of the second and fourth quadrants ($k = 1, 3$). Simply averaging a pair of forward traces or a pair of reverse traces of the scanner action is sufficient to determine the steady state line profile. Thus, these pairings are acting effectively in antiphase for every pixel along the scanline, cancelling any AC modulation from the difference frequency exactly. The next equation in (18.8) gives the result for the tangent of the overall phase, effectively as a ratio of quadrant differences multiplied by a spatially dependent factor. This multiplication factor is encoding the spatial variation of the automatic instrumental phase adjustment occurring in this experiment given the sinusoidal scanner action. It will give increasing error propagation issues as $x \to 1$. The last equation in (18.8) gives the result for the modulated amplitude line profile and is in the form of a sum of squares of differences (quadrant differences). The overall multiplier on this equation now gets increasingly large as we approach both ends of the scanline.

Therefore, just two different combinations of two quadrant differences suffice to determine the time resolved parameters. As in any frequency domain experiment, the phase delay and normalised demodulation of the fluorescence emission is always relative to the values for scattered light from either the same object or a related reference sample whose properties are known. So ideally one could perform the above experiment monitoring both scattered light and fluorescence giving phase delay and modulated amplitude scanline data for both cases. One then differences the net phase delay data (fluorescence-scattered light) or ratios the normalised modulated amplitudes (fluorescence/scattered light) to obtain the final expressions $\varphi(x), M(x)$ to be used for lifetime estimation $\tau(x)$. As noted before, the original parent modulation frequency $(\omega_0 + \Delta\omega)$ is used in such lifetime calculations. If the fluorescence decay behavior is more complex than a single exponentially decaying component, then the above experiment could be repeated at a series of modulation frequencies, whilst maintaining the same frequency difference $\Delta\omega$ and r value.

18.3
Simulations

As an illustration of the method described above, it is instructive to simulate specific 1D lifetime patterns $\tau(x)$ for particular values of modulation, sampling and scanning frequencies. The aim of such simulations is to see whether the lifetime profiles recovered from Eqs. 18.8 match the input model simulated. One example of such simulations will now be briefly summarised.

Taking frequency values that will be used in practical experimental work described later in this paper, we take as our $\tau(x)$ model a set of Gaussian shaped peaks where the peak positions are equi-spaced in space and linearly increasing in lifetime as we move from $x = 0$ to $x = 1$ along the scan line. In addition we gradually increase the spatial width of the peaks as we move towards $x = 1$ and add a small amount of random noise on the calculated fluorescence profiles. Fig. 18.3 shows such a lifetime model and the phase delay and demodulation profiles that result. The simulated parameters were: lifetime range of 2 to 10 nsec across the Gaussian peak set, scanner frequency of 598 Hz, sampling frequency of 25 MHz, excitation modulation frequency of 25.000299 MHz. We note that the difference frequency is 299 Hz which is half that of the scanner frequency i.e., $r = 0.5$ case. A spatially constant DC component is also present in the simulation but not graphed. The division between scanner passes is indicated in Fig. 18.3 and subsequent graphs by the dotted vertical lines, i.e., $k = 0,1,2$ three of such quadrants shown. Fig. 18.3(a) shows the input model as lifetime values versus normalised scan distance x (top left) and in particular how this pattern is visited in reverse for $k = 1$. Fig. 18.3(b) shows the phase delay versus scan time at the modulation frequency specified above (top right) and Fig. 18.3(c) the normalised demodulation values versus scan time (bottom left). Because of the continual varying velocity of sinusoidal scanners along their scan line, the features in the middle of the scan are compressed in time relative to the edges for sample features that were originally equally spaced in distance. This data then allows us to calculate the resultant fluorescence intensities at a set of equi-spaced intervals in time corresponding to the sampling frequency chosen. Fig. 18.3(d) shows such a set of samples as a function of scan time (bottom right). We note the slow sinusoidal envelope which will repeat after 4 quadrants exactly, this is the difference frequency of 299 Hz which now appears due to the deliberate sampling process. Computerised digital sampling and real-world signal sampling may differ in their degree of ideality in approaching a Dirac comb, but the sampling theorem and aliasing phenomena apply equally to both. Superimposed on this are the sample intensity changes caused by the spatially variable set of phase delays and demodulations.

If we now isolate each quadrant and plot back against scan distance, we will obtain the data of Fig. 18.3(e),(f),(g) for $s(0, x)$, $s(1, x)$, $s(2, x)$, respectively. We can see that the data of quadrants 1 (top left) and 2 (top right) are not the same or not just related by a reflection or sign change, whereas quadrants 1 and 3 (bottom panel) are indeed related in such a way. This confirms the antiphase relationship of quadrants 1 and 3, which can be used to estimate the steady state line profile (remember simulated constant DC, normalised level of 1.0).

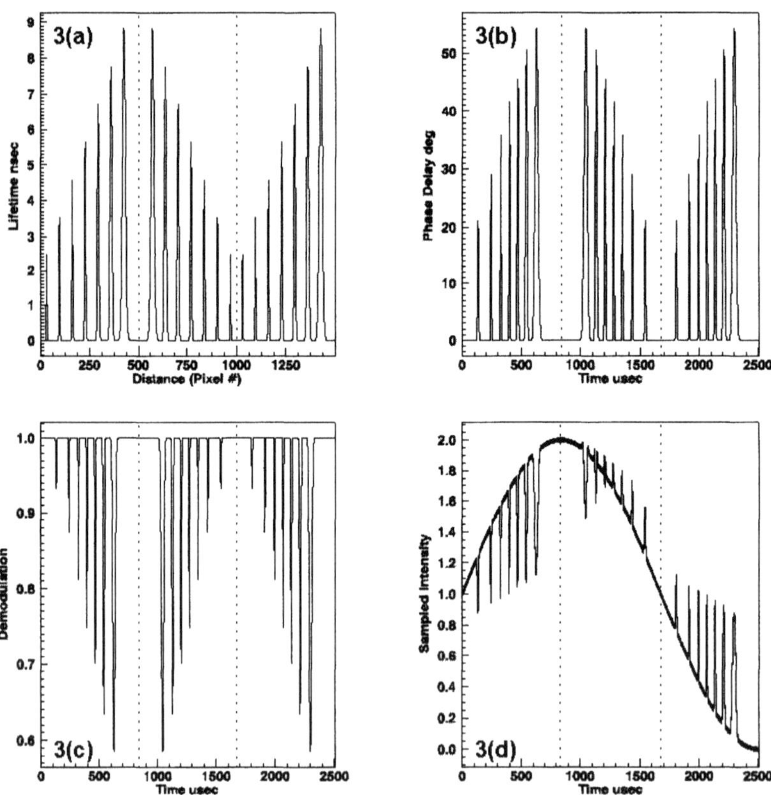

Fig. 18.3. (a), (b), (c), (d)

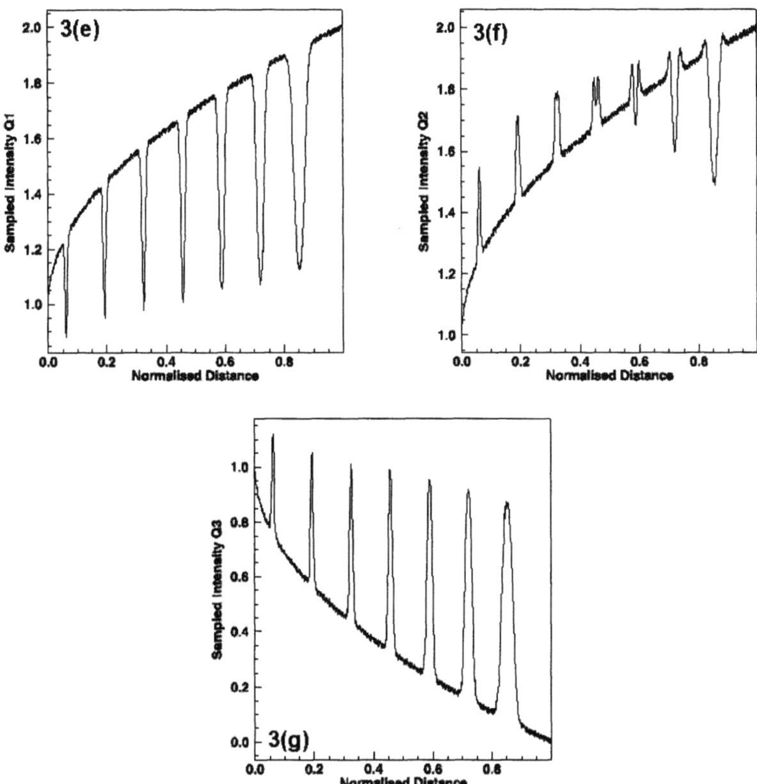

Fig. 18.3. (e), (f), (g)

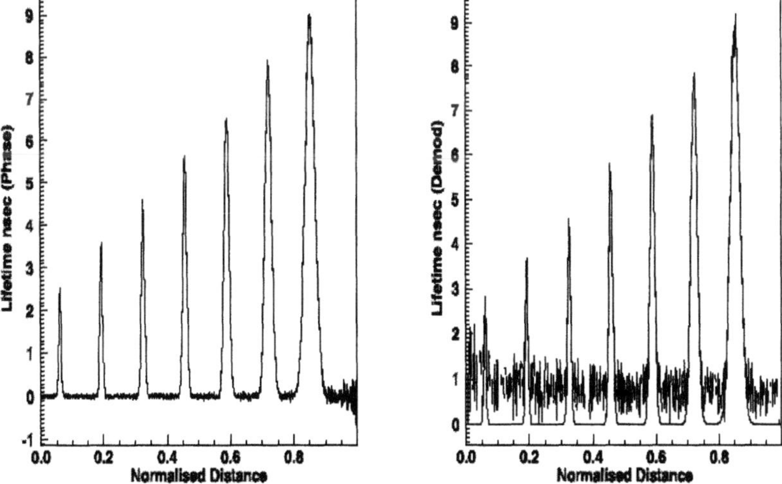

Fig. 18.4. (a) left, (b) right

Fig. 18.4(a),(b) shows the calculated lifetime profiles by using either the phase (left) or demodulation (right) route of Eq. 18.8. These calculated curves are the noisier of the two curves depicted in each graph, the smooth curve is the simulated input model $\tau(x)$. Exact agreement results if no random noise is used in the simulation. We note satisfactory recovery of the model lifetime profile $\tau(x)$, with the phase route, as is usual with frequency-domain approaches, being a bit more robust compared to demodulation in the presence of noise. As pointed out earlier, the phase estimations are expected to degrade in the presence of noise as we approach $x = 1$ because of the instrumental phase multiplier form. Arbitrary DC spatial distributions in addition to $\tau(x)$ are also recoverable in such simulations (data not shown).

Such simulations emphasise the general point about the method that it succeeds by engineering a continuous phase adjustment at a frequency which not only is synchronised to the scanner action, but critically is fractionally related in value to that frequency. At every pixel along the scan line, three basic unknowns need to be estimated, which requires a minimum of three passes of the scanner through the same set of pixels. The interaction between excitation modulation and sampling frequencies generates a difference frequency which allows four such passes (in the case of $r = 0.5$) to occur. The separate passes have different but known instrumental phase adjustments occurring at a given pixel and so the required parameters can be extracted successfully. Any method that generates a difference frequency, which has these two properties linking them to the scanner frequency, should also work, e.g., directly heterodyned PMT detectors.

18.4
Experimental Set-up

A proof-of-principle experimental rig was constructed to test the method. See Fig. 18.5 for a schematic of this apparatus. A Lexel model 85 CW argon-ion laser operating at 488 nm was externally modulated with an acousto-optic modulator (IntraAction model 125, MOD) driven from an analog driver, about 30 MHz bandwidth, adjusted for about 70% modulation depth. The 1st order modulated beam was isolated with a pinhole and reflected towards the scanner via an Olympus microscope dichroic filter block (U-MNIB). The beam is then focussed through the scanner onto the sample through a 500 mm focal length lens. The resonant scanner itself was a nominally 600 Hz model IDS600 from GSI Lumonics driven from an AX730 amplifier. The maximum optical peak to peak excursion for this scanner is 60 deg, the mirror aperture is 9 mm. The AX730 driver can either operate off internal or external sinewave signals, it was always used on external mode for the current work. The purpose of the AX730 is to control the amplitude of the scan compensating for temperature drifts, etc. It also outputs a TTL sync timing signal useful for identifying start and end of scan line and a sensor sinewave output pro-

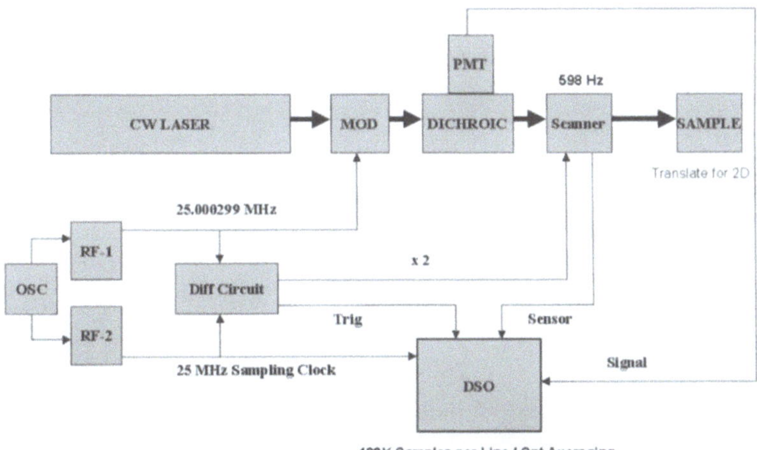

Fig. 18.5. Experimental set-up

portional to the scanner excursion. The scanned laser line length is adjusted on the sample through control of the input sinewave amplitude to the AX730 amplifier. The macroscopic samples to be scanned were placed on a 1D linear translation table which is driven by a stepper motor/controller (Bentham model SMD3B-IEEE), thus translating the object through the scan line in order to obtain 2D images. Because of the long focal length of the scanning lens and the shallow nature of the samples used, no significant defocusing of the beam occurs from edge to middle of scan line. The difference in optical path length between edge and middle is also not sufficient to set up significant phase delay differences in scattered light reference scans because of the modest modulation frequencies involved.

The scanning mirror at any one time of the scan is also acting as the light collection element, descanning a small proportion of the emitted fluorescence propagating back towards the dichroic filter block. Now the longer wavelength fluorescence transmits through the block and is focussed onto a photomultiplier tube (PMT, side-window model 9781R Thorn-EMI, 2 nsec risetime). The pre-amplifier used was a Stanford Research model SR440, gain x25. This amplified signal was sampled, digitised and averaged using a LeCroy digital storage oscilloscope (DSO, model 9384TM, external sampling clock mode, 100K samples taken per displayed line giving at least three scanner quadrant coverage, external trigger from difference circuitry (see Fig. 18.6), start of records adjusted so that $t = 0$ represents the actual start of a scan line). The high frequency generator side of the experiment comprised two RF generators (Marconi 2023 RF-1 and Hewlett Packard HP8656B RF-2, both using an external 10 MHz reference clock supplied by a Stanford Research SC-10 oscillator). The Marconi generator was used to drive the modulator (25.000299 MHz, 0dBm), the other generator drives the sampling side of the experiment at 25 MHz. Each of these two RF drive signals were split to act also as inputs to a homemade difference generator circuit which mixes the two in-

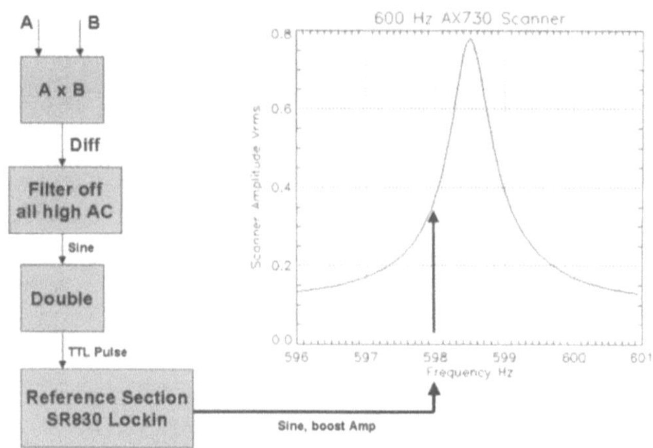

Fig. 18.6. Difference circuitry scanner drive

puts and removes the sum frequency to leave a clean difference sinewave output. This is converted to a TTL square wave and doubled exactly in frequency. The circuitry can handle up to an upper limit of 350 MHz for the combined sum frequency. The frequency doubled TTL output then acts as an external reference for the sinewave generator side of a DSP lock-in amplifier (Stanford Research SR830). The output of this device acts as the final driving signal for the scanner amplifier.

Fig. 18.6 shows a schematic of the difference circuitry, inset is a graph of the resonance frequency response of the actual scanner used. The peak is at 598.5 Hz, however the resolution of the high frequencies on the RF generator side is 1 Hz at best, so given that we need to double the difference frequency, the best that can be achieved is 598 Hz in this case. RF generators with 0.1 Hz resolution would increase the ability to tune to the exact resonance peak of the scanners. To compensate for this off-peak situation, the SR830 output amplitude is increased to give adequate scan line lengths on the sample. The overall strategy is unusual in that the low frequency scanner is driven from the high frequency side of the experiment through the difference circuitry.

A suitable test sample was a series of concentrations of a typical water-soluble fluorescer in MilliQ-grade water, tetramethyl BODIPY disulfonate (Molecular Probes D-3238). Such a sample should give steady-state intensity changes but no substantial lifetime changes over the concentration range studied (0–50 μM). This is contrasted with a series of samples of the same fluorescer at constant concentration (50 μM) but with a controlled variation of added quencher (KI, 0–240 mM). This series should also show a decrease in intensity with increased concentration of quencher, but now because of a reduction in fluorescence lifetime caused by collisional dynamic quenching. This same basic experiment but at altered frequencies was previously used as a test sample for the traditional homodyne 2D image intensifier FLIM route [20]. Both types of solutions were mounted in individual 400 μm pathlength capillary microslides (Camlab Cambridge), these placed on a stiff black card background, this single unified sample then placed on the scanning table and measured in a light tight enclosure. Each raw sampled record, consisting of 100 K points covering more than three quadrants of the scanner action, was averaged 100 times in the DSO to reduce detector noise, external sampling frequency of 25 MHz exactly. The DSO was triggered from the difference circuit signal before doubling, and the record aligned so that the first sampled point corresponded to the exact start of a forward pass of the scanner. The sync and sensor outputs of the scanner amplifier were used to determine this point with the aid of fluorescing white card samples which were placed so as to leave just the edges of the scanline off the card areas, giving a strong dark/bright transition in the sampled signals. This trigger alignment is critical, as linearising of the resonant scanner action (time to distance inter- conversions) is performed in software in the current implementation. We find slow changes in this trigger alignment over many hours presumably due to ambient temperature changes causing off-resonance shifts of the scanner, so the trigger position is checked carefully before each run. The average scattered light phase delay and demodulation reference values were estimated by scanning a sub-nsec emitter (eosin) under the same conditions, this is taken as substantially equivalent to scattered light given the low modulation frequency of the current experiment.

18.5
Results

Fig. 18.7 shows the steady state image that results from the BODIPY experiment using the first of Eqs. 18.8, i.e., averaging the first and third quadrants. The image is 500x550 pixels, the 500 pixels being interpolated from the much larger sampled record per quadrant. The left half of the image is the concentration series of microslides, the right half is the quenching series. The number labels for each microslide indicate either the fluorescer or quencher concentration. The individual microslides have not all filled to the same extent and there are occasional bubbles/unfilled areas clearly evident in some. A slight lensing effect is evident, compare centers to edges. Nevertheless, it is clear that we have the expected increase in steady state signal in the concentration series and the expected decrease in the quencher series.

Fig. 18.8 shows the corresponding phase image i.e., estimated via the second of Eqs. 18.8. A 3 % intensity threshold has been used on this image, i.e., all pixels with steady state intensity less than 3 % of the peak global image value have been set to zero phase. We see from the associated colourbar that the total phase variation over the sample is about 40 deg. It is clear that essentially the same phase values are seen throughout the concentration series whereas there is an increase in absolute phase as the quencher concentration increases. The reference phase for this experiment was estimated via the eosin reference sample at 141 deg, so this is

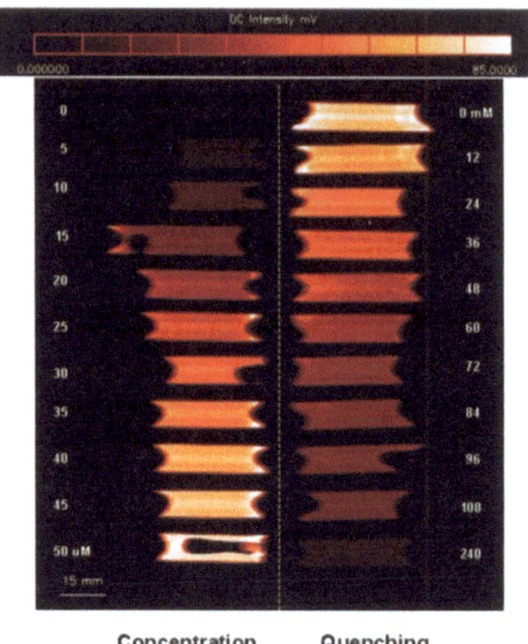

Fig. 18.7. Steady-state intensity image

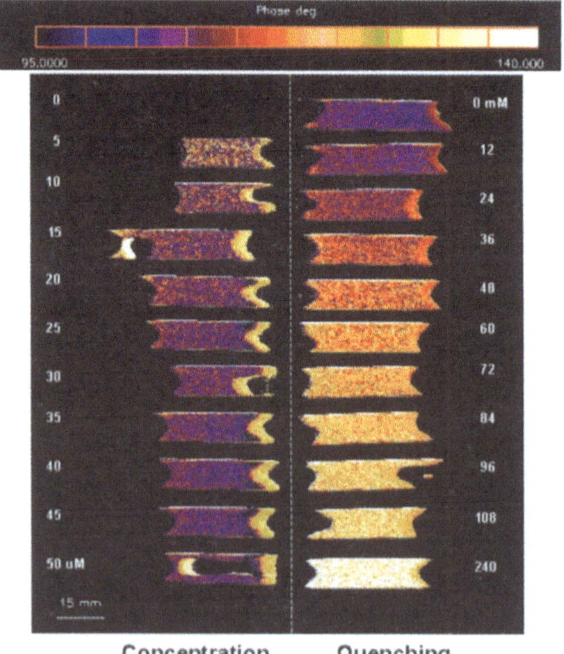

Fig. 18.8. Phase image

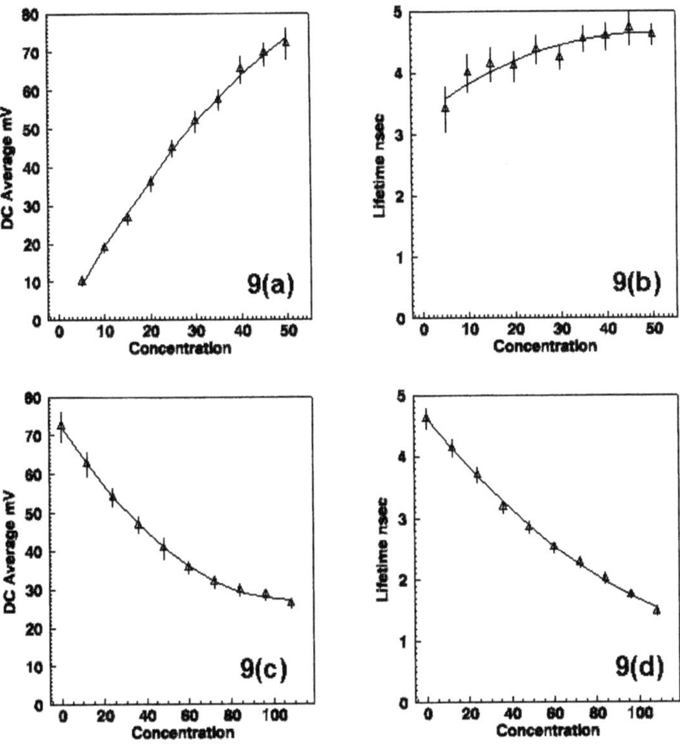

Fig. 18.9. (a), (b), (c), (d)

equivalent to a decreasing phase delay in the quencher series, or a shortened lifetime as the quencher concentration increases.

Fig. 18.9 collects the trends together for the two series using the mean values
estimated from placing regions-of-interest (ROI) over each filled microslide area,
top panels (a), (b) are the concentration series results, bottom panels (c), (d) the
quenching series results. The error bars on each point are derived by appropriate
error propagation rules from the standard deviation estimates of each ROI. The
lifetime is seen to be relatively constant over the concentration series but shows a
clear 3 nsec fall over the quencher range studied. Low concentrations of fluorescer
have relatively high noise, an indication that an improved detector should really be
used. The estimated unquenched lifetime is about 4.6 nsec under these conditions.
This value was checked by performing a traditional frequency-domain cuvette experiment on a homemade multifrequency rig, see Fig. 18.10 for demodulation (a)
and phase delay (b) versus modulation frequency data measured. The solid curves
are a single-exponential fit to the data and give us an unquenched lifetime estimate
of about 4.5 nsec by both phase and demodulation. This is satisfyingly close to the
value seen in the scanning imaging experiment. We conclude that both absolute
and relative lifetime trends can indeed be correctly measured by the scanning lifetime method described.

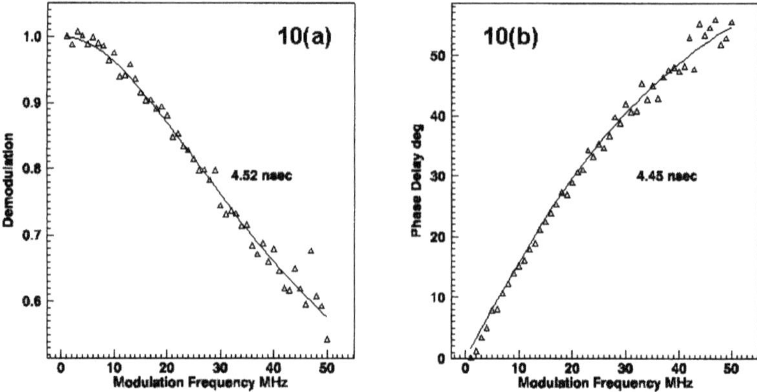

Fig. 18.10. Non-imaging lifetime comparison

18.6
Conclusions

The method described in this paper is one example of how to fold the standard frequency-domain fluorescence lifetime approach with inexpensive continuously oscillating scanning elements in order to achieve a lifetime imaging property. The excitation is continuously modulated at one frequency, the scanner also executes continuously but at a different frequency. These two tones are mixed together in such a way as to achieve a necessary down-shifting of the lifetime data synchronised to the scanner action. The way this achieved in this work is to introduce a third tone, via sampling the data at a third specific frequency. This same function could also be implemented in other ways, e.g., mixing at the detector stage.

As the excitation beam scans along a line, we have a continuous instrumental phase shift occurring which is arranged to repeat only after multiple passes of the scanner have occurred through a given pixel. This allows each pixel along the scanline to be multiply sampled at a small number of known instrumental phase values, and this is sufficient to extract steady state and lifetime data in parallel. The equations necessary to extract this data are simple in form and involve numerical differences for the sampled fluorescence intensities for a given pixel calculated between specific passes of the scanner.

It is interesting to note that the overall form of the equations derived somewhat resembles that used in calculating microscopic optical sections by grid-projection techniques [21, 22]. In such methods a series of microscope images are acquired using a spatial sinusoidal illumination mask at usually three positions, or phases, of the grid. In that case the modulated amplitude image constructed by sums of squares of differences at a given pixel across the image set, exhibits good optical sectioning properties. In our case such modulation parameter images show lifetime contrast from quite similar equations. The same basic Fourier analysis applied in our case in time and in the structured light approaches in space, results in a similar form of final equations.

The experimental rig described above proved sufficient for proof-of-principle purposes. It is almost the simplest possible optical system to perform the measurements, i.e., modulated light source, scanner, optical filters, detector. No imaging devices such as gated intensifiers and cameras are necessary. Such a simple scanning approach easily facilitates both macroscopic and microscopic imaging applications. The overall performance could be improved markedly by using more sophisticated detectors (cooled and higher gain PMTs, possibly heterodyned as noted earlier). The use of electro-optic modulators would permit easy extension to higher modulation frequencies and hence provide access to shorter lifetime ranges. Telecentric scanning lenses could be incorporated to ensure the excitation beam is optimally normal to all sample locations along a scan line. Resonant galvanometer scanners up to 8 kHz frequency are commercially available and could be used; they form the line scanning engine in the latest commercial real-time two-photon microscopes for example. As the scanning frequency increases, more sophisticated trigger alignment mechanisms than used in this paper may become necessary to implement the linearisation function [23]. Parallel scattered light reference images could also be acquired by adding additional optical filters and detection paths. As indicated earlier, conventional galvanometer scanners could act as the second dimension generator rather than the simple sample translation method used in this work. However, even with a very basic system as described in this paper, it is possible to successfully image fluorescence lifetimes with results comparable to more expensive 2D routes.

An alternative route to either detector heterodyning or synchronised sampling to achieve the necessary demodulation of the high frequency signals, would be to use an RF lock-in amplifier, e.g., Stanford Research model SR844. This device has previously been used to perform non-imaging frequency-domain fluorescence lifetime detection [24]. However, it should be noted that such a device even in its fastest state as a raw demodulator generating analog in-phase and in-quadrature outputs, is too slow to adequately track even the slowest resonant scanner models (15 μsec maximum update rate). If future types of such instrumentation had faster update rates enabled, then they could be used to directly generate demodulated data as a function of scanner position with no need for explicit difference frequency generation as in the current work.

Acknowledgements. I wish to thank my colleagues in the Instrument Workshops at Port Sunlight, in particular Steve Cowan for expert design and construction of the difference circuitry, and Pete Carroll for continual help with mechanical mounting and assembly of instrument components.

References

1. Wang XF, Herman B (1996) Fluorescence imaging spectroscopy and microscopy. J Wiley, New York, pp 273–374
2. Lakowicz JR (1999) Principles of fluorescence spectroscopy. Kluwer, New York
3. Ghiggino KP, Harris MR, Spizzirri PG (1992) Fluorescence lifetime measurements using a novel fiber-optic laser scanning confocal microscope. Rev Sci Instrum 63(5): 2999–3002

4.   Dowling K, Dayel MJ, Lever MJ, French PMW, Hares JD, Dymoke-Bradshaw AKL
     (1998) Fluorescence lifetime imaging with picosecond resolution for biomedical appli-
     cations. Optics Letters 23(10):810–812
5.   Lakowicz JR, Berndt KW (1991) Lifetime-selective fluorescence imaging using an rf
     phase-sensitive camera. Rev Sci Instrum 62(7):1727–1734
6.   Clegg RM, Feddersen B, Gratton E, Jovin TM (1992) Time resolved imaging fluores-
     cence microscopy. SPIE 1640:448–460
7.   Morgan CG, Mitchell AC, Murray JG (1992) In-situ fluorescence analysis using nano-
     second decay time imaging. Trends Anal Chem 11(1):32–41
8.   Gadella TWJ, Jovin TM, Clegg RM (1993) Fluorescence lifetime imaging microscopy
     (FLIM): spatial resolution of microstructures on the nanosecond time scale. Biophys
     Chem 48:221–239
9.   Verveer PJ, Squire A, Bastiaens PIH (2000) Global analysis of fluorescence lifetime
     imaging microscopy data. Biophys J 78:2127–2137
10. Neumann M, Herten DP, Dietrich A, Wolfrum J, Sauer M (2000) Capillary array scan-
     ner for time-resolved detection and identification of fluorescently labelled DNA frag-
     ments. J Chromatography A 871:299–310
11. Lassiter SJ, Stryjewski W, Legendre BL, Erdman R, Wahl M, Wurm J, Peterson R,
     Middendorf L, Soper SA (2000) Time-resolved fluorescence imaging of slab gels for
     lifetime base-calling in DNA sequencing applications. Anal Chem 72:5373–5382
12. Piston DW, Sandison DR, Webb WW (1992) Time-resolved fluorescence imaging and
     background rejection by two-photon excitation in laser scanning microscopy. SPIE
     1640:379–390
13. Dong CY, So PTC, French T, Gratton E (1995) Fluorescence lifetime imaging by
     asynchronous pump-probe microscopy. Biophys J 69:2234–2242
14. Buurman EP, Sanders R, Draaijer A, Gerritsen HC, van Veen JJF, Houpt PM, Levine
     YK (1992) Fluorescence lifetime imaging using a confocal laser scanning microscope.
     Scanning 14:155–159
15. Vroom JM, de Grauw KJ, Gerritsen HC, Bradshaw DJ, Marsh PD, Watson GK, Bir-
     mingham JJ, Allison C (1999) Depth penetration and detection of pH gradients in
     biofilms by two-photon excitation microscopy. Appl Env Micro 65(8):3502–3511
16. Abramowitz M, Stegun IA (1965) Handbook of mathematical functions. Dover, New
     York, p 80, formula 4.4.10
17. Haykin S (1983) Communication Systems. J Wiley, New York, pp 364–371
18. Blackman RB, Tukey JW (1959) The measurement of power spectra from the point of
     view of communications engineering. Dover, New York, pp 31–33
19. Boas ML (1983) Mathematical methods in the physical sciences. J Wiley, New York,
     pp 81–95
20. Birmingham JJ (1997) Frequency-domain lifetime imaging methods at Unilever Re-
     search. J Fluorescence 7(1):45–54
21. Wilson T, Neil MAA, Juskaitis R (1998) Real-time three-dimensional imaging of mac-
     roscopic structures. J Microscopy 191(2):116–118
22. Neil MAA, Squire A, Juskaitis R, Bastiaens PIH, Wilson T (2000) Wide-field optically
     sectioning fluorescence microscopy with laser illumination. J Microscopy 197(1):1–4
23. Tweed DG (1984) Resonant scanner linearization techniques. SPIE 498:161–168
24. Harris P, Sipior J, Ram N, Carter GM, Rao G (1999) Rev Sci Instrum 70(2):1535–
     1539

Spectral Imaging of Single CdSe/ZnS Quantum Dots Employing Spectrally- and Time-resolved Confocal Microscopy

W.G.J.H.M. VAN SARK, P.L.T.M. FREDERIX, M.A.H. ASSELBERGS, D.J. VAN DEN HEUVEL, A. MEIJERINK, AND H. C. GERRITSEN

A spectrally- and time-resolved study of single CdSe/ZnS quantum dots (QDs) is presented. To this end a versatile, high sensitivy spectrograph is coupled to a confocal laser-scanning microscope. The spectrograph is built in-house and is especially developed for use in fluorescence microscopy. The high sensitivity is achieved by using a prism for the dispersion of light in combination with a state-of-the-art back-illuminated charge-coupled device (CCD) camera. The detection efficiency of the spectrograph, including the CCD camera, amounts to 0.77 ±0.05 at 633 nm. Full emission spectra with a 1–5 nm spectral resolution can be recorded at a maximum rate of 800 spectra per second. The spectrograph can easily be fiber-coupled to any confocal laser-scanning microscope.

The spectral characteristics of the QDs are studied in two different ambients: air and nitrogen. For QDs in ambient air, a clear 30 nm blue-shift in the emission wavelength is observed, before the luminescence stops after about 2–3 minutes due to photobleaching. In a nitrogen atmosphere, the blue shift is absent while photobleaching occurs after much longer times, i.e., 10–15 minutes. These observations are explained by photo-induced oxidation. The CdSe surface is oxidized during illumination in the presence of oxygen. This effectively results in shrinkage of the CdSe core diameter by almost 1 nm, and consequently in a blue shift. The faster fading of the luminescence in air suggests that photo-induced oxidation results in the formation of non-radiative recombination centers at the CdSe/CdSeO$_x$ interface. In a nitrogen atmosphere photo-induced oxidation is prevented by the absence of oxygen. Additionally, a higher initial light output for CdSe/ZnS QDs in air is observed. This can be explained by a reduction of the lifetime of the long-lived defect states of CdSe QDs by oxygen.

19.1
Introduction

Nanometer-sized semiconductor quantum dots (QDs) nowadays are extensively studied [1–4]. The change in the electronic structure as a function of QD-size is most evident from the variation of their emission wavelength [2, 5–7], and is the basis for many new applications in physics, chemistry, and biology, see, e.g., [8]. Due to the confinement of electrons and holes in the nanocrystallites the energy-level scheme resembles that of an atom, with many discrete energy levels. The separation between energy levels increases as the particle size decreases. As an example, the radius of the bulk Bohr exciton amounts to 5.6 nm for CdSe [6], hence nanocrystallites with a size smaller than the Bohr radius, will show quantum confinement effects. Dabbousi et al. have shown that CdSe QDs with a diameter of 2.3 nm fluoresce at 470 nm, and QDs with a diameter of 5.5 nm fluoresce at 610 nm [5]. Capping of CdSe QDs with a few monolayers of ZnS increases the quantum efficiency considerably to values exceeding 50% [9, 10].

The tunability of the optical properties of QDs by changing their size also makes them difficult to study. A prerequisite for the study of QDs is the ability to produce them with a narrow size distribution (< 5%). Variations of size and shape within ensemble samples result in extensive inhomogeneous broadening of absorption and emission spectra. Measuring spectra of *single* QDs can circumvent these effects. Since Blanton et al. [11] and Nirmal et al. [12] showed that it is possible to measure *single* QD emission spectra, the study of the luminescence of single semiconductor QDs has revealed many interesting properties that cannot be observed for an ensemble of QDs. Phenomena of fundamental interest such as spectral diffusion and blinking (on–off behavior [13, 14]) are also important for potential applications of single quantum dots as luminescent labels in biological systems [10, 15–17].

Most studies on spectral diffusion have been performed at cryogenic temperatures. Random spectral diffusion at low temperatures has been related to blinking and therefore the ionization of quantum dots due to Auger processes [12–14, 18–20]. As a result of the ionization and subsequent recombination processes, the charge distribution around the QD changes, resulting in a spectral (Stark) shift of the emission [20, 21]. A clear relation between blinking (explained by photo-ionization) and the occurrence of a spectral jump has been established. At room temperature less information is available on spectral diffusion of single QD luminescence. A non-random blue shift of about 10–15 nm has been reported [12, 22], which was attributed to photo-oxidation of the quantum dot.

The measurement of *single* QD emission spectra is just one example of the many applications in fluorescence microscopy that require multiple wavelength band detection. Other examples include the simultaneous imaging of multiple probes in morphological studies, the quantification of ion concentrations using the ratio of the fluorescence signal in two emission bands imaging [23], the measurement of nanometer co-localization by means of Förster Resonance Energy Transfer (FRET) [24] and the very recently observed FRET between QDs and organic dyes [25, 26].

A common and simple way to implement the detection of multiple wavelength bands is the use of multiple emission filters. However, in general the bandwidth of the emission filters is comparatively broad and the number of wavelength bands used in the imaging experiment is limited. The number of wavelength bands that can be detected (simultaneously) limits the number of parameters that can be monitored. Importantly, the broad emission bands of organic fluorescent probes often overlap. This complicates the separation of multiple spectra. Moreover, the broad detection bandwidth and limited number of detection channels do not allow assessment of the shapes of the emission spectra. Therefore, no direct indication of unexpected spectral shifts or the presence of artifacts such as auto-fluorescence or scattered excitation light is present in the images. This may result in the misinterpretation of the observed intensities.

The measurement of complete emission spectra under the microscope can be used to solve these problems [27–34]. However, the acquisition of reliable emission spectra requires the detection of a large number of photons. This restricts both the acquisition rate of the images and the number of images that can be recorded before a fluorescent probe is photobleached.

In this chapter we describe a highly sensitive spectrograph that is optimized for use in (scanning) fluorescence microscopy. The high sensitivity is achieved by using a prism for the dispersion in combination with a back illuminated CCD camera. Complete emission spectra with a 1–5 nm resolution can be recorded with millisecond dwell times. The spectrograph is connected to a commercial confocal laser-scanning microscope (CLSM). Several of its unique features will be elucidated in a time-resolved study at room-temperature of emission spectra of *single* CdSe/ZnS QDs.

19.2
Experimental

19.2.1
Spectrograph-CLSM Set-up

19.2.1.1
Description

The spectrograph was designed and constructed in-house. Two important design parameters were the wavelength range and the spectral resolution. As the spectrograph was intended for use in biological studies, they were set to 450–750 nm, and 1–5 nm, respectively. Furthermore, reasonable spectral imaging acquisition times require short integration times per spectrum. In a scanning microscope, dwell times per pixel of up to a few milliseconds were considered acceptable. Another important consideration was the detection efficiency, as in fluorescent microscopy only a limited number of fluorescent probes are present per pixel, i.e., ranging from one to several hundreds. This limits the number of available photons. Also photobleaching effects will reduce the number of photons. Finally, in order to record a reliable emission spectrum a comparably large amount of photons are to be detected. Further detailed design considerations can be found elsewhere [35, 36]. On the basis of these considerations a back-illuminated CCD camera was chosen as the detector, and an antireflection-coated prism as the dispersive element.

For flexibility reasons the entrance of the spectrograph was equipped with a standard multimode fiber adaptor. Light emerging from the fiber (or directly coupled light) was collimated by a 100 mm F/2.5 achromatic lens (Melles-Griot). Next, the light was dispersed by the prism and focused on the CCD camera by an identical lens. Note that the fiber end was projected on the camera without magnification.

An equilateral SF10-glass prism (Linos), operating at minimum deviation conditions, dispersed the light. The minimum deviation wavelength was chosen at 550 nm. For this wavelength the angle of incidence at the entrance surface equaled the angle of emergence at the exit surface. For the prism employed here, this yielded an entrance angle of 60.1 degrees. The interfaces of the prism were coated with a single layer MgF_2-coating, to enhance the transmission. The lengths of the legs of the prism (60 mm) limited the numerical aperture of the spectrograph. The usable prism width was slightly (i.e., about 1 mm) diminished by the fact that the dispersed beam was wider than the beam before dispersion. From the focal length of the collimating lens and the length of the legs of the prism a maximum numerical aperture of 0.14 was found for the beam of light entering the spectrograph.

The spectrograph was equipped with a Peltier cooled, back-illuminated CCD camera (Princeton Instruments NTE/CCD-1340, 16 bit ST133 controller, readout noise 6e⁻ at 1 MHz ADC). The CCD chip was 1340 pixels long in the dispersion direction (horizontal) and 100 pixels high (vertical) and had 20×20 μm^2 square pixels. The spectral measurements were carried out on a small sub-area of the CCD chip of less than 10 pixels high and between 50 and 400 pixels long. The pixels in the direction perpendicular to the dispersion direction were always hardware binned. Moreover, hardware binning can be employed in the dispersion direction as well, albeit at the price of a loss of spectral resolution. To gain speed, the sub-area was positioned at the corner of the CCD-chip closest to the readout amplifier. To facilitate alignment of the CCD-chip it was mounted on two orthogonal translation stages. In order to select another wavelength region the camera was simply translated with respect to the prism.

The experiments presented here were all recorded using a confocal laser-scanning microscope (CLSM, Nikon PCM2000, in combination with a Nikon Optiphot 2 microscope). The CSLM scan head was equipped with a standard single-mode fiber adaptor for excitation and two multimode fiber adapters for coupling the emission light to the (remote) detectors. The first of the two detection channels was coupled to the spectrograph. A schematic diagram of the set-up is shown in Fig. 19.1.

A 50 μm core diameter fiber patch cord equipped with standard connectors on both ends (Thorlabs, FG-050-GLA) was used to interface the spectrograph to the CLSM. Because of the comparatively large fiber core diameter of 50 μm, the alignment of the fiber was not critical. The fiber adapter at the CLSM was manually adjusted to optimize the fluorescence signal. Furthermore, the numerical aper-

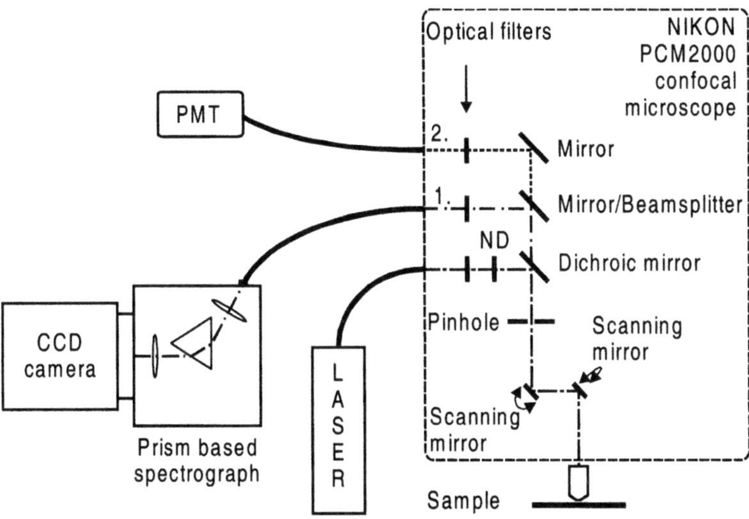

Fig. 19.1. Schematic overview of the experimental set-up. The excitation light from the mixed gas Argon-Krypton laser is guided to the CLSM by a single mode fiber. An excitation filter and a neutral density filter (ND) can be placed in the excitation beam. Each of the two emission paths can be equipped with a longpass or bandpass emission filter. Detection channel 1 of the CLSM is connected to the spectrograph and channel 2 is connected to a PMT. Channel 1 and 2 are selected by the insertion of a mirror or beamsplitter in the emission path, respectively. The lenses of the CLSM are omitted from the picture for clarity

ture of the light emerging from the CLSM ($<$ 0.12) and that of the spectrograph (0.14) were reasonably matched. This ensures efficient coupling of the two devices.

The pixel clock pulses generated by the CLSM were used to synchronize the CLSM and the CCD camera controller. The CLSM was used for acquiring 2D spectral images (2D-mode) or for time resolved spectral analysis of small volume elements (time-mode). The wavelength, the confocal pinhole size, the objective magnification and the numerical aperture of the objective determine the spatial resolution. The measurements discussed below, were carried out with a 60× oil immersion objective (Nikon PlanApo 1.4 NA) and with the larger of two available pinholes (50 μm). The axial and lateral resolutions were estimated to be approximately 1 μm and 0.3 μm, respectively.

The Argon-Krypton mixed gas laser (Spectra Physics, 2060-10SA) used here provides a large number of excitation wavelengths (400–650 nm). A single mode fiber guided the laser light to the CLSM. The excitation light was reflected towards the sample by the built-in dichroic mirror of the confocal microscope (Nikon). Typical laser powers at the sample ranged from 1–100 μW.

19.2.1.2
Performance

The main factors determining the spectral resolution of the device are the fiber-core diameter, the dispersion of the prism, the focal lengths of the lenses and the pixel size of the CCD camera. In addition, the resolution also depends on the number of pixels binned in the dispersion direction. Here, a fiber with a 50 μm fiber-core diameter is used at the input of the spectrograph. It is projected on the CCD without magnification. The projection of the fiber-core on the CCD is larger than the pixel size; i.e., monochromatic light spreads out over more than one pixel. In the following, results are presented which have been taken from Frederix et al. [35]. Further details can be found there as well.

Scattered laser light at 476.5 nm, 514.5 nm and 632.8 nm yielded peaks with a full width at half maximum (FWHM) of 2.2 ± 0.1 pixels in the dispersion direction after binning vertically over 8 pixels. The FWHM is close to the ratio of 2.5 between the fiber-core diameter and the pixel size.

The non-linear spectral bandwidth per camera pixel was calculated using the angular dispersion of the prism [37], the focal length of the second lens, the camera pixel size. This pixel bandwidth is shown in Fig. 19.2. The spectrograph is operated at a fixed geometry of the prism and lenses. Therefore the shape of the wavelength calibration curve is constant. The calibration of the spectrograph was accomplished by fitting the measured CCD pixel positions of a number of calibration lines (3 or more) to the theoretical expression of the calibration curve. An additional scaling factor was included in the fit to account for errors in the magnification of the spectrograph. Calibration measurements were carried out with laser lines or with a calibration lamp (HgAr lamp, Oriel), depending on the wavelength range of interest. The peak positions were determined using the Winspec-software of the CCD camera (Princeton Instruments). Typically, the calibration accuracy was better than 1 nm over the whole wavelength range.

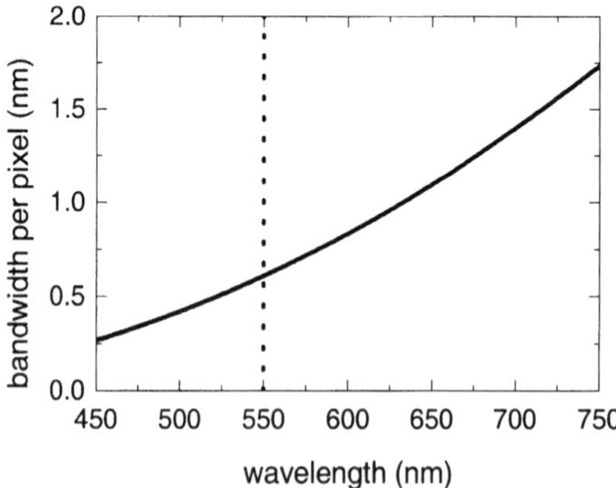

Fig. 19.2. Pixel bandwidth of the spectrograph, plotted against the wavelength. The vertical dotted line at 550 nm denotes the minimum deviation wavelength of the prism

In general, the spectral resolution at a specific wavelength can be simply calculated by multiplying the number of binned pixels with the bandwidth per pixel at that wavelength. However, the projection of the fiber-core on the detector of 2.2 pixels sets a lower limit on the maximum achievable resolution. At wavelengths of 450, 550, and 750 nm a spectral resolution of 1.1, 2.5, and 7.0 nm, respectively, is found using 4-pixel binning in the dispersion direction.

The throughput of the spectrograph (prism plus two lenses) was found to be 0.85 ±0.05 at 633 nm, and agreed well with the calculations. The quantum efficiency of the CCD camera (η) at this wavelength is approximately 90%. The overall detection efficiency of the spectrograph plus camera at this wavelength is the product of these two values and amounts to 0.77±0.05.

The shape of the overall detection efficiency curve, now including the fiber, is determined using a calibrated tungsten band lamp. The (absolute) overall detection efficiency curve is found by normalizing this curve to 0.77 at 633 nm. The throughput of the fiber is about 90% and constant within 2.5% in the wavelength range from 450–750 nm.

The CCD camera has a maximum rate of approximately 800 spectra (of 20–100 points/spectrum) per second at a $N_{RD} = 6$ e$^-$. Therefore, pixel dwell times in excess of 1.2 ms were employed. In 2D-mode, a 160×160 points image with 1.2 ms dwell time is recorded in approximately 40 seconds. Due to the time lost during the retrace of the mirror, the acquisition time is somewhat longer than the number of spectra multiplied by the dwell time.

The internal timing of the camera can be modified to increase the maximum spectral rate, at the price of extra noise. Alternatively a slow ADC (100 kHz) can be employed to reduce the N_{RD} to 3.5 e$^-$. However, this decreases the maximum spectral rate.

Measurements were carried out using two PCs both running Windows 98 (Microsoft). The first PC controls the CLSM and the second controls the CCD camera. The Microsoft automation server and automation manager were used to synchronize the start of the recording of spectra by the CCD camera and of the scanning. To this end the operating software of the CLSM (EZ2000 Nikon, Coord automatisering) was slightly modified. Acquisition of the spectra was carried out using the Winspec-software that controls the CCD camera and raw data are saved to disk.

For each point in a 2D-image a whole emission spectrum was recorded. Consequently, 3D data sets were created (x, y, λ). Data processing and visualization of the 3D data sets were carried out with a program written for use in IDL (Creaso). The program subtracts a (user-defined) background from the raw spectra and corrects for the detection efficiency of the set-up. The wavelength calibration was performed as explained above. Where possible, the average spectrum from an area in the image with no obvious fluorescence was employed as a background. If such an area was not available, a background spectrum from a comparable (reference) specimen was used. The wavelength dependency of the detection efficiency of the complete set-up was corrected for with a so-called 'flat-field' spectrum. The flat-field spectra were determined by using a calibrated tungsten band lamp. The band lamp spectrum was assumed to be identical to the spectrum of a black body radiator of the same temperature. The flat-field spectrum was calculated from the ratio of the band lamp spectrum and the spectrum of the black body radiator. The flat-field spectrum also includes the variable bandwidth per CCD pixel. A more detailed description of the data correction can be found in [36]. Intensity images can be viewed at each recorded wavelength band. Furthermore, emission spectra from selected pixels, or regions of interest, can be viewed and exported for further analysis.

The data sets in time-mode are 2D (t, λ). The processing of these data sets was done in a similar way as described above. The background was taken from time intervals with no obvious fluorescence. If this was not available, a background spectrum from a comparable (reference) specimen was used.

19.2.2
QD Synthesis and Characterization

The ZnS-capped CdSe nanoparticles were synthesized by using a TOP/TOPO method similar to the one described by Hines and Guyot-Sionnest [9, 38, 39]. Two batches of overcoated dots were synthesized. For batch 1 about 4 monolayers of ZnS were grown over the CdSe core, while for batch 2 the thickness amounted to about 7 monolayers. The number of monolayers is based on the amounts of precursors used in the synthesis.

The synthesis was performed in a glovebox filled with nitrogen. Stock solutions of Cd, Zn, S and Se in TOP were prepared. For the Cd/TOP stock solution 1.6 g (0.011 mol) dimethylcadmium was dissolved in 15.5 mL TOP. The Zn/TOP stock solution was prepared by dissolution of 1.23 g (0.01 mol) diethylzinc in 9.0 mL TOP. 1.3 g Se (0.016 mol) was dissolved in 16.0 mL TOP and 2.0 mL (0.01 mol) $(TMS)_2S$ was dissolved in 8.0 mL TOP to obtain the Se/TOP and S/TOP stock solutions, respectively. Cd/Se/TOP and Zn/S/TOP stock solutions were freshly prepared for every synthesis. The Cd/Se/TOP stock solution was prepared by diluting 0.4 mL Cd/TOP and 0.4 mL Se/TOP in 2.0 mL TOP. Dissolution of 1.6 mL S/TOP and 1.12 mL Zn/TOP in 8.28 mL TOP gave the Zn/S/TOP stock solution.

The synthesis was performed by the following method. 25 g TOPO was heated to 300°C and kept at this temperature for half an hour to degas and dry the TOPO. The temperature was raised to 370°C. The heater was then removed and the temperature dropped. At 360°C 1.4 mL Cd/Se/TOP stock solution (containing 0.13 mmol Cd and 0.20 mmol Se) was injected rapidly. The reaction mixture was allowed to cool to 300°C and at this temperature 5.5 mL of the Zn/S/TOP stock solution (0.62 mmol Zn, 0.88 mmol S) was added in five portions at approximately 20 s intervals (batch 1). For batch 2, 13.73 mL of the Zn/S/TOP was added in five portions. After this injection the reaction mixture was allowed to cool down to 100°C and was kept at this temperature for one hour. The nanocrystals were purified by precipitation with anhydrous methanol. The precipitate was collected by centrifuging (4000 rpm, 5 min). The precipitate was then washed three times with anhydrous methanol. The nanocrystals were dispersed in doubly distilled chloroform.

Absorption spectra were recorded using a double beam Perkin-Elmer Lambda 16 UV/VIS spectrophotometer. Electron microscopy was performed with a Philips CM300UT-FEG electron microscope operating at 300 kV. Emission and excitation spectra were recorded with a SPEX Fluorolog spectrofluorometer equipped with two double grating 0.22 m SPEX 1680 monochromators and a 450 W Xenon lamp as excitation source. The emission was detected with a cooled Hamamatsu R928 photomultiplier.

To determine the luminescence quantum efficiency of a sample the integrated emission intensity was compared to that of a reference solution with known quantum efficiency, i.e., sulphorhodamine 101 in ethanol with 90% quantum efficiency. If necessary the solutions were diluted in order to have an absorbance that is in the regime where the emission intensity scales linearly with the number of absorbed photons.

For single particle luminescence measurements, small droplets of a strongly diluted QD stock solution were deposited, spread out, and dried on cover glass slides. The final density was approximately 0.1 dot/μm^2. The slides were prepared, mounted and sealed in a flowchamber in nitrogen. Prior to and during the experiments nitrogen or air was flushed through the flowchamber.

Fast spectral imaging of single dots was performed employing the above described spectrograph-CLSM set-up. The 468 nm line of the Ar-Kr CW laser was used for excitation (power $\approx$ 20 kW/cm^2). The photomultiplier of the CLSM (channel 2 in Fig. 19.1) was first used to locate a single QD, applying a 590/60 bandpass filter (center transmission wavelength is at 590 nm, FWHM of the band is 60 nm). Subsequently, the laser beam was parked at the position of single QDs and spectra (range 500–700 nm) were recorded with a 6 ms dwell time using the spectrograph (channel 1 in Fig. 19.1). Four pixels were binned in the dispersion direction, yielding a spectral resolution of about 3 nm at 580 nm. A 480 nm longpass filter (Ba480, Nikon) was inserted in the detection path to block scattered laser light.

The luminescence of QDs was followed until the luminescence had been bleached away. Data are corrected for the background signal from the system and sample as described above [35, 36]. Data are further corrected for the detection sensitivity of the set-up using a calibrated tungsten bandlamp. The spectra are fitted by a Lorentzian peak function [2] for quantification of the integrated intensity, the peak position, and the peak width.

19.3
Results and Discussion

A typical absorption spectrum of a sample of CdSe/ZnS quantum dots is shown in Fig. 19.3. Due to quantum size effects the absorption spectra of the CdSe/ZnS nanoparticles has shifted to the blue compared to the absorption spectrum of bulk CdSe (absorption onset around 730 nm [40]). The structure in the absorption band is a result of the formation of discrete energy levels caused by quantum size effects. The insert in Fig. 19.3 shows a histogram of the size distribution of batch 1

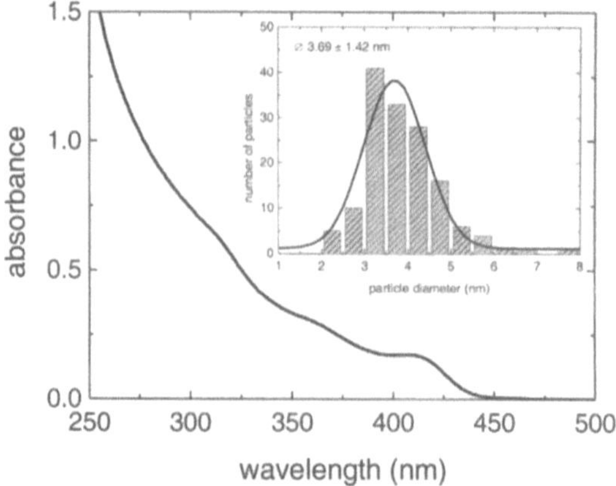

Fig. 19.3. UV-VIS absorption spectrum of CdSe QDs overgrown with 4 monolayers of ZnS (batch 1) as determined from HR-TEM photographs. The insert shows a histogram of the particle size distribution, revealing a ~50% polydispersity

obtained from HR-TEM pictures [39]. It was not possible to distinguish between the CdSe core and the ZnS shell on the HR-TEM pictures. For the histogram the size of 147 quantum dots was determined. The width of the size distribution of batch 1 is about 50%. The average particle diameter of the CdSe/ZnS core shell QDs is about 3.7 nm. For batch 2 similar results were obtained.

By spectral imaging of the CdSe/ZnS QDs with the spectrograph-CLSM set-up individual QDs and their emission spectra could easily be resolved. In Fig. 19.4 the spectra of a few of these QDs of batch 1 are shown, with emission wavelengths of 538, 562, 601, 619, and 640 nm, and FWHMs between 13 and 18 nm.

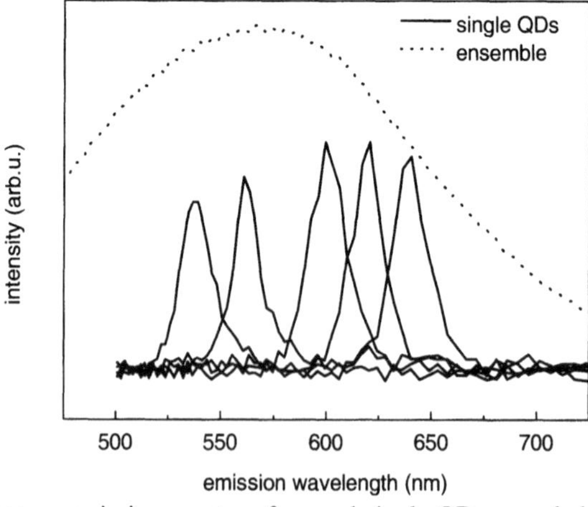

Fig. 19.4. Fluorescence emission spectra of several single QDs recorded with the spectrograph-CLSM set-up, compared to the spectrum of an ensemble of QDs. The emission wavelengths of the single QDs are 538, 562, 601, 619, and 640 nm (excitation wavelength 468 nm)

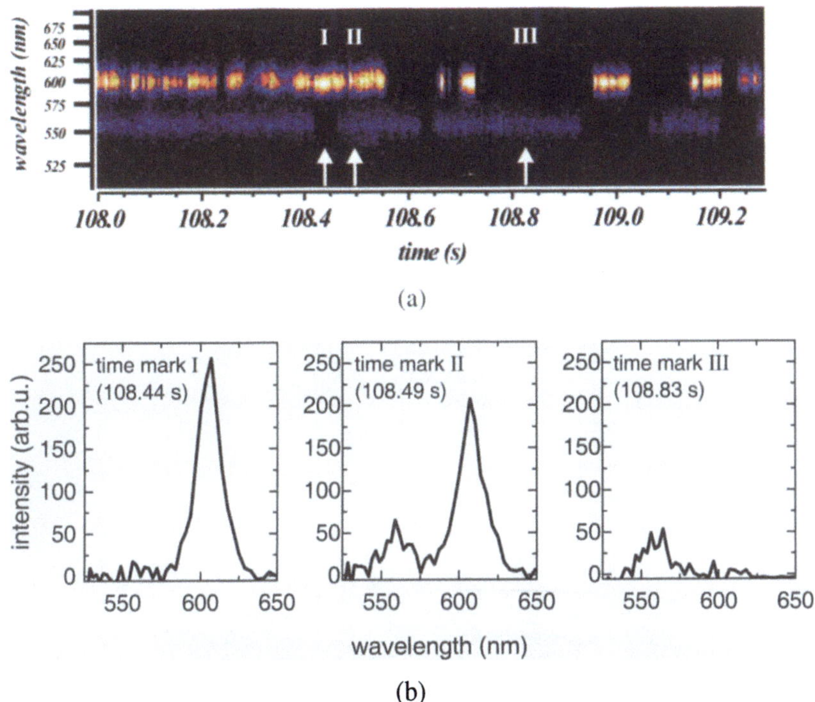

Fig. 19.5. a Spectrally resolved time trace (2.3 s) of two different QDs present in the detection volume of the spectrograph-CLSM set-up. The QDs can be separated due to their different emission wavelength, i.e., 605 and 560 nm, respectively. Complete spectra at times marked with I, II, and III are shown in **b**. Note that the intensity is a stretched false color representation

The corresponding diameters are 3.5, 4.1, 5.3, 6.1, and 7.1 nm, respectively, as calculated using the relation between diameter and emission wavelength [5]. The emission spectrum of the ensemble is shown for comparison. The value of the linewidth is in agreement with the homogeneous broadening at room temperature by phonon dephasing processes [41]. As a result of quantum confinement effects every single quantum dot has its own characteristic emission wavelength, which is determined by the size of the quantum dot. Due to the large size distribution the emission spectrum of an ensemble of CdSe/ZnS quantum dots is extensively broadened compared to the emission spectrum of a single quantum dot.

An example of a spectrally resolved time-trace is shown in Fig. 19.5a. Apparently and, we must stress, exceptionally, two different quantum dots were present in the detection volume of the spectrograph-CLSM set-up, clearly separated due to their different emission wavelengths, i.e., 605 and 560 nm, respectively. Spectra at time marks I, II, and III are shown in Fig. 19.5b. Clearly, the emission intensity of the two QDs differs. In addition, the blinking behavior is apparent, and also, both QDs blink independently.

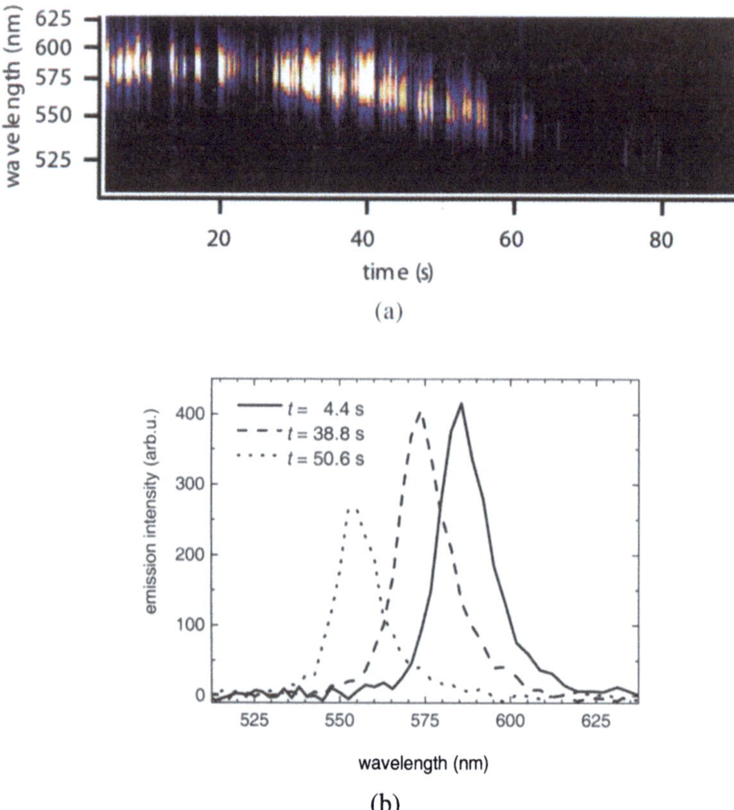

(a)

(b)

Fig. 19.6. Spectrally resolved time trace **a** of a CdSe/ZnS QD of batch 1 in ambient air present in the detection volume of the spectrograph-CLSM set-up (excitation at 468 nm, room temperature). Note that the intensity is a stretched false color representation. Emission spectra at different illumination times are shown in **b**

Luminescence spectra of individual CdSe/ZnS quantum dots were measured for a large number of quantum dots. Clear differences are observed between the time evolutions of the emission spectra of the quantum dots, even for dots of the same batch in the same atmosphere. To obtain reliable information on the influence of the atmosphere and the thickness of the capping layer on the spectral diffusion and on/off behavior, the emission spectra of 41 different dots were followed in time. In Fig. 19.6a a typical spectrally resolved time trace (duration 80 s) is presented of a CdSe/ZnS QD of batch 1 in *air*. Initially this QD emits at 585 nm (for about 20 s), and then the emission wavelength starts to shift to the blue. After a blue shift of about 40 nm the emission of the QD is fully photo-bleached. A blow-up of a smaller part of the time trace (not shown) shows blinking of the QD, see [42]. Figure 19.6b shows spectra of the dot collected (6 ms integration) at different times. It is clear that the two spectra recorded at the later times are blue shifted with respect to the initial wavelength. The FWHM of the peaks is about 15 nm. The result of fitting of the emission maxima is shown in Fig. 19.7a, where the

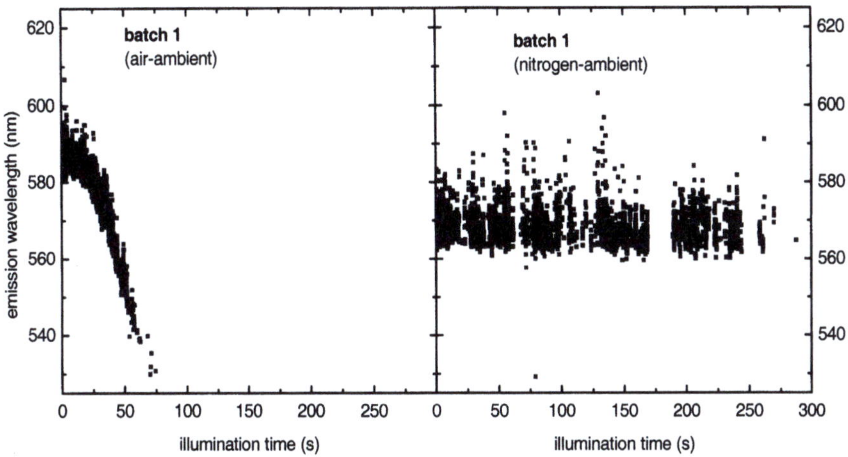

Fig. 19.7. Emission wavelength as a function of time of the CdSe/ZnS QD of Fig. 19.6, i.e., batch 1 in air (**a**) and of a CdSe/ZnS QD of batch 1 in a nitrogen atmosphere (**b**)

emission wavelength of a CdSe/ZnS QD from batch 1 in *air* is depicted as a function of time. Similar results are obtained for other quantum dots of batch 1 and 2 in *air*, with a blue shift of 29 ±10 nm and 29 ±17 nm, respectively, before the luminescence disappears. In some cases an initial fast blue shift is observed, while for most QDs a behavior similar to that depicted in Fig. 19.7a is observed [36]. For QDs of batch 1, the luminescence intensity decreases in time and no luminescence is observed after typically 2.5 minutes. There are large differences between quenching times for individual dots, in agreement with previously reported results [12]. Due to the increased thickness of the passivating ZnS layer for QDs from batch 2, the time scales for the blue shift and the photobleaching are significantly longer (bleaching occurs on average after 3.5 minutes).

The time-evolution of the fitted emission wavelength for a single CdSe/ZnS QD of batch 1 in *nitrogen* atmosphere is depicted in Fig. 19.7b. Clearly, no blue shift is observed. The emission wavelength of this QD varies around an average value of about 570 nm (Fig. 19.7b). A spectral variation of about 10 nm is observed. This is in agreement with previous measurements, which show that over longer time periods random spectral diffusion can result in shifts of up to 10 nm [43,44]. *Contrary* to the situation in *air*, there is only random spectral diffusion in time and no shift to shorter wavelengths is observed. The time periods for which the dots show emission (until photobleaching occurs) are significantly longer for the QDs in *nitrogen* than in *air* (about 10 minutes on average for both batches).

The observed differences between the time evolution of the emission spectra of single quantum dots in *air* and in *nitrogen* provide convincing evidence that the observed blue shift of the emission in *air* is due to photo-oxidation of CdSe [45]. From the blue shift and the well-known relation between bandgap and the diameter of CdSe particles [2, 5–7] it can be calculated that the effective CdSe core diameter decreases from about 5 nm to about 4 nm before the dot is completely bleached. A change in particle diameter of 1 nm corresponds to photo-oxidation of almost two layers of CdSe from the surface [12]. Oxidation of CdSe nanocrystals is well known. Even under ambient conditions (without intense excitation) surface oxidation of CdSe nanoparticles has been reported [46]. The main oxidation product was suggested to be SeO_2. In the case of CdS, illumination of CdS nanocrystals leads to the photo-oxidation products Cd^{2+} and SO_4^{2-} [47–49]. For photo-oxidation of CdSe evidence for the formation of $CdSeO_x$ ($x = 2, 3$) has been found [5, 46, 47]. Here, we did not analyze the photo-oxidation products. For photo-oxidation to take place, oxygen has to diffuse through the passivating ZnS layer that has been grown on the CdSe nanocrystals. The observation that photo-oxidation occurs, indicates that the ZnS layer is not a closed epitaxial layer, but rather a layer with grain boundaries, presumably at places where ZnS islands, which started to grow at different locations on the CdSe nanocrystal, meet. At these boundaries oxygen can diffuse to the CdSe core inside the ZnS shell. For the thicker shell (batch 2) the oxidation rate is reduced due to the slower diffusion of oxygen to the CdSe core through a thicker ZnS shell.

As a result of the photo-oxidation at the CdSe surface, quenching states are expected to be formed at the $CdSe/CdSeO_x$ interface. The formation of surface quenching states causes a decrease of the number of photons emitted. In the single QD emission spectra a gradual decrease in light output is indeed observed as the emission shifts to shorter wavelengths. Finally, the luminescence disappears and the QD has bleached. The occurrence of photo-oxidation for QDs can explain the shorter bleaching times observed for QDs in *air*. Also in a *nitrogen* atmosphere photobleaching occurs, albeit after much longer times. In view of the high laser power (20 kW/cm^2) photobleaching is not unexpected. Few materials are stable against photodegradation under the presently used laser power. The nature of the photo-induced quenching states is not clear. The efficiency of the photo-induced formation of quenching states in *nitrogen* is much lower than for photo-oxidation observed in *air*. Possibly, a high-energy bi-exciton state in a single dot has enough energy to rearrange or break bonds at the CdSe/ZnS interface, which gives rise to non-radiative recombination channels and finally leads to bleaching.

The blinking behavior of the ZnS coated CdSe QDs can also be studied from the spectral time-traces as shown in Fig. 19.6. The emission intensity was used a means to distinguish between an 'on' and an 'off' state of the QD. A threshold set just above the background was determined to discriminate between 'on' and 'off' state. Blinking can be quantified by thus determining the 'off' and 'on' time distributions. Blinking behavior of ZnS coated CdSe QDs has been studied in detail by Kuno et al. [13]. Evidence was provided for an inverse power law behavior of the off-time intervals, i.e., $P(t_{off}) \propto t_{off}^{-(1 + \alpha)}$. A good agreement between the experimentally observed off-time distribution over nine decades in probability density and five decades in time was observed, with $(1 + \alpha) = 1.6 \pm 0.2$. Since the process that leads to the off state is assigned to photo-ionization (by an Auger process) of the QD, the off-time intervals are related to recombination of the ejected charge carrier with the ionized dot. As a result the off-time interval distribution is expected to be excitation intensity independent, which has indeed been observed [13]. The most probable explanation for the inverse power law behavior is a distribution of tunneling rates for recombination processes.

Analysis of the distribution of 'off'-time intervals for all our QDs shows that the probability density of 'off'-time intervals can indeed be described by an inverse power law. For our QDs we find an average value of $(1 + \alpha) = 1.4 \pm 0.2$, close to the value found by Kuno et al. [13] and to earlier work by our group [50]. This value is the average of an analysis for 41 QDs, both in air and in nitrogen atmosphere. Due to the limited statistics at present a significant difference between the values for $(1 + \alpha)$ in air and nitrogen has not been observed. A difference is expected as the shrinking of the CdSe core and the formation of a $CdSeO_x$ at the interface is likely to influence the tunneling rate for recombination of an ejected charge carrier and an ionized dot. Further research is in progress to study differences in the blinking of QDs in air and nitrogen [38].

The potential application of single quantum dots as luminescent labels (e.g., in biological systems) is based on the high stability in combination with a relative narrow emission band and a large 'Stokes shift' [10, 15]. The total number of photons emitted by a single QD until bleaching occurs is an important number. For single dye molecules (e.g.,, rhodamine [51–53]) the highest number of photons emitted before bleaching is around 10^6 at room temperature. Based on an overall efficiency of about 5% for the system (i.e., one out of every twenty emitted photons is counted), the number of photons emitted by a single QD before bleaching occurs, is typically 2×10^7, with numbers exceeding 10^8 for the more robust quantum dots. These numbers are more than an order of magnitude higher than for single dye molecules, which reflects the higher stability of inorganic chromophores such as semiconductor QDs.

Surprisingly, the total number of photons emitted before bleaching occurs is not much less for QDs in air than for QDs in nitrogen. The shorter lifetime of QDs in air is compensated for by a higher initial photon count in air. The initial emission intensity is about two times higher for QDs in air (about 3000 counts in air vs. 1500 counts in nitrogen during the 6 ms integration period for the brightest QDs). At present further experiments are conducted to provide more precise information on the difference in initial light output. A possible explanation for the higher initial light output in air is quenching of QD defect luminescence by oxygen. It is well known that in addition to the fast (nanosecond) exciton emission, also relatively long-lived (microsecond) defect emission can occur in QDs [47, 54]. Due to the long lifetime, the fast photon absorption and emission process is interrupted, until the long-lived excited state has returned (via radiative or non-radiative relaxation) to the ground state. If oxygen can quench the defect luminescence, a higher exciton emission yield is expected by reduction of the time spent in the 'dark state'. For CdSe we have not found studies on the role of oxygen on the efficiency of the defect emission. For other II–VI semiconductors like CdS and ZnS it has been established that oxygen can quench the defect-related emission [47, 54]. If the same occurs for CdSe QDs, this can explain the higher initial light output observed for CdSe QDs in air. The presently applied excitation power (20 kW/cm^2) yields initial intensities of 1000–3000 counts in 6 ms. Assuming a 5% collection efficiency, this corresponds to time intervals of 100–300 ns between emission of photons. In this situation relaxation to a trap state with a microsecond lifetime will reduce the number of photons emitted and a fast return to the ground state by non-radiative relaxation enhances the photon output. A similar situation has been reported for dye molecules where higher fluorescence light yields were measured by quenching of the triplet state luminescence [55].

19.4
Conclusion

In this chapter, we have described a highly sensitive home-built spectrograph that is optimized for use in (scanning) fluorescence microscopy. It has been designed such that it can be easily retrofitted to any commercially available CLSM. At present the time-resolution is limited by the CCD-camera and amounts to 1.2 ms. Fast spectral imaging is possible up to a rate of 800 spectra per second. This is of great importance in biological studies, where one wants to follow protein-protein reaction/interaction kinetics on a sub-second timescale.

As an example of the versatility in use of the spectrograph-CLSM set-up time-resolved luminescence experiments on single CdSe/ZnS core-shell quantum dots were shown. A large blue shift in the emission wavelength of about 30 nm was observed for QDs in ambient air. In contrast, this shift was not observed for QDs in a nitrogen atmosphere. Moreover, the bleaching time and initial emission intensity of single QDs were influenced by the presence of oxygen. The results were explained by photo-induced oxidation of the CdSe crystallites.

Acknowledgements. We thank Cees van der Oord for additional programming, Ageeth Bol and Joost van Lingen for QD synthesis. Further, we gratefully acknowledge the Netherlands Technology Foundation (STW) and the Netherlands Council for Earth and Life Sciences (ALW) of the Netherlands Organization for Scientific Research (NWO) for financial support.

References

1. Alivisatos AP (1996) Perspectives on the physical chemistry of semiconductor nanocrystals. J Phys Chem 100:13226–13239
2. Gaponenko SV (1998) Optical properties of semiconductor nanocrystals. Cambridge University Press, Cambridge, U.K.
3. Efros AlL, Rosen M (2000) The electronic structure of semiconductor nanocrystals. Annu Rev Mater Sci 30:475–521
4. Yoffe AD (2001) Semiconductor quantum dots and related systems: electronic, optical, luminescence and related properties of low dimensional systems. Adv Phys 50:1–208
5. Dabbousi BO, Rodriguez-Viejo J, Mikulec FV, Heine JR, Mattoussi H, Ober R, Jensen KF, Bawendi MG (1997) (CdSe)ZnS core-shell quantum dots: synthesis and characterization of a size series of highly luminescent nanocrystallites. J Phys Chem B 101: 9463–9475

6. Empedocles S, Bawendi MG (1999) Spectroscopy of single CdSe nanocrystallites. Acc Chem Res 32:389–396

7. Mikulec FV, Kuno M, Bennati M, Hall DA, Griffin RG, Bawendi MG (2000) Organometallic synthesis and spectroscopic characterization of manganese-doped CdSe nanocrystals. J Am Chem Soc 122:2532–2540

8. Nanotech, special issue. (2001) Scientific American 285 (3)

9. Hines MA, Guyot-Sionnest P (1996) Synthesis and characterization of strongly luminescent ZnS-capped CdSe nanocrystals. J Phys Chem 100:468–471

10. Bruchez Jr. M, Moronne M, Gin P, Weiss S, Alivisatos AP (1998) Semiconductor nanocrystals as fluorescent biological labels. Science 281:2013–2016

11. Blanton SA, Dehestani A, Lin PC, Guyot-Sionnest P (1994) Photoluminescence of single semiconductor nanocrystallites by two-photon excitation microscopy. Chem Phys Lett 229:317–322

12. Nirmal M, Dabbousi BO, Bawendi M, Macklin JJ, Trautmann JK, Harris TD, Brus LE (1996) Fluorescence intermittency in single cadmium selenide nanocrystals. Nature 383:802–804

13. Kuno M, Fromm DP, Hamann HF, Gallagher A, Nesbitt DJ (2000) Nonexponential "blinking" kinetics of single CdSe quantum dots: a universal power law behavior. J Chem Phys 112:3117–3120

14. Kuno M, Fromm DP, Hamann HF, Gallagher A, Nesbitt DJ (2001) "On"/"off" fluorescence intermittency of single semiconductor quantum dots. J Chem Phys 115: 1028–1040

15. Chan WCW, Nie S (1998) Quantum dot bioconjugates for ultrasensitive nonisotopic detection. Science 281:2016–2018

16. Dahan M, Laurence T, Pinaud F, Chemla DS, Alivisatos AP, Sauer M, Weiss S (2001) Time-gated biological imaging by use of colloidal quantum dots. Opt Lett 26:825–827

17. Gerion D, Pinaud F, Williams SC, Parak WJ, Zanchet D, Weiss S, Alivisatos AP (2001) Synthesis and properties of biocompatible water-soluble silica-coated CdSe/ZnS semiconductor quantum dots. J Phys Chem B 105:8861–8871

18. Efros AlL, Rosen M (1997) Random telegraph signal in the photoluminescence intensity of a single quantum dot. Phys Rev Lett 78:1110–1113

19. Banin U, Bruchez M, Alivisatos AP, Ha T, Weiss S, Chemla DS (1999) Evidence for a thermal contribution to emission intermittency in single CdSe/CdS core/shell nanocrystals. J Chem Phys 110:1195–1201

20. Neuhauser RG, Shimizu KT, Woo WK, Empedocles SA, Bawendi MG (2000) Correlation between fluorescence intermittency and spectral diffusion in single semiconductor quantum dots. Phys Rev Lett 85:3301–3304

21. Empedocles SA, Bawendi MG (1997) Quantum-confined stark effect in single CdSe nanocrystallite quantum dots. Science 278:2114–2117

22. Cordero SR, Carson PJ, Estabrook RA, Strouse GF, Buratto SK (2000) Photo-activated luminescence of CdSe quantum dot monolayers. J Phys Chem B 104:12137–12142

23. Tsien RY, Poenie M (1986) Fluorescence ratio imaging: a new window into intracellular ionic signaling. Trends Biochem Sci 11:450–455

24. Matyus L (1992) New trends in photobiology. J Photochem Photobiol B:Biol 12:323–337

25. Willard DM, Carillo LL, Jung J, Van Orden A (2001) CdSe-ZnS quantum dots as resonance energy transfer donors in a model protein-protein binding essay. Nano Lett 1:469–474

26. Gordon GW, Berry G, Liang XH, Levine B, Herman B (1998) Quantitative fluorescence resonance energy transfer measurements using fluorescence microscopy. Biophys J 74:2702–2713
27. Feofanov A, Sharonov S, Valisa P, Da Silva E, Nabiev I, Manfait M (1995) A new confocal stigmatic spectrometer for micro-Raman and microfluorescence spectral imaging analysis: design and application. Rev Sci Instrum 66:3146–3158
28. Millot J-M, Pingret L, Angiboust J-F, Bonhomme A, Pinon J-M, Manfait M (1995) Quantitative determination of free calcium in subcellular compartments, as probed by Indo-1 and confocal microspectrofluorometry. Cell Calc 17:354–366
29. Martínez-Zaguilán R, Gurulé MW, Lynch M (1996) Simultaneous measurement of intracellular pH and Ca^{2+} in insulin-secreting cells by spectral imaging microscopy. Am J Physiol 270:C1438–C1446
30. Vereb G, Jares-Erijman E, Selvin PR, Jovin TM (1998) Temporally and spectrally resolved imaging microscopy of lanthanide chelates. Biophys J 74:2210–2222
31. Favard C, Valisa P, Egret-Charlier M, Sharanov S, Herben C, Manfait M, Da Silva E, Vigny P (1999) A new UV-visible confocal laser scanning microspectrofluorometer designed for spectral cellular imaging. Biospectrosc 5:101–115
32. Rigacci L, Alterini R, Bernabei PA, Ferrini PR, Agati G, Fusi F, Monici M (2000) Multispectral imaging autofluorescence microscopy for the analysis of lymph-node tissues. Photochem Photobiol 71:737–742
33. Lacoste TD, Michalet X, Pinaud F, Chemla DS, Alivisatos AP, Weiss S (2000) Ultra-high-resolution multicolor colocalization of single fluorescent probes. Proc Natl Acad Sci USA 97:9461–9466
34. Tsurui H, Nishimura H, Hattori S, Hirose S, Okumura K, Shirai T (2000) Seven-color fluorescence imaging of tissue sample based on fourier spectroscopy and singular value decomposition. J Histochem Cytochem 48:653–662
35. Frederix PLTM, Asselbergs MAH, Van Sark WGJHM, Van den Heuvel DJ, Hamelink W, De Beer EL, Gerritsen HC (2001) High sensitivity spectrograph for use in fluorescence microscopy. Appl Spec 55:1005–1012
36. Frederix PLTM (2001) Spectral analysis in microscopy: A study of FRET and single quantum dot luminescence. Ph.D. Thesis, Utrecht University
37. Born M, Wolf E (1999) Principles of optics. Cambridge University Press, Cambridge, U.K.
38. Van Sark WGJHM, Frederix PLTM, Bol AA, Van den Heuvel DJ, Gerritsen HC, Meijerink A (2002) Blinking, blueing, and bleaching of single CdSe/ZnS quantum dots. Phys Chem Chem Phys (submitted)
39. Bol AA (2001) Luminescence of doped semiconductor quantum dots. Ph.D. Thesis, Utrecht University
40. Yen WM, Shionoya S (1999) Phosphor handbook. CRC Press, Boca Raton, FL
41. Li X-Q, Arakawa Y (1999) Optical linewidths in an individual quantum dot. Phys Rev B 60:1915–1920
42. Van Sark WGJHM, Frederix PLTM, Van den Heuvel DJ, Gerritsen HC, Bol AA, Van Lingen JNJ, De Mello Donegá C, Meijerink A (2001) Photo-oxidation and photobleaching of single CdSe/ZnS quantum dots probed by room-temperature time-resolved spectroscopy. J Phys Chem B 105:8281–8284
43. Blanton SA, Hines MA, Guyot-Sionnest P (1996) Photoluminescence wandering in single CdSe nanocrystals. Appl Phys Lett 69:3905–3907

44. Empedocles SA, Neuhauser R, Shimizu K, Bawendi MG (1999) Photoluminescence from single semiconductor nanostructures. Adv Mater 11:1243–1256
45. Van Sark WGJHM, Frederix PLTM, Van den Heuvel DJ, Bol AA, Van Lingen JNJ, De Mello Donegá C, Gerritsen HC, Meijerink A (2001) Time-resolved fluorescence spectroscopy study on the photo-physical behaviour of quantum dots. J Fluorescence 12:69–76
46. Bowen Katari JE, Colvin VL, Alivisatos AP (1994) X-ray photoelectron spectroscopy of CdSe nanocrystals with applications to studies of the nanocrystal surface. J Phys Chem 98:4109–4117
47. Henglein A (1988) Mechanism of reactions on colloidal microelectrodes and size quantification effects. Top Curr Chem 143:113–180
48. Dunstan DE, Hagfeldt A, Almgren M, Siegbahn HOG, Mukhtar E (1990) Importance of surface reactions in the photochemistry of ZnS colloids. J Phys Chem 94:6797–6804
49. Spanhel L, Haase M, Weller H, Henglein A (1987) Photochemistry of colloidal semi-conductors. 20. Surface modification and stability of strong luminescing CdS particles. J Am Chem Soc 109:5649–5655
50. Van Sark WGJHM, Frederix PLTM, Van den Heuvel DJ, Asselbergs MAH, Senf I, Gerritsen HC (2000) Fast imaging of single molecules and nanoparticles by wide-field microscopy and spectrally resolved confocal microscopy. Single Mol 1:291–298
51. Rosenthal I (1978) Photochemical properties of rhodamine 6G in solution. Opt Comm 24:164–166
52. Huston AL, Reimann CT (1991) Photochemical bleaching of adsorbed rhodamine 6G as a probe of binding geometries on a fused silica surface. Chem Phys 149:401–407
53. Schmidt Th, Schütz GJ, Baumgartner W, Gruber HJ, Schindler H (1996) Imaging of single molecule diffusion. Proc Natl Acad Sci USA 93:2926–2929
54. Chestnoy N, Harris TD, Hull R, Brus LE (1986) Luminescence and photophysics of CdS semiconductor clusters: the nature of the emitting electronic state. J Phys Chem 90:3393–3399
55. Song L, Varma CAGO, Verhoeven JW, Tanke HJ (1996) Influence of the triplet ex-cited state on the photobleaching kinetics of fluorescein in microscopy. Biophys J 70:2959–2968

Imaging of Oxidative Stress in Plant Cells by Quantitative Fluorescence Microscopy and Spectroscopy

J. W. BORST, M. A. USKOVA, N. V. VISSER, AND A. J. W. G. VISSER

In this chapter the application of two different fluorescent probes is described that are able to interrogate lipid peroxidation in whole plant cells and hydrogen peroxide production in plant cell suspensions.

20.1
Introduction

The production of reactive oxygen species (ROS) plays an important role in the plant defense mechanism and is one of the reactions of plants to pathogen infection or elicitor production (elicitors are pathogen-produced signaling compounds) [1, 2]. The actual involvement of oxygen radicals in plant diseases has been difficult to assess, since an array of protective enzymes as well as antioxidants are involved. In addition, lipid peroxidation may also be initiated by production of ROS in response to pathogen invasion [3–7]. Generally, peroxidation of unsaturated lipids plays an important role in cell membrane properties, signal transduction pathways, apoptosis and the deterioration of foods.

Two phases of responses to pathogens exist in plant tissues and suspension-cultured cells. First, there is a fast response (the so-called oxidative burst) occurring within minutes after elicitor treatment, which is followed after a few hours by a second phase of ROS production and is accompanied by specific gene activation. Several enzymatic systems have been proposed to generate ROS. [1, 8–10] (see Fig. 20.1 for a schematic view). One of them involves the activation of a large protein: NADPH oxidase. This enzyme facilitates reduction of oxygen to a superoxide radical through the oxidation of cytosolic NADPH. It consists of a membrane-bound flavocytochrome b and a larger glycoprotein subunit and requires a number of cytosolic factors for enzymatic activity. Because the oxidant O_2^- (superoxide radical) is poorly diffusible across membranes, it is most likely released at the surface of the membrane, where O_2^- is spontaneously or enzymatically (by superoxide dismutase) dismutated into H_2O_2. Hydrogen peroxide participates in many signal transduction pathways and serves as a substrate for extracellular peroxidases located in the apoplastic region [10]. In most plant systems H_2O_2 accumulation can be detected. Another oxidant produced includes the highly toxic OH (hydroxyl radical). The oxygen compounds may have direct antimicrobial activity and have a role in other defense mechanisms including lipid peroxidation, oxida-

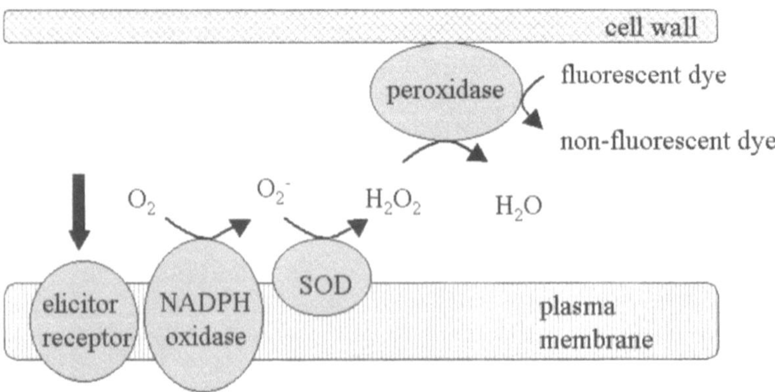

Fig. 20.1. Schematic view of the enzymes involved in oxidative burst

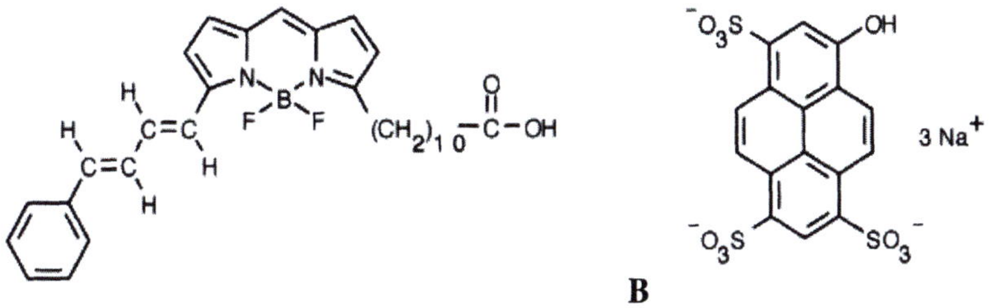

Fig. 20.2. **A** Chemical structures of 4,4-difluoro-5-(4-phenyl-1,3-butadienyl)-4-bora-3a,4a-
-diaza-*s*-indacene-3-undecanoic acid (BP-C11), and **B** 8-hydroxypyrene-1,3,6-trisulfonic
acid, trisodium salt (pyranine)

tive cross-linking of cell walls and hypersensitive response [9]. The oxidative
burst includes the activation of several other enzymes such as phospholipases and
lipoxygenases and is governed by the phosphorylation–dephosphorylation poise
because the phosphatase 2A inhibitor cantharidin induces ROS production in several plant species while the protein-serine kinase inhibitor K252A suppresses it [1,
11, 12].

In this chapter attention is focused on two aspects of ROS detection, namely
lipid oxidation and hydrogen peroxide production. Fluorescent probes were used
that report on oxidation and can look at the effect of specific inhibitors allowing
interrogating the mechanism of oxidative burst in plant cells. In order to track lipid
oxidation a new fluorescent oxidation probe, abbreviated as BP-C11 (Fig. 20.2A),
was used [13–15]. BP-C11 represents a long-chain fatty acid with a fluorophore
composed of a boron dipyromethene difluoride extended with a phenyl moiety by
a diene bond. The probe is sensitive to the presence of ROS, which are likely to
oxidise the diene bond. The BP-C11 probe has been used previously for oxidation
assays in model systems of membrane bilayers [16] and animal cells (rat fibroblasts) [14, 15]. In Fig. 20.3 an example of the oxidation of BP-C11 in DOPC
vesicles is shown. It can be clearly seen that oxidation with cumene hydroperoxide
and hemin as oxidant–catalyst pair causes a shift from "red" to "green" fluorescence. The beneficial property of BP-C11 upon oxidation is allowing 2-channel
tracking fluorescence imaging microscopy in live cells [14]. It was further found
that the oxidation of BP-C11 proceeds at a comparable rate as arachidonic acid
[15]. Utilizing all the advantages of the BP-C11 probe we have developed a fluorescent assay for on-line localization and quantification of lipid oxidation in living
tobacco plant cells. A second line of approach is followed by use of the fluorescent peroxidase substrate pyranine (Fig. 20.2B) for studying the effect of elicitors,
inhibitors and ROS scavengers on H_2O_2 production in suspensions of tobacco
cells. Upon oxidation the probe becomes nonfluorescent. Pyranine has been
widely used as a pH-sensitive probe [17], but has also found applications as a sensor for peroxidases [10]. Because of its high charge pyranine will not pass the
plasma membrane and is confined to the periplasmic cellular space.

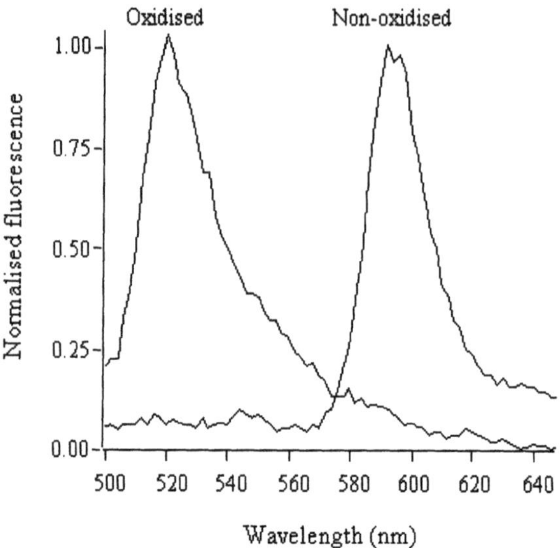

Fig. 20.3. Fluorescence spectra of BP-C11 in DOPC vesicles. Oxidation was induced by adding 150 μM cumene hydroperoxide (oxidant) and 0.5 μM hemin (catalyst). Non oxidized BP-C11 has a fluorescence maximum at 591 nm, while the maximum of the oxidized BP-C11 spectrum is at 520 nm

20.2
Experimental Procedures

20.2.1
Tobacco BY-2 Cells

The tobacco cell line *Nicotiana tabacum* cv. Bright Yellow-2 (abbreviated as BY--2) possesses unique features such as very large growth rates and high homogeneity. The cells lack chlorophyll yielding therefore relatively low background fluorescence [18].

20.2.2
Preparation of Cell Suspension

BY-2 suspension culture was grown in a liquid medium containing Murashige and Skoog salts (without vitamins), supplemented with 30 g sucrose, 0.255 g KH_2PO_4, 0.1 g myo-inositol, 1 mg thiamine-HCl and 0.2 mg 2,4-dichlorophenoxyacetic acid per liter (pH 5.8) [19]. The cell culture was maintained by subculturing 1 mL of cells into 50 mL of fresh medium weekly and kept on a rotary shaker (130 rpm) in the dark at 22°C. For experiments, 3–4 days-old cells were collected by centrifugation for 5 minutes at 130 g and resuspended in fresh medium to a concentration of 0.01 packed cell volume (pcv). The cells are in the log-phase at 2–5 days after subculturing.

20.2.3
Protoplast Isolation

For protoplast isolation 1 mL of BY-2 cells was transferred into 50 mL of fresh medium and grown for 2–3 days before use. The cells were collected by centrifugation to obtain a 2.5 pcv value and immersed in 10 mL of enzymatic solution (1% cellulase, 0.1% pectolyase, 0.4 M mannitol, pH 5.8), transferred to a 9-cm petri-dish and incubated for 2–3 hours at 25°C on a rotary shaker with low speed in the dark. Prior to further purification, protoplasts were checked with light microscopy to ensure that all cell walls had been removed. The protoplasts were purified over 63 μm nylon filter and transferred to 50 mL plastic tube, collected by centrifugation at 130 g, washed three times with 0.4 M mannitol solution (pH 5.8) and resuspended in protoplast culture medium (BY-2 medium with addition of 0.4 M mannitol). The BY-2 protoplasts were used for experiments on the same day of isolation.

20.2.4
Labeling Cells with BP-C11

BY-2 cells of 3-days-old were diluted with fresh growth medium to 0.01 pcv and incubated with 1 μM BP-C11, taken from an ethanol stock solution, for 1 hour in a petri-dish at room temperature on a shaker at low speed in the dark. Cells were washed twice with medium prior to use.

20.2.5
Confocal Microscopy

The cells were examined under a LSM 510 confocal laser scanning microscope (Carl Zeiss). Samples (100–300 μL) were placed into the 8-well chamber holder of the microscope and treated with the desired agents. The fluorescence of BP-C11 was acquired using 488 and 543 nm laser-line excitation and emission was collected in two channels: 505–550 nm for the oxidised form (channel 2) and 560–615 nm for the nonoxidized form (channel 1) of BP-C11. Integration of the fluorescence intensity of the desired image area was performed using LSM 510 software.

20.2.6
Steady-state Fluorescence

Steady-state fluorescence spectra or intensities were obtained using a Spex Fluorolog 3.2.2 spectrofluorometer. The assay for H_2O_2 production was performed in BY-2 cells using pyranine. Basically, 1 µL from a stock solution of 1 mg/mL pyranine in water was added directly to 1 mL of a prepared cell suspension in a quartz cuvette having a magnetic stirring bar approximately 1 minute prior to elicitor addition. The cells were continuously stirred at low speed to avoid mechanical disruption and cell sedimentation. The excitation wavelength was 405 nm and the emission was monitored at 512 nm.

20.3
Results and Discussion

20.3.1
Measurements and Imaging of Oxidative Stress in Tobacco BY-2 Cells

The susceptibility of the probe towards reactive oxygen is comparable to that of endogenous fatty acids. Various radicals, such as for example, oxyl, peroxyl, or hydroxyl radicals, can initiate oxidation of BP-C11 but not superoxide, nitric oxide or hydroperoxide [15]. Because of its lipophilic character BP-C11 incorporates easily into plasma and organelle membranes. After five minutes of incubation, only the fluorescence of the nonoxidized fraction of the probe in the plasma membrane is visible (data not shown). After 1 hour most of the probe was taken up by the cells staining also organellar membranes. It is remarkable, that the incorporated probe did not show toxicity towards the cells and was resistant to self-oxidation, degradation by cellular enzymes or bleaching during the course of an experiment requiring typically 1 hour. An incubation time of 60 minutes was chosen for further experiments.

In order to investigate whether the incorporated probe into BY-2 cells is susceptible to oxidation, we exposed the cells to cumene hydroperoxide and hemin as an oxidant-catalyst pair (Fig. 20.4). Within one hour a significant fraction of the probe is oxidised (Fig. 20.4B). The degree of oxidation depends, among others, on the concentrations of oxidative agents used and the age of the cells (data not shown). Similar experiments were done with BY-2 protoplasts. The absence of the cell wall did not greatly affect the lipid oxidation rate, which led us to conclude that the cell wall was permeable to the chosen oxidants. The experiments were quantified in terms of average fluorescence intensity changes in a selected region of interest in the image. Examples of selected regions of interest (ROI) are shown in Fig. 20.4.

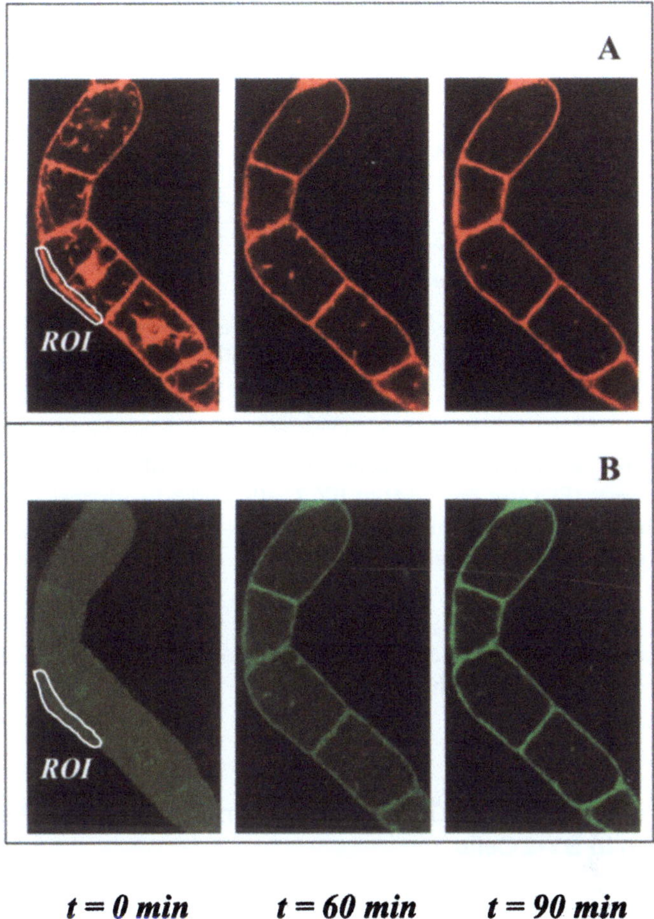

Fig. 20.4. Time course of oxidation of BP-C11 in BY-2 cells induced by cumene hydroperoxide (1.5 mM) and hemin (5 µM) by confocal laser scanning microscopy. **A** represents channel 1 (nonoxidized), while **B** represents channel 2 (oxidized). A region of interest (ROI) has been indicated

In Fig. 20.5 the average fluorescence intensities in membrane ROI in BY-2 cells and protoplasts are plotted against time. During the exposure to oxidative agents, the average fluorescence of the nonoxidized fraction of the probe decreases, while the fluorescence of the oxidized fraction becomes more intense. Within 1 hour approximately 20% of the probe underwent oxidation. The observed lipid oxidation rate in BY-2 cells is much smaller than reported previously in animal cells, in which about 70% oxidation of probe was detected within 30 minutes using a 20-fold lower concentration of oxidative agents [14]. This difference is probably due to the strong defense response of higher plants [1]. Another

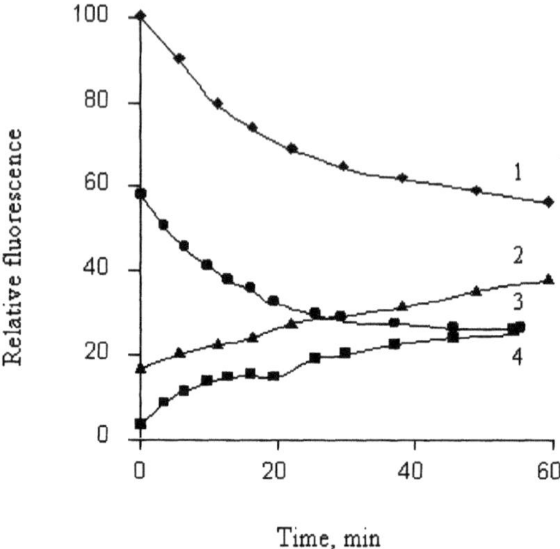

Fig. 20.5. Time traces of the average fluorescence intensity of BP-C11 probe in selected regions of interests (ROI, see Fig. 20.4) of BY-2 cells (curve 1 – channel 1, nonoxidized; curve 2 – channel 2, oxidized) and protoplasts (curve 3 – channel 1, nonoxidized; curve 4 – channel 2, oxidized) during incubation with 1.5 mM cumene hydroperoxide and 5 μM hemin

factor to consider is that the composition and morphology of the membranes in plant cells are different from those in animal cells.

20.3.2
Effect of Inhibitors and ROS Scavengers on the Hydrogen Peroxide Production in BY-2 Cells

Oxidative burst is accompanied by the activation of several enzymes such as phospholipases and lipoxygenases. The phosphorylation–dephosphorylation balance also governs oxidative burst, because the protein phosphatase 2A inhibitor cantharidin induces ROS production in several plant species, while the protein-serine kinase inhibitor K252A suppresses it [1, 11, 12]. The production of ROS can be measured by hydrogen peroxide production in cell suspensions induced by the elicitor cantharidin [20]. Activation of the oxidative burst also involves perturbations of cytosolic calcium levels [1, 21, 22]. Calcium serves as an activator of many enzymes, including phospholipase A_2, and is an important second messenger in signal transduction. The nature of the enzymatic system responsible for generation of H_2O_2 in the tobacco BY-2 cell suspension was studied using pyranine, which can act as a fluorescent peroxidase substrate becoming nonfluorescent after oxidation. The probe is located in the extracellular space and serves as a probe for cell wall peroxidases. The treatment of the BY-2 cell suspension culture with cantharidin leads to an increase of hydrogen peroxide production (see Fig. 20.6A). An oxidative burst occurred within 30 minutes following a brief lag pe-

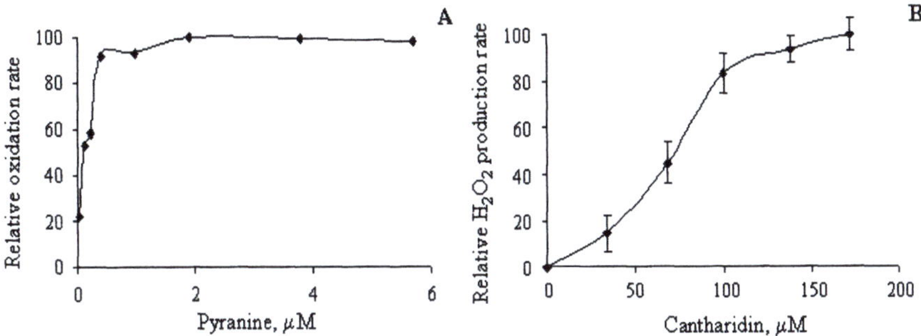

Fig. 6. Effect of pyranine concentration (**A**) and elicitor concentration (**B**) on the relative oxidation rate of the fluorescent probe

riod. The rate of H_2O_2 production depends on the concentration of elicitor (see Fig. 20.6B) and the growth stage of the cells (logarithmic phase). For inhibition experiments a concentration of 100 µM cantharidin and 2 µM pyranine was used. The inhibitor effect was calculated based on oxidative quenching of the pyranine fluorescence.

A few enzymatic systems have been proposed to generate active oxygen species during the elicited oxidative burst [1, 8–10]. One of them involves activation of NADPH oxidase [1, 23], but also superoxide dismutase and cell wall peroxidases play a role (see Fig. 20.1). In our studies we tested a number of specific inhibitors of these enzymes for their ability to inhibit production of hydrogen peroxide in BY-2 cells induced by treatment with cantharidin. The molecular compounds, their common effects and results of inhibition studies are summarized in Table 20.1. Activators of phospholipase A_2, calcium and mastoparan, did not stimulate hydrogen peroxide production, although mastoparan may induce an oxidative burst involving other oxidative species. We tested α-tocopherol (vitamin E) and ascorbic acid (vitamin C) in a near-physiological concentration range for their influence on the elicited response. The effect of α-tocopherol was not strong (in any tested concentration), although, because of its poor solubility, there is some doubt as to the actual concentration coming into contact with cells, whilst ascorbic acid completely abrogated the response at 10 mM. The temptation is to speculate that most reactions involved in the elicited cascade do not occur in the membrane region (vitamin E locates in hydrophobic areas), but in the apoplastic water phase where vitamin C can scavenge the produced oxygen species. Several inhibitors of NADPH oxidase and peroxidase result in almost complete inhibition and indicate participation of these enzymes in a cascade reaction. Removal of the cell walls diminishes the oxidation rate consistent with peroxidases being attached to the cell walls. The results obtained support the mechanism of cantharidin-elicited hydrogen peroxide production that has been proposed for other plant cells.

Table 20.1. Effect of inhibitors and ROS scavengers on H_2O_2 production by BY-2 cell suspension induced by cantharidin

Addition to BY2 suspension culture	Common effects	Pyranine oxidation rate, % of 100 μM cantharidin
None or 2%(*v/v*) DMSO		0
1 mM Calcium	Stimulates phospholipase A_2 activity	0
200 μM Mastoparan	Stimulates phospholipase A_2 activity	0
100 μM Cantharidin (A)	Inhibitor of protein phosphatase 2A	100 ± 2
A + catalase	H_2O_2 scavenger	19 ± 4
A + superoxide dismutase	Catalyst of conversion of O_2^- into H_2O_2	91 ± 1
A + peroxidase		146 ± 2
A + 40 μM α-tocopherol	Lipid-soluble antioxidant, reacts directly with superoxide, singlet oxygen	80 ± 16
A + 10 mM ascorbic acid	water-soluble antioxidant	2 ± 0.2
A + 25 μM diphenylene iodonium	Flavin analog inhibitor of plant and mammalian NADPH oxidase	0.3 ± 0.4
A + 0.1 mM *p*-hydroxymercurybenzoic acid	Inhibitor of mammalian NADPH oxidase	0.7 ± 0.7
A + 100 μM iodoacetate	Membrane-permeative thiol reagent	5 ± 1
A + 100 μM *N*-pyrene maleimide	Membrane-permeative thiol reagent	0
A + 2 mM salicylhydroxamic acid	Plant peroxidase inhibitor	5 ± 2
A + 10 mM imidazole	Inhibitor of plant NADPH oxidase	15 ± 3
Protoplasts	Removed cell walls and attached peroxidases	11 ± 4

20.4
Conclusions

The lipophilic probe BP-C11 has the advantageous property that the fluorescence changes color upon oxidation and is therefore very appropriate to be used in live cells as a probe for lipid oxidation as demonstrated here. Pyranine can be easily oxidized by hydrogen peroxide. This probe can then be conveniently applied in suspension cells to monitor hydrogen peroxide levels for screening activation or inhibition of the pathways leading to H_2O_2 formation. In the future pyranine can be viewed with two-photon confocal laser scanning microscopy opening the way for *in situ* detection.

Acknowledgements. This research was supported by EU-FAIR project 97-3228. Marsha Uskova was supported by the Netherlands Organisation of Scientific Research (NWO) grant 047-007-005 providing scientific cooperation between The Netherlands and Russian Federation.

References

1. Lamb C, Dixon RA (1997) The oxidative burst in plant disease resistance. Annu Rev Plant Physiol Plant Mol Biol 48:251–275
2. Ebel J, Mithoefer A (1998) Early events in the elicitation of plant defence. Planta 206: 335–348
3. Keppler LD, Novacky A (1987) The initiation of membrane lipid peroxidation during bacteria-induced hypersensitive reaction. Physiol Mol Plant Pathology 30:233–245
4. Keppler LD, Baker CJ (1989) O_2^- initiated lipid peroxidation in bacteria-induced hypersensitive reaction in tobacco cell suspensions. Phytopathology 79:555–562
5. Rusterucci C, Stallaert V, Milat M-L, Pugin A, Ricci P, Blein J-P (1996) Relationship between active oxygen species, lipid peroxidation, necrosis, and phytoalexin production induced by elicitins in *Nicotiana*. Plant Physiol 111:885–891
6. Rogers KR, Albert F, Anderson AJ (1988) Lipid peroxidation is a consequence of elicitor activity. Plant Physiol 86:547–553
7. Adam A, Farkas T, Somlyai G, Hevesi M, Kiraly Z (1989) Consequence of O_2^- generation during a bacterially induced hypersensitive reaction in tobacco: deterioration of membrane lipids. Physiol Mol Plant Pathology 34:13–26
8. Auh C-K, Murphy TM (1995) Plasma membrane redox enzyme is involved in the synthesis of O_2^- and H_2O_2 by *Phytophthora* elicitor-stimulated rose cells. Plant Physiol 107:1241–1247
9. Mithofer A, Daxberger A, Fromhold-Threu D, Ebel J (1997) Involvement of an NAD(P)H oxidase in the elicitor-inducible oxidative burst of soybean. Phytochemistry 45:1101–1107
10. Apostol I, Heinstein PF, Low PS (1989) Rapid stimulation of an oxidative burst during elicitation of cultured plant cells. Plant Physiol 90:109–116
11. Levine A, Tenhaken R, Dixon R, Lamb C (1994) H_2O_2 from oxidative burst orchestrates the plant hypersensitive disease resistance response. Cell 79:583–593
12. Li Y-M, Casida JE (1992) Cantharidin-binding protein: Identification as protein phosphatase 2A. Proc Natl Acad Sci USA 89:11867–11870
13. Haugland RP (1996) Handbook of fluorescent probes and research chemicals, 6th ed., Eugene, OR.
14. Pap EHW, Drummen GPC, Winter VJ, Kooij TWA, Rijken P, Wirtz KWA, Op den Kamp JAF, Hage WJ, Post JA (1999) Ratio-fluorescence microscopy of lipid oxidation in living cells using C11-BODIPY[581/591] FEBS Lett 453:278–282
15. Pap EHW Drummen, GPC, Post JA, Rijken PJ, Wirtz KWA (2000) Fluorescent fatty acid to monitor reactive oxygen species in single cells. Methods Enzymol 319:603–612

16. Borst JW, Visser NV, Kouptsova O, Visser AJWG (2000) Oxidation of unsaturated phospholipids in membrane bilayer mixtures is accompanied by membrane fluidity changes. Biochim Biophys Acta 1487:61–73

17. Oja V, Savchenko G, Jakob B, Heber U (1999) pH and buffer capacities of apoplastic and cytoplasmic cell compartments in leaves. Planta 209:239–249

18. Nagata T, Nemoto Y, Hasezawa S (1992) Tobacco BY-2 cell line as the "HeLa" cell in the cell biology of higher plants. Int Rev Cytol 132:1–30

19. Linsmaier E.M, Skoog F (1965) Organic growth factor requirements of tobacco tissue cultures. Physiol. Plant 18:100–127.

20. Van Gestelen P, Ledeganck P, Wynant I, Caubergs R J, Asard H (1998) The cantharidin-induced oxidative burst in tobacco BY-2 cell suspension cultures. Protoplasma 205,83–92

21. Takahashi K, Isobe M, Muto S (1998) Mastoparan induces an increase in cytosolic calcium ion concentration and subsequent activation of protein kinases in tobacco suspension culture cells. Biochim Biophys Acta 1401:339–346

22. Jones DL, Kochian LV, Gilroy S (1998) Aluminium induces a decrease in cytosolic calcium concentration in BY-2 tobacco cell cultures. Plant Physiol 116:81–89

23. Dwyer SC, Legendre L, Low PS, Leto TL (1996) Plant and human neutrophil oxidative burst complexes contain immunologically related proteins. Biochim Biophys Acta 1289:231–237

The Biomedical Use of Rescaling Procedures in Optical Biopsy and Optical Molecular Imaging

O. MINET, J. BEUTHAN, K. LICHA, AND C. MAHNKE

Using fluorescence imaging in biomedicine, the analysis of laser-induced autofluorescences shows comparable strong intensity distortion for endogenous NADH in the UV and synthesized markers in the NIR range due to tissue optics. Rescaling, taking into account biochemical and biooptical methods, results in the chromophore profile in the observed tissue region. One of the outstanding advantages of near infrared so-called optical molecular imaging (OMI) is the high-detection sensitivity and the capability to brightly modulate the fluorescence injectable markers, e.g., by specific molecular interaction with tumor-specific enzymes. For the *in vivo* tests of the dyes an experimental NIR imager was used. NIR fluorescence of the entire body of small animals can be imaged. For first experiments an undifferentiated superficial tumor of mouse thigh was used. Corrections due to tissue optics must take care of a stronger scattering of the light in the NIR range in comparison with the UV fluorescence, like in optical biopsy. For example, the diameter of the fluorescent volume is apparently larger for the same reason. Therefore, the established rescaling from the UV adapted to the NIR range is important for the interpretation of fluorescence pictures in biomedicine.

21.1
Introduction

A central issue of oncologic therapy is the correct evaluation of malign tumor extension and local metastatic spread. Based on the macroscopic evaluation of the tumor extension the surgeon will resect the tumor with a certain safety margin in the healthy tissue. That safety margin often lets the surgery become a mutilating intervention, which refers, e.g., to larynx or brain tumors. The fluorescence technology may be a valuable tool to optimize the safety margin, i.e., to make it as small as possible and as large as necessary. The possibilities and restrictions of this method are as follows:

Potentials
- Minimally invasive
- Early and *in situ* diagnosis
- Surface diagnosis
- Evaluation of radicalism / dignity
- Cost-favorable
- Simultaneous PDT is possible

Restrictions
- Low specificity
- Work in progress for quantitative statements
- Severe individual variations
- Metabolism-dependent
- Low penetration depths in the UV range

From a physical point of view, the quantitative evaluation of laser-induced fluorescence in strongly scattering media remains a challenge. The fluorophores are embedded in this medium and the fluorescence intensity values are changing with different optical properties of the medium. Since the first investigations of fluorescence in biomedicine [1–3] the quantitative interpretation of the signal becomes more and more important [4–7] and the possibilities of the diagnostics in medicine grow rapidly with such tools.

21.2
Method of Rescaling

Biomedical applications in fluorescence diagnostics mainly deal with the excitation and detection of fluorescent light at macroscopic level. Laser light in the infrared spectral range of low intensity excites fluorescence in tissue still at a depth of some millimeters. On its way to the detector, the fluorescent light passes through the tissue again, but now it has another, larger wavelength than the excitation radiation and is, therefore, scattered and absorbed to different extent than the incident light. It is obvious that the intensity of the fluorescence radiation is chan-

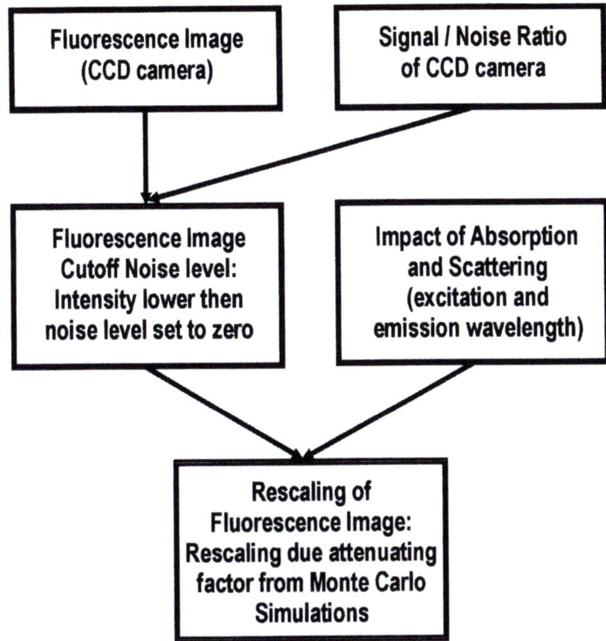

Fig. 21.1. Scheme of the rescaling procedure

ged by these optical processes. This influence can be described and thus optimized, e.g., by rescaling the measured intensity.

The rescaling procedure consists of two steps (Fig. 21.1). First, the signal/noise ratio of the CCD camera is set off in the black and white bitmap mainly by cutting off those signals, which are below the noise limit compared to the maximum intensity. In another step, tissue optics knowledge is used for rescaling fluorescence intensities by means of a Monte Carlo simulation (MCS) because analytical and approximative solutions of the transport equation break down even for less complicated geometries.

Besides these technical aspects like shifting of the spectra due to the instrumental set-up (see for example [7]), the physical basis of rescaling is a careful analysis of the light transport in biological, so turbid, media (see for example [8, 9]). At first the optical parameters, i.e., the absorption and scattering coefficient μ_a and μ_s, respectively, and the anisotropy factor of the scattering g, were determined by the double-integrating-sphere technique [10]. Then, these parameters were used for Monte Carlo simulations of the fluorescence intensity and backscattering intensity. The simulation permits to separate the influence of all parameters as follows. All but one of the involved parameters were fixed to their mean values.

Exemplarily, a closer look at the excitation of fluorescences in the UV range is taken. The lines in Fig. 21.2 show the natural variation of μ_a and μ_s for the excitation wavelength in human liver tumor and the resulting variation of the fluorescence signal, respectively. The variation is remarkably high, up to 70%.

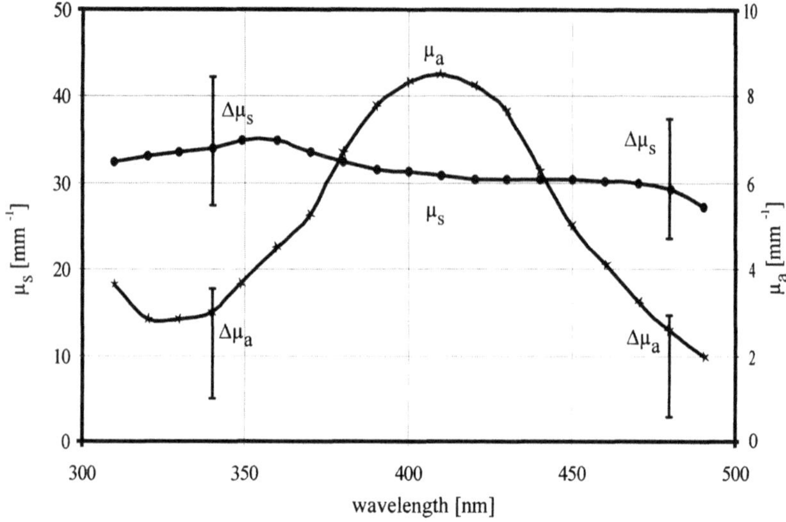

Fig. 21.2. Variation of the absorption and scattering coefficient for human liver tumor [11]

Monte Carlo simulations demonstrate the overwhelming influence of μ_a on the fluorescence intensity [12–14]. Its dependence on other optical tissue parameters (μ_s for excitation and μ_a and μ_s for the fluorescence wavelength) is much weaker and therefore neglected. In a first approximation the anisotropy factor g can be assumed to be almost constant. According to our Monte Carlo simulations the backscattering intensity also depends essentially on the same μ_a. Therefore, a simultaneous measurement of the backscattering intensity provides the data for rescaling of the fluorescence intensity with respect to the deviations of the given μ_a from its mean value.

Besides NADH, especially a number of cytokeratines are excited at 340 nm, too. The use of time-resolved techniques analysing the fluorescence decay seems to be a promising tool to separate these different contributions to the signal [14]. A mathematical model explains the connection between the rescaled fluorescence signal and the NADH concentration in the tissue region under investigation [14, 11].

21.3
Biomedical Examples

21.3.1
Optical Biopsy in the UV Range Using Endogenous Chromophores

"Optical Biopsy" is a valuable tool in biomedical research and medical diagnosis detecting the relative distribution of endogenous chromophores. In the case of the reduced form of nicotinamide adenine dinucleotide (NADH), the fluorescence signal reveals the activity status of the observed cells, because NADH comprises

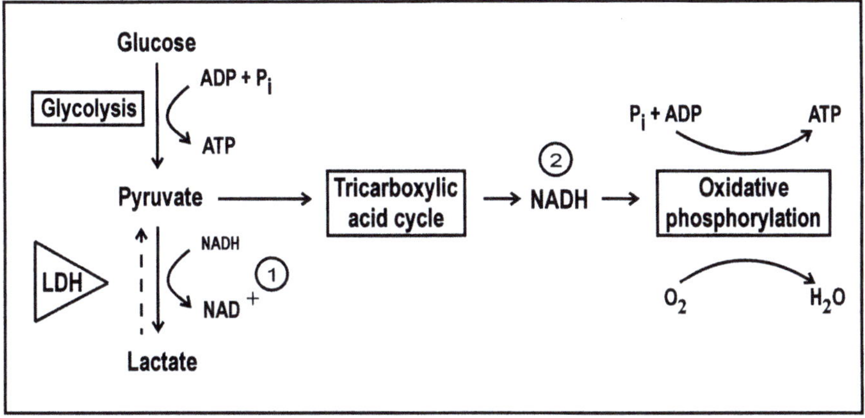

Fig. 21.3. Biochemical pathways of NADH in the cellular respiration

some essential features of cellular energy metabolism (see, for example [15, 16]). Mainly, this coenzyme takes part in the respiratory cycle (Fig. 21.3), as was discovered by Otto Warburg in 1931.

21.3.1.1
Experiment

The experimental set-up for two-dimensional fluorescence imaging is based on early investigations taking line scans using an optical fiber [17, 18].

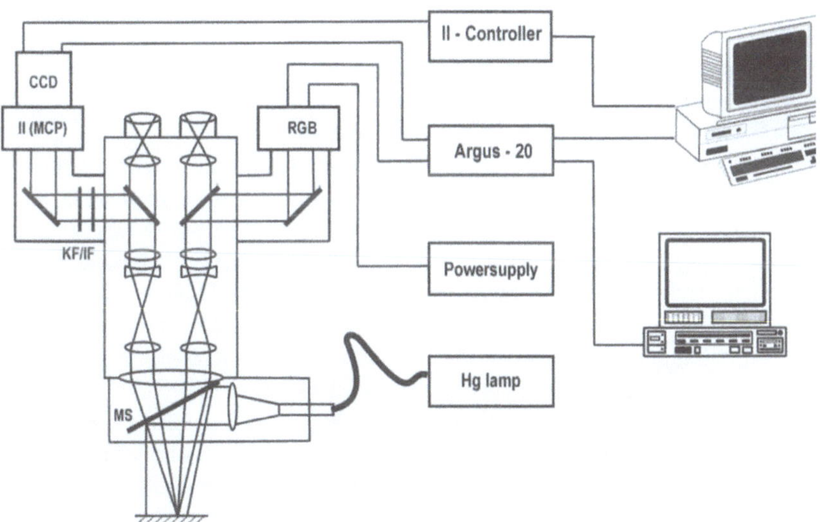

Fig. 21.4. Experimental set-up for two-dimensional fluorescence imaging [19]. **II** Image intensifier (Micro channel plate); **FLL** Liquid core wave guide; **MS** Monochromatic mirror (HR 365 nm); **KF** Edge filter; **IF** Interference filter (band-pass 460 nm)

21.3.1.2
Results

As target tissue we used human tissue from of a tumor of the base of the tongue (squamous cell carcinoma) *in vivo*. As an example for a two-dimensional fluorescence mapping a native picture of a tumor of the base of a human tongue is shown (Fig. 21.5). Right next to it is the corresponding fluorescence and the rescaled image. The rescaling is made by using the optical parameters in relationship to the tissue region of interest as discussed in Sect. 21.2.

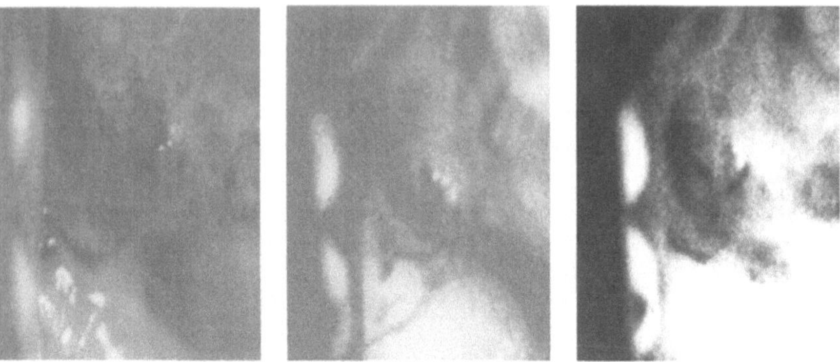

Fig. 21.5. Example for the rescaling method in optical biopsy of a base tumor of a human tongue *in vivo*, *left*: native image, *middle*: fluorescence image, *right*: rescaled image

21.3.2
Optical Molecular Imaging in the NIR Range Using Exogenous Contrast Agents

Approaches using autofluorescence are increasingly replaced by contrast-enhanced optical imaging methods using drugs such as indocyanine green (ICG) or 5--ALA in order to enhance the fluorescence intensity within tissues and improve the diagnostic specificity. A variety of novel dyes has recently been synthesized and characterized regarding its potential to detect tumors and other diseases. Generally, such dyes have to be nontoxic and absorb in the near infrared (NIR) to ensure deep photon penetration. The class of cyanine dyes has shown greatest potential since many derivatives out of this structural family are already known and have proven to be efficient fluorescent labels. For *in vivo* diagnostic applications, different pharmacological and molecular principles to achieve tumor-specific signal enhancement can be followed. For example, fluorescence-tagged receptor-specific peptides have been used to impart molecular specificity for the detection of tumor-specific receptor expression [20, 21]. A more simple approach involved the synthesis of indotricarbocyanines with hydrophilic chemical substitution. As described in [22], different structures comprising two different amino-sugar derivatives, D-glucamine and D-glucosamine, were synthesized and characterized

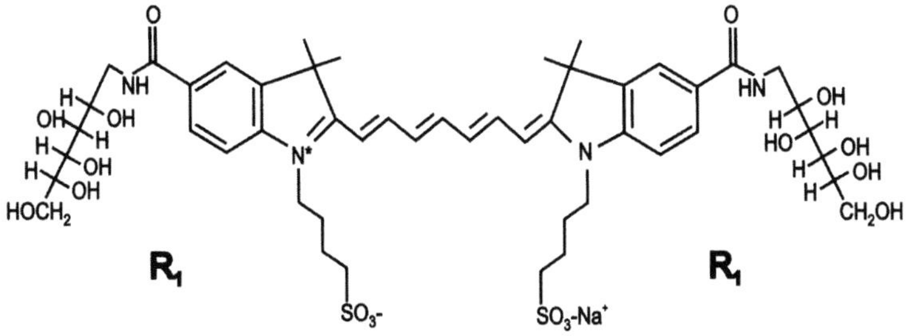

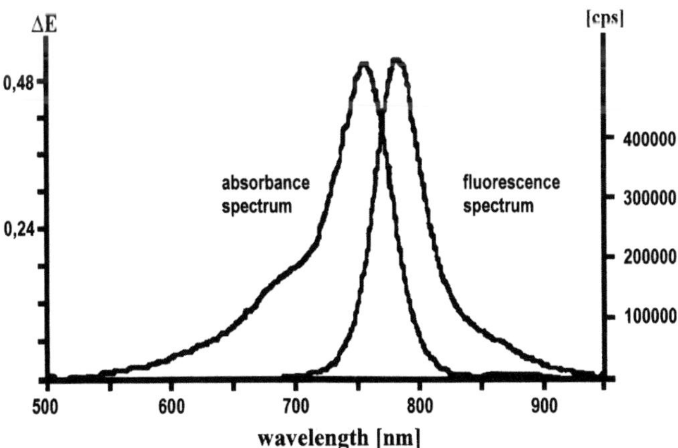

Fig. 21.6. Chemical structure of NIR96010 (bis-1,1'-(4-sulfobutyl)indotricarbocyanine-5,5'-dicarboxylic acid diglucamide, monosodium salt)

for their physicochemical and pharmacological properties. The plasma protein binding affinity decreases in agreement with increasing hydrophilicity. The most hydrophilic compound, NIR96010 (bis-1,1'-(4-sulfobutyl)indotricarbocyanine-5,5'-dicarboxylic acid diglucamide, monosodium salt, chemical structure see Fig. 21.6), was thoroughly studied in different animal models.

NIR96010 has an absorption maximum at 754 nm, a fluorescence maximum at 782 nm and a fluorescence quantum yield of 7–12% in physiological environment. In Fig. 21.7 the absorption and fluorescence emission spectrum is depicted. At doses in the range of 0.5–2 µmol/kg it shows improved properties regarding its ability to generate increased tumor-to-normal tissue concentration ratios, compared to the known drug indocyanine green (ICG) [22].

An additional asset was observed in fluorescence imaging studies when looking at late times after intravenous injection. While ICG is strongly protein-bound, thus distributed in the intravascular space and rapidly cleared from the intravascular space by the liver, NIR96010 extravasates into the extracellular compartment. As a consequence, a significantly descreased tissue clearance was apparently leading to a preferential retention in tumor tissues and thus a subsequently improving contrast, which shows highest values at 24 h after injection of the compound [23, 24].

Fig. 21.7. Absorbance and fluorescence spectrum of NIR96010 in bovine plasma (concentration 2 µM)

This behavior makes the compound suitable as fluorescent probe for studying imaging methods and rescaling principles throughout a variety of *in vivo* models.

One of the outstanding advantages of near infrared OMI using injectable fluorescence probes is the fact that the fluorescence signal of an organic dye molecule can potentially be affected by its local environment. For instance, energy transfer, fluorescence quenching or dequenching processes can be employed to report molecular or physiological conditions. Weissleder et al. [25] have introduced a novel approach which employes so-called fluorescence-quenched, enzyme-activatable polymers which carry cyanine dye molecules in close distance to each other so that an efficient quenching of fluorescence emission results. Upon enzymatic cleavage and liberation of dye molecules by tumor-specific enzymes, a bright fluorescence recovers in areas where enzymes are active, thus permitting the optical detection of specific protein expression, e.g., proteolytic enzymes, *in vivo*.

The different design approaches for optical contrast agents have in common that resulting images have to be analyzed regarding its diagnostic accuracy. One dye is presented here as an example.

21.3.2.1
Experiment

For the *in vivo* tests of the new dyes a NIR Imager was designed. The imager consists of two laser systems and a cooled CCD-camera (Hamamatsu). The first laser system is a 742 nm cw diode laser (1.5 W, CeramOptec) with an interference filter for 740 nm (*filter 1*). The NIR fluorescence at 800 nm of an entire mouse body can be imaged with a specific interference filter (*filter 3*), as shown in Fig. 21.8.

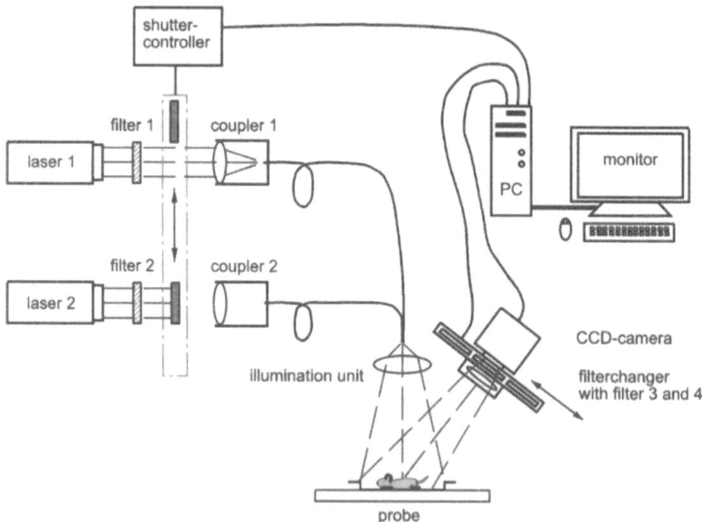

Fig. 21.8. Experimental set-up for OMI. *Laser 1*: excitation wavelength $\lambda = 742$ nm, *filter 1*: transmission wavelength $\lambda = 742$ nm, *filter 3*: fluorescence wavelength $\lambda = 800$ nm. *Laser 2* (example): excitation wavelength $\lambda = 488$ nm, *filter 2*: transmission wavelength $\lambda = 488$ nm, *filter 4*: fluorescence wavelength $\lambda = 509$ nm

The second laser can be used for different purposes. It is possible, e.g., to excite EGFP (CLONTECH) with an argon laser that is adjusted and filtered at 488 nm (*filter 2*). The fluorescence light of this laser excitation can be imaged with another interference filter for 509 nm (*filter 4*).

21.3.2.2
Results

The *in vivo* imaging potential was studied in nude mice bearing F9 teratocarcinoma. After intravenous injection of the dye, a typical contrast performance was observed with highest contrast at approximately 20 h. The contrast signal ratio (tumor/normal tissue) was determined to be about 10 from the image depicted in Fig. 21.9. As described above, this high contrast results from a tumor retention of a small amount of NIR96010 accompanied by a renal clearance of the majority of injected dose. Compared to previous studies using other tumor models [23, 24] a higher contrast was observed.

To improve the image of the concentration of the NIR-fluorophore in the mouse, a new robust rescaling principle was developed as a first approach. For tissue optics related corrections it is to be considered that the light scattering in the NIR range is stronger compared to the UV fluorescence used in optical biopsy. Nevertheless, the crucial point is the knowledge of the optical parameters μ_a, μ_s and g (anisotropy) and the use of Monte Carlo methods. The diameter of the fluorescent volume is apparently larger for the same reason. Furthermore, the signal/noise ratio can be improved by suppressing parasitic fluorescences. The following image (Fig. 21.10) shows the first result in rescaling the above fluorescence image of the mouse thigh. The effects of rescaling can be clearly seen by comparison of the line scans taken in the middle of the tumor of the fluorescence images.

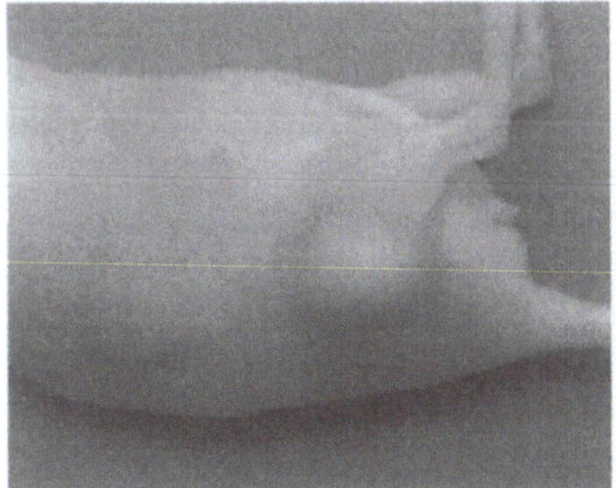

Fig. 21.9. Native (*left*) and fluorescence image (*right*) with line scan in the middle of the superficial tumor of the mouse thigh (nude mouse, F9 teratocarcinoma of approx. 8 mm × 8 mm in size) at 20 h after intravenous injection of NIR96010 at a dose of 2 µmol dye/kg body weight

Fig. 21.10. Fluorescence image from Fig. 21.9 (*right*) after rescaling procedure with line scan

21.4
Discussion and Conclusions

With the modified rescaling principle for the NIR-spectral region rescaled images of NIR-fluophores can be achieved. The concentration profile of NIR-fluophores in the border region of tumors can be achieved just like the NADH concentration profile in optical biopsy and can be used as a first step of the three-dimensional reconstruction of the distribution of the fluorophore [26]. This aspect along with the tumor specific quenching of the fluorophore enhances the border and the local resolution and therefore permits a chance to analyse even early stage tumors.

NIR96010 has been described as a simple and efficient optical contrast agent, which permits tumor detection in animals through early time contrast enhancement based on a more distinct tumor perfusion [22] as well as by a certain retention in tumor tissue leading to improved contrasts up to 24 h after administration [23, 24]. Therefore, this dye was selected as a first model compound in our studies.

The OMI procedure presented herein, however, includes only three of the four characteristic features defined by Weissleder [27]. While the biocompatible suitable ligands, targeting strategies and the dedicated imaging system have already been realized, the amplification strategies are still missing. Nevertheless, the dye used in this procedure can be called enhancer. But the described rescaling is part of the imaging system and remains unchanged in terms of its physical and mathematical principles. The image blurred by photon migration is to be rescaled according to the respective tissue optical parameters. Various methods are suitable for rescaling. The suggested methodical combination of a cut off according to the signal/noise ratio and the intensity scaling for setting off the partial elimination of the signal in the tissue is an applicable first approximation.

In a total of 30 measurements in subcutaneous tumors, a rescaling error of 12% was determined. At a tumor growth process to 10 mm in diameter this corresponds to a resulting surgical unreliability of approximately 1 mm. These statements, of course, have to be considered strictly conditioned. Further investigations on the influences of the marker concentration, the saturation behaviour and the respective quantum yield will follow. These findings may serve to improve the presented approximation.

References

1. Chance B, Schoener B, Oshino R, Itshak F, Nakase Y (1979) Oxidation-reduction ratio studies of mitichondria in freeze-trapped samples. J Biol Chem 254:4766–4771
2. Renault G, Raynal E, Sinet M, Berthier JP, Godard B, Cornillault J (1982) A laser fluorimeter for direct cardiac metabolism monitoring. Opt Laser Technol 14:143–148
3. Renault G, Sinet M, Muffat-Joly M, Fourati T, Polianski J, Meric P, Weiser M, Pocidalo J (1984) Evaluation in situ du métabolisme tissulaire par fluorimétrie laser. La Presse Médicale 13:2381–2385
4. Richards-Kortum, R, Rava R, Fitzmaurice M, Tong L, Ratliff N, Kramer J (1989) A one-layer model of laser-induced fluorescence for diagnosis in human tissue: applications to artheroscleroris. IEEE Trans Biomed Eng 36:1222–1232
5. Gandjbakhche A, Gannot I (1996) Quantitative fluorescent imaging of specific markers of diseased tissue. IEEE Sel Topics QE 2:914–921
6. Pogue BW, Hasan T (1996) Fluorophore quantition in tissue-simulating media with confocal detection. IEEE Sel Topics QE 2:959–964
7. Zellweger M, Goujon D, Conde R, Forrer M, van den Bergh, H, Wagnières G (2001) Absolute autofluorescence spectra of human healthy, metaplastic, and early cancerous bronchial tissue in vivo. Appl Opt 40:3784–3791
8. Ishimaru I (1978) Wave propagation and scattering in random media. Academic Press, New York
9. Tuchin V (1994) Selected papers on tissue optics. SPIE MS 102, Bellingham
10. Roggan A, Minet O, Schröder C, Müller G (1993) Measurements of optical tissue properties using integrating sphere technique. In: Müller G et al. (eds), Medical optical tomography. SPIE IS 11, Bellingham, pp 149–165
11. Beuthan J, Minet O (1999) Fluorescence diagnosis in the border zone of liver tumors. In: Rettig W et al. (eds), Applied fluorescence in chemistry, biology and medicine. Springer, Berlin, pp 537–551
12. Beuthan J, Bocher T, Minet O, Roggan A, Schmitt I, Weber A, Müller G (1994) Investigations concerning the determination of NADH concentrations using optical biopsy. Proc. SPIE 2135:147–156
13. Beuthan J, Minet O, Müller G (1998) Optical biopsy of cytokeratin and NADH in the tumor border zone. Ann NY Acad Sci 838:150–170
14. Beuthan J, Weber A, Minet O, Hagemann R, Roggan A, Schmitt I, Müller G, Germer C, Albrecht D, Bocher T (1994) Untersuchungen zur NADH-Konzentrationsbestimmung mittels optischer Biopsie. Lasermedizin 10:57
15. Chance B, Legallais V, Schoener B (1962) Metabolically linked changes in fluorescence emission spectra of cortex of rat brain, kidney and adrenal. Nature 195:1073–1075
16. Lehninger AL (1993) Principles of biochemistry. Worth Publishers, New York

17. Beuthan J, Zur C, Hofmann H (1990) Quantitative (in vivo) NADH-Messung – ein methodisch klinischer Ansatz zur biologischen Äquivalentdosimetrie. Adv Laser Medicine 5:253–260
18. Beuthan J, Minet O, Müller G (1993) Observations of the fluorescence response of the coenzyme NADH in biological samples. Opt Lett 18:1098–1099
19. Bocher T, Beuthan J, Minet O, Naber R, Müller G (1995) Frequency domain techniques for 2-dimensional mapping of optical tissue properties. Proc. SPIE 2626:283–294
20. Becker A, Hessenius C, Licha K, Ebert B, Sukowski U, Semmler W, Wiedenmann B, Grötzinger C (2001) Receptor-targeted optical imaging of tumors with near-infrared fluorescent ligands. Nature Biotech 19:327
21. Bugaj JE, Achilefu S, Dorshow RB, Rajagopalan R (2001) Novel fluorescent contrast agents for optical imaging of in vivo tumors based on a receptor-targeted dye-peptide conjugate platform. J Biomed Opt 6:122–133
22. Licha K, Riefke B, Ntziachristos V, Becker A, Chance B, Semmler W (2000) Hydrophilic cyanine dyes as contrast agents for near-infrared tumor imaging: synthesis, photophysical properties and spectroscopic in vivo characterization. Photochem Photobiol 72:392–398
23. Riefke B, Licha K, Nolte D, Ebert B, Rinneberg H, Semmler W (1996) In vivo characterization of cyanine dyes as contrast agents for near-infrared imaging. Proc. SPIE 2927:199–208
24. Ebert B, Sukowski U, Grosenick D, Wabnitz H, Moesta KT, Licha K, Becker A, Semmler W, Schlag PM, Rinneberg H (2001) Near-infrared fluorescent dyes for enhanced contrast in optical mammography: phantom experiments. J Biomed Opt 6:134–140
25. Weissleder R, Tung C, Mahmood U, Bognanov A (1999) In vivo imaging of tumors with protease-activated near-infrared fluorescent probes. Nature Biotech 17:375–378
26. Minet O (1995) Zur Bestimmung der räumlichen Verteilung von Fluoreszenzzentren in streuenden Medien. Fortschritte in der Lasermedizin 12:98
27. Weissleder, R (1999) Molecular imaging: exploring the next frontier. Radiology 212:609–614

CHAPTER 22

Looking into a Living Cell

M. VAN BORREN, N. R. BRADY, J. RAVELSLOOT, AND H. V. WESTERHOFF

Microscopic detection of fluorescent dyes is a powerful tool to monitor dynamics of intracellular parameters, in the living cell. In contrast to green fluorescent proteins (GFPs), which require expertise in molecular biology, the ease at which fluorescent dyes can be used makes them appealing for a larger audience of biologists. In this overview we will highlight certain methodologies and considerations when imaging a selection of cell-permeable fluorescent dyes and endogenous fluorescent molecules. We will do this with an emphasis on pH and mitochondrial energetics in cardiomyocytes.

22.1
Introduction

Genomes having been sequenced, blueprints of life and living cells are available at least in one sense. To attain the goal of biology of understanding life, much more than the blueprints are required. The functioning of macromolecules cannot be calculated from their primary sequence, neither reliably nor quantitatively. The nonlinear nature of the interactions of the macromolecules, which is largely responsible for the essence of life, can only be understood on the basis of reliable and quantitative information obtained at the operating point of a living organism [1].

Fluorescence techniques, combined with advances in microscopy, data collection and analytical methods offer the investigator a non-intrusive means to study molecular events that underlie cell functioning, *in the living cell*. Fluorescent microscopy can be divided into two main categories: (1) studies using fluorescent dyes, and (2) studies using genetically-encoded and -targeted green fluorescent protein variants (GFPs) [2]. Here we will only discuss the use of fluorescent dyes, either extraneous or endogenous.

We shall focus on the practicalities of fluorescence microscopy when examining the role of Ca^{2+} signalling, intracellular pH (pH_i)-regulation, mitochondrial energetics and reactive oxygen species (ROS) in heart failure, ischemia, and apoptosis in isolated ventricular cardiomyocytes. Examples from our work will illustrate some of these techniques. Photo-toxicity and loading techniques will also be discussed.

22.2
How to Choose your Fluorescent Indicator

Many of the available fluorescent probes have limitations in specificity, responsiveness and reliability. In this section we shall go through some considerations pertinent to the selection of the most suitable dye.

22.2.1
Process of Interest

Certainly the intracellular event of interest is the first criterion that determines which dye to choose. There are ion specific dyes (e.g., Na^+, Ca^{2+} and H^+), dyes that are sensitive for ROS, and autofluorescent endogenous molecules (e.g., NADH) that reflect the metabolic status of the cell. There are two main types of dyes available that function as indicators of membrane potentials (plasma and inner mitochondrial membranes): (1) the rhodamine-based dyes, whose change in transmembrane concentrations can reflect changes in transmembrane electric potential difference, and (2) electrochromic probes that integrate into the membrane and undergo a spectral shift in response to changes of the electric field in the membrane. In addition, organelles (e.g., nucleus, mitochondria, lysosomes) can be loaded specifically with dyes, e.g., to study morphological changes. For most of these objects more than one dye is available.

22.2.2
Ratiometric Dyes

A ratiometric dye is either a dual excitation or a dual emission probe. The ratio of the fluorescence obtained at two wavelengths is an indirect measure of the event of interest. The advantage of a ratiometric dye is that neither the bleaching, variable loading, nor the leakage of the dye should compromise the method. The obtained ratios can be converted to quantitative assessments of the phenomenon of interest provided that the latter can be manipulated independently so as to allow calibration of the dye response. Most ion specific dyes can be calibrated *in vitro,* by titration. *In vivo (in situ)* calibration is more difficult, due to the inability to inhibit active transport across the membrane of the specific ion of interest. This is especially true for Ca^{2+} and Na^+. For *in situ* calibration, incubations with a series of bathing solutions containing different concentrations of the ion of interest together with the appropriate ionophores are required. *In situ* calibration is preferred since a dye inside cells might behave differently due to dependence of its affinity for the probed molecule on its microenvironment. An example is given of an *in vivo* calibration by the high K^+ nigericin technique [3] of the pH-sensitive dye SNARF (Fig. 22.1).

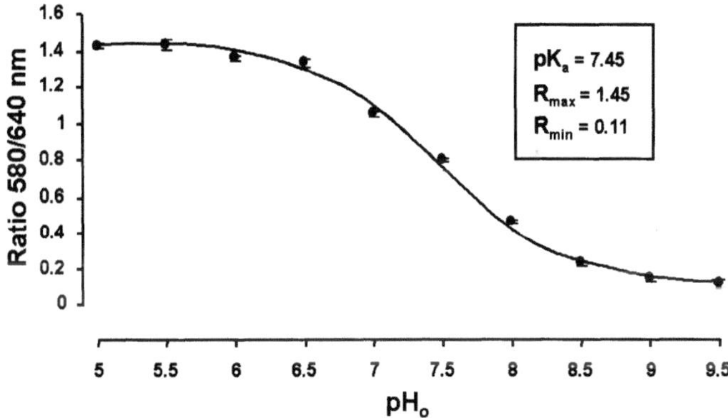

Fig. 22.1. *In situ* calibration-curve of SNARF-AM: Rabbit (3–6 months) ventricular cardiomyocytes were isolated as described [4]. Cells were loaded with 10 µM SNARF-AM (Molecular Probes) for 10 minutes, allowed to settle down on a poly-L-lysin coated glass coverslip, and examined using an inverted, Nikon Diaphot, wide-field fluorescence microscope (40X objective). After addition of 10 µM of the H^+/K^+ ionophore nigericin, the cells were superfused with several high K^+ solutions at various external pH (pH_o) values. When the external and internal K^+ free concentrations are equal, pH_i will be the same as pH_o. The calibration curve was obtained by plotting the ratios (580/640 nm) against the corresponding pH_o. We fitted a Henderson-Hasselbalch equation through these data (line), which revealed a maximum ratio of 1.45, a minimum ratio of 0.11 and a pK_a of 7.45

22.2.3
Buffering Power

A factor to consider when using fluorescent ion indicators, such as for Ca^{2+}, Na^+, and H^+, is the indicator's intrinsic buffering power (β). Dyes not only have their fluorescent properties altered upon ion binding, but they can effectively retain an amount of added ions, thereby blunting changes in free ion concentration. High concentrations of the dye can therefore diminish changes in free ion concentrations and thus interfere with the biological process under study [5].

22.2.4
Photo-toxicity

Depending on the type of dye employed, fluorescence excitation (the imaging process) can contribute considerably to subcellular production of ROS. Fluorescent molecules, including dyes [6] and GFPs [7], can act as photosensitizers, where a non-radiative transition of the excited fluorophore to the ground state occurs. This can result in spin exchange with molecular oxygen bringing the latter in the more reactive singlet state or in the transfer of the electron to molecular, triplet-state oxygen, producing superoxide anion [8]. The photosensitization properties of many dyes have been characterized [6]. See Fig. 22.4 for an example of the effects of photo-toxicity on mitochondrial energetics.

In addition, UV excitation, as required when imaging certain fluorescent dyes (e.g., Indo-1) or NAD(P)H can also have damaging effects on the living cell. UV excitation can elicit undesirable intracellular responses, such as DNA damage [9], ROS generation and even programmed cell death: apoptosis [10]. However, if a UV-excitable dye is necessary the total exposure time must be minimized.

22.3
Considerations Concerning the Experimental Approach

The most suitable dye having been selected, there are also considerations that concern the experimental set-up, dye-loading techniques and the imaging apparatus (microscope, excitation source, detector type and sensitivity). In addition one must consider the inverse relationship between spatial and temporal resolution.

22.3.1
Selective Loading

Many fluorescent dyes comprise a membrane-permeable acetoxymethyl (AM) ester group. Upon entrance, endogenous non-specific esterases hydrolyse the lipophilic AM group, releasing the free acid form of the dye, which is then trapped inside of the cell. This is a simple and less invasive approach.

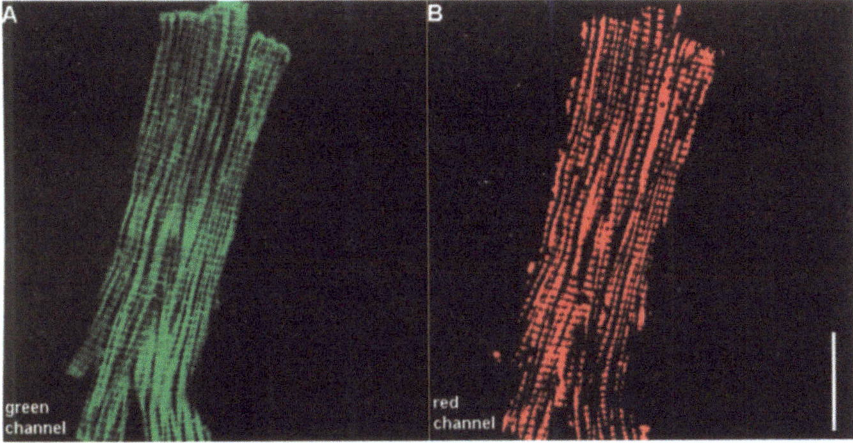

Fig. 22.2. Cytosolic loading of calcein-AM: Rabbit (3–6 months) ventricular cardiomyo-cytes were isolated as described [4]. Cells were dual-loaded 30 minutes with 0.7 µM TMRM and 10 µM calcein-AM (Molecular Probes) in 25 mM Hepes-buffered (pH 7.4) M199 medium (supplemented with 1% bovine serum albumin (BSA)) at 37°C, thereby re-sulting in a cytosolic distribution of calcein initially in excess over that in mitochondria. Cells were centrifuged to remove supernatant, mixed with 0.5% agarose (~36°C; M199/1% BSA), and then deposited on glass-well plates (MatTek) with M199/1% BSA. These fig-ures were obtained using a Leica TCS-4D laser scanning confocal microscope (Plan-Neofluar 63X/1.4 N.A. oil immersion objective, Krypton/Argon laser). The confocal pin-holes were configured to obtain images of 1 µm in the axial dimension. **A** Calcein fluores-cence was collected in the green channel (488 nm exc.; BP 520–560 em.). **B** Simultane-ously, TMRM fluorescence was collected in the red channel (568 nm exc.; LP 580 em.). Scale bar = 20 µm

The AM group has the additional advantage of allowing selective subcellular targeting, i. e. into either the cytosol or into subcellular organelles, as a function of the loading temperature. For example, at 4°C endogenous esterases are inactive and therefore the fluorescent AM dye partitions to both the cytosol and organelles. Following cold loading, the temperature is increased to 37°C and the esterases, once again active, trap the dyes in the organelles by hydrolysing them to a mem-brane-permeant derivative. Following a prolonged (4–6 hours) incubation at 37°C the cytosolic dyes tend to be transported by anionic channels in the plasma mem-brane to the extracellular medium [11, 12], the organelles maintaining a significant concentration [13]. Conversely, loading at 37°C is used to label the cytosolic compartments in excess of the organelles. Fig. 22.2 demonstrates cytosolic load-ing of calcein-AM. Tetramethylrhodamine methyl ester (TMRM) accumulates electrophoretically to the mitochondrial matrix (discussed below) and mitochon-dria appear as longitudinally aligned arrays (Fig. 22.2B). Calcein fluorescence is most intense in regions that are not labelled by TMRM, i.e., the cytosol and nuclei (Fig. 22.2A). Consequently, the two images are virtually negative images of one another.

22.3.2
Additional Loading Techniques

When there are no membrane-permeable dyes for the event of interest, there are still other possibilities to internalise the dye, such as application through a pipette, osmotic shock, or lipid vesicles. A single injection of a dye into the cytoplasm can be performed through a sharp micropipette, with the hazard of damaging the cell. If electrophysiological recordings are required, a dye can be dialysed into the cytoplasm through a patch-pipette. In this case important intracellular constituents may be diluted. Both techniques have the advantage that the amount of dye loading can be controlled, although, in terms of interference with the cell's membrane integrity, they are inferior to AM loading.

22.3.3
The Imaging System

A wide-field microscope coupled with a fluorescence excitation lamp and a photomultiplier tube (PMT) suffices to determine ionic concentrations over time, when spatial resolution is not an important factor. For increased spatial resolution, in the case of subcellular signalling, gradients, compartmentalization, organelles, etc., a wide-field fluorescence microscope coupled to a cooled charge coupled device (CCD), or a confocal fluorescence microscope (CFM) can be used. A CFM offers the additional advantage of increased resolution, which, via the rejection of out-of-focus background fluorescence, allows the investigator to obtain precise 2D and 3D subcellular information [14]. Figs. 22.2 and 22.4 are examples of confocal imaging.

22.4
Examples of Fluorescence Microscopy in Living Cells

22.4.1
Detection of the Ca^{2+} Ion

Calcium is a ubiquitous messenger, involved in fertilization, muscle contraction, cell death, etc. Ca^{2+} homeostasis also plays an important role in aspects of heart failure, such as ischemia [15]. The versatility of Ca^{2+} as a signalling molecule arises from modulations in its frequency (rate of progression through the cell) or in its amplitude (either in terms of concentration or in terms of space) [16]. Increased CCD sensitivity and increased laser scanning rate have enhanced the ability to record millisecond images, mitigating the temporal problem of imaging calcium signalling. The variety of indicators with differing dissociation constants (K_d, low to high affinities) allows for assessment of the role of calcium in the control of muscle concentration, which is an issue of renewed interest [17]. The detectable concentration range will be about $0.1 \times K_d$ to $10 \times K_d$ [18]. The fluorescence of the dyes increases upon Ca^{2+} binding, and as the concentration of intracellular free calcium increases, so does the fluorescence of the dye, such as Fluo3 or 4. Measuring the emission ratio using two different excitation wavelengths, such as Fura-2 or Indo-1, enables a more accurate determination of Ca^{2+} concentrations [5].

22.4.2
Determination of Intracellular pH

Many processes in the cell depend on the intracellular pH (pH_i). Accordingly, pH_i is highly regulated and kept constant near neutral values. Under pathophysiological conditions this equilibrium is disturbed. For example, when cells are exposed to ischemic conditions due to the lack of oxygen, as occurs during ischemic heart disease, lactic acid is produced and pH_i drops [15]. Acid/base transporters in the plasma membrane enable the cell to recover from acidosis except under pathological conditions. We are trying to elucidate the underlying mechanisms using the most common dye for measuring pH_i, carboxy-seminaphthorhodafluor-1-acetoxymethyl ester (SNARF-AM). SNARF-AM is readily loaded into the cell at 37°C and one can start measuring after 10 minutes of loading. As SNARF is relatively insensitive to bleaching; pH_i can be measured accurately for hours. SNARF is excited at 515 nm and the fluorescence of the acidic form (580 nm) is divided by the fluorescence of the alkaline form (640 nm). The obtained ratio is then converted to pH values. For calibration purposes, we use the high-K^+/nigericin technique as described by Thomas et al. [3].

To study the activity of the acid extruders (Na^+/H^+ exchanger, Na^+/HCO_3^- cotransporter) and acid loaders (Cl^-/OH^- and Cl^-/HCO_3^- exchangers), the steady-state pH_i has to be perturbed. An ammonium (a membrane permeant base) or an acetic acid (a membrane permeant acid) prepulse, respectively, is used to acid or alkaline load the cells. The recovery rate from the acidosis or alkalosis is an indirect measure of the transport rate through acid extruders or acid loaders, respectively, as well as adjusted metabolic rates. When these activities are multiplied with the corresponding intracellular buffer capacity (β_i) the real proton flux through the transporter is obtained. The activity of the acid/base transporters, expressed as the amount of acid or base extruded or loaded per second, is called the proton flux ($JH^+ = \beta_i * (dpH_i / dt)$). We determined the relationship between β_i and pH_i in rabbit ventricular cardiomyocytes. This approach to measure β_i is called the "stepwise reduction in extracellular NH_3/NH_4^+ approach" [19]. Contributing to the apparent internal buffer capacity is the internal CO_2/ bicarbonate equilibrium. A typical pH_i trace is shown in Fig. 22.3.

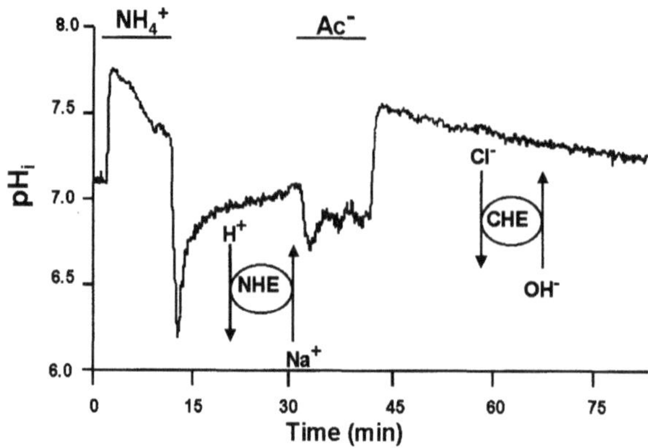

Fig. 22.3. Typical pH$_i$ recording using SNARF. Rabbit (3–6 months) ventricular cardio-myocytes were isolated as described [4]. Cells were loaded with 10 µM SNARF-AM for 10 minutes, allowed to settle down on a poly-L-lysin coated glass coverslip, and examined using an inverted, Nikon Diaphot, wide-field fluorescence microscope (40X objective). Cells were acutely acid loaded using the ammonium (NH$_4^+$) prepulse technique, whereas an acute alkalosis was introduced by the acetate (Ac$^-$) prepulse technique. Wash in of NH$_4$Cl caused first a rapid alkalosis due to the rapid entry of the membrane-permeable NH$_3$ and a subsequent slow recovery from the alkalosis due to NH$_4^+$ entry through potassium channels. Washout of extracellular NH$_4$Cl caused a rapid acidosis due to the rapid diffusion of intracellular NH$_3$ out of the cell. Under Hepes-buffered conditions cardiomyocytes are able the recover fast from an acidosis using the Na$^+$/H$^+$ exchanger. On the other hand, exposing the cells to Na-acetate caused first a rapid acidosis due to diffusion of the membrane permeable HAc, whereas pH$_i$ slowly recovers due to acid extrusion. Washout of the cells caused an alkalosis due to rapid diffusion of HAc out of the cells. Cells were able to recover from this alkalosis using the slow Cl$^-$/OH$^-$ exchanger

22.4.3
Mitochondrial Energetics

In addition to monitoring temporal and spatial dynamic changes in ion concentrations, one can use fluorescence microscopy to study events at the level of organelles. The mitochondria are of interest in studies of Ca^{2+} signalling, acting as intracellular Ca^{2+} modulators [20], as well as participating in both apoptotic and necrotic cell death [21]. The mitochondrial membrane potential (i.e., the electric potential difference across the inner mitochondrial membrane, $\Delta\Psi$m) is an important indicator of the functional state of mitochondria and of the energetic state of the entire cell.

Rhodamine-based dyes are among the commonly used probes of mitochondrial membrane potential, and of these the most commonly used are the rhodamine derivatives tetramethylrhodamine methyl ester (TMRM) and tetramethylrhodamine ethyl ester (TMRE). TMRM is arguably the best choice due to its (1) lower phototoxicity [6] and (2) lowest inhibition of mitochondrial respiration [22]. Based on a plasma membrane potential of –60 mV they will accumulate about 10x in the cytosol. Due to the highly negative membrane potential across the inner-mitochondrial membrane, they should accumulate a further 1000x into the mitochondria [23]. A corollary is that the medium probe concentration should not exceed 1 µM.

When cells are equilibrated with their incubation medium, then the partitioning of the fluorophore between two aqueous phases separated by a membrane should be described by the Nernst equation. If the changes in signal intensities can be related quantitatively to changes to the dye concentration, $\Delta\Psi$m can be calculated [24]. When using either TMRM or TRME, depolarization is assessed from the changes of mitochondrial TMRM or TMRE fluorescence. The exit of TMRE due to depolarization results first in an increase of fluorescence, presumably due to the reversal of auto-quenching [20]. Furthermore, when using TMRE (unpublished results) or TMRM, the loss of mitochondrial fluorescence can also be a marker for depolarization [25]. However, the ambiguity of the quenching versus unquenching properties of the rhodamine derivatives can cause confusion, and furthermore, depolarized mitochondria are no longer detectable.

These problems have been solved by the development of a novel technique using fluorescence resonance energy transfer (FRET) to detect changes in mitochondrial membrane potential [26] (Fig. 22.4). MitoTracker Green FM (MTG), the donor molecule, is a blue-excited, green-fluorescing, cell-permeant fluorophore and has been shown to concentrate to the mitochondrial matrix, reportedly due to the covalent binding between its chloromethyl moiety and matrix thiol groups [18]. TMRM, the acceptor molecule, is a red-fluorescing, lipophilic, cationic probe, which accumulates electrophoretically in the cell as discussed above. When the two dyes are within 10 nm, upon excitation at 488 nm (blue light) the green fluorescence of MTG is quenched and TMRM fluoresces red [26]. FRET between MTG and TMRM permits solely the excitation of TMRM localized in mitochondrial matrix. Fig. 22.4A demonstrates that upon excitation at 488 nm the green MTG is quenched and TMRM fluoresces red: the mitochondria have a high membrane potential. This FRET technique, combined with laser scanning confocal microscopy for high spatial resolution, evidences the architecture of cardiomyocyte mitochondria as a highly ordered array, each mitochondrion measuring about 1 µm^2. In Fig. 22.4B, heterogeneous depolarizations among the mitochondrial population within a single cardiomyocyte were induced by photosensitization, as described by Zorov et al. [25]. The depolarizations were doubly evidenced using the FRET technique: the loss of $\Delta\Psi$m caused both the loss of red fluorescence (TMRM) as well as the gain (unquenching) of green fluorescence (MTG) among a subset of mitochondria, indicating reversal of the FRET interaction. Practically, this FRET technique permits a more informative and precise localization of changes in $\Delta\Psi$m than do TMRM or TMRE fluorescence alone.

Photo-toxicity-induced ROS production is an omnipresent problematic factor when imaging mitochondrial membrane potential. Since mitochondria are ROS excitable organelles [21, 25, 27]; the photosensitizing effect of potentiometric dyes can activate mitochondria to locally produce more ROS. As mitochondria are important in calcium signalling [20] and cell death, any disruption of their functional state by the applied fluorescent probes could affect the system under study and compromise the experimental results and the interpretation thereof.

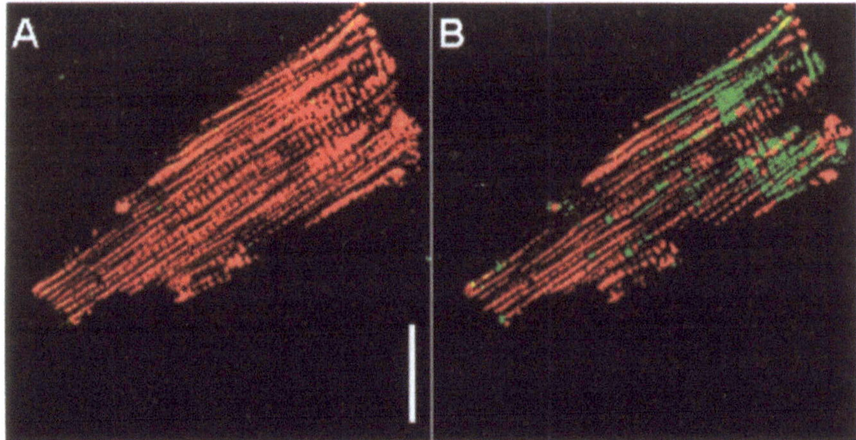

Fig. 22.4. Photosensitization of TMRM induces depolarization of the mitochondria, as visualized by fluorescence resonance energy transfer (FRET) between MTG and TMRM: Rabbit (3–6 months) ventricular cardiomyocytes were isolated as described [4]. Cells were dual-loaded for 30 minutes with 0.3 µM MitoTracker Green (MTG) and 0.7 µM TMRM (Molecular Probes) in 25 mM Hepes-buffered M199 (pH 7.4) medium supplemented with 1% bovine serum albumin (BSA)) at room temperature (~25°C). Cells were centrifuged to remove supernatant, mixed with 0.5% agarose (~36°C; M199/1% BSA), and then deposited on glass-well plates (MatTek) with M199/1% BSA. These figures were obtained using a Leica TCS-4D laser scanning confocal microscope (Plan-Neofluar 63X/1.4 N.A. oil immersion objective, Krypton/Argon laser). The confocal pinholes were configured to obtain images of 1 µm in the axial dimension. The FRET interaction between the MTG and the electrophoretically-accumulating TMRM was used to image the energetic state of the mitochondria, via excitation of MTG (488 nm) and collection of red (red channel; LP 580 em.) and green (green channel; BP 520–560 nm) fluorescence. Figs. A and B are overlays of red and green channels. **A** Red fluorescence should indicate a FRET interaction between MTG and TMRM, and thereby polarized mitochondria. After image A was generated the cell was subjected to photo-oxidative stress in the form of intense, prolonged excitation of TMRM at 568 nm. **B** The gain, or unquenching, of green fluorescence is taken to indicate loss of TMRM from the mitochondria, thereby mitochondrial depolarization. Scale bar = 20 µm. 2.5 minutes between images

22.4.4
Reactive Oxygen Species (ROS)

ROS are of considerable physiological interest, regulating or participating in such cellular processes as apoptosis and necrosis following reperfusion injury and heart failure [21].

The most common indicator of ROS production is dichlorodihydrofluorescein diacetate (DCFH2-DA), which is oxidized to the fluorescent molecule 2',7'-dichlorofluorescein (DCF) by the reactive oxygen species H_2O_2 or OH [28]. However, its specificity has recently been questioned [29].

BODIPY C11$^{(581/591)}$, which contains a lipophilic moiety making it distribute into membranes, appears to be an additional ROS indicator and novel marker of lipid peroxidation. Upon its oxidation by ROS (e.g., H_2O_2 and OH·) BODIPY C11$^{(581/591)}$ undergoes a red-to-green shift in fluorescence. BODIPY C11$^{(581/591)}$ is a ratiometric dye, i.e., ratioing the green and red emissions one can become less vulnerable to subcellular heterogeneity in dye loading and obtain more quantitative data for comparing separate experiments [30].

22.4.5
Autofluorescence: Marker of Redox State

Finally, it is of interest to observe the metabolic status during ischemia in cardiomyocytes, since the lack of O_2 may deplete the ATP pool. NAD(P)H, endogenous, UV excitable, autofluorescent molecules provide markers for the cellular redox state, although their detection by UV excitation is damaging to the cell. An alternative redox marker, the FADH$_2$-linked alpha-lipoamide dehydrogenase, is excitable at visible wavelengths. This mitochondrial protein shows inverse fluorescence behavior with NAD(P)H. It is therefore a suitable indicator of the mitochondrial redox state [31].

22.5
Conclusions

At the level of the single cell, fluorescent microscopy provides a complex, yet powerful and non-intrusive, quantitative or qualitative approach for many unanswered biological questions. One may well argue that, thanks to fluorescent dyes, green fluorescent proteins and fluorescent microscopes capable of resolutions below 100 nm [32], a renaissance in fluorescent cell biology is underway.

Acknowledgements. We thank J. Bourier and others at the Amsterdam Medical Centrum (University of Amsterdam) for the gift of the isolated rabbit ventricular cardiomyocytes. Confocal imaging was possible thanks to Prof. dr. J. Lankelma and H. Dekker.

References

1. Westerhoff HV (2001) Metab Eng 3:207
2. Zacharias DA, Baird GS, Tsien RY (2000) Curr Opin Neurobiol 10:416
3. Thomas JA, Buchsbaum RN, Zimniak A, Racker E (1979) Biochemistry 18:2210
4. Veldkamp MW, de Jonge B, van Ginneken AC (1999) Cardiovasc Res 42:424
5. Takahashi A, Camacho P, Lechleiter JD, Herman B (1999) Phys Rev. 79:1089
6. Minamikawa T, Sriratana A, Williams D, Bowser D, Hill J, Nagley P (1999) J Cell Sci 112:2419

7. Zhang C, Sriratana A, Minamikawa T, Nagley P (1998) Biochem Biophys Res Commun 242:390
8. Matroule JY, Piette J (2000) Antioxid Redox Signal 2:301
9. Griffiths HR, Mistry P, Herbert KE, Lunec J (1998) Crit Rev Clin Lab Sci 35:189
10. Godar DE (1999) J Invest Dermatol 112:3
11. Di Virgilio F, Fasolato C, Steinberg TH (1988) Biochem J 256:959
12. McDonough PM, Button DC (1989) Cell Calcium 10:171
13. Trollinger DR, Cascio WE, Lemasters JJ (2000) Biophys J 79:39
14. Pawley J (1994) Handbook of Confocal Microscopy 2nd ed, Plenum Press, New York
15. Halestrap AP, Wang X, Poole RC, Jackson VN, Price NT (1997) Am J Cardiol 80:17A
16. Thomas AP, Bird GS, Hajnoczky G, Robb-Gaspers LD, Putney JW Jr. (1996) Faseb J 10:1505
17. Jeneson JA, Westerhoff HV, Kushmerick MJ (2000) Am J Physiol Cell Physiol 279: C813
18. Haugland R (2001) Section 12.2: Probes for Mitochondria in Handbook of Fluorescent Probes and Research Products, Eugene, OR, USA
19. Boyarsky G, Ganz MB, Sterzel RB, Boron WF (1988) Am J Physiol 255:C844
20. Boitier E, Rea R, Duchen MR (1999) J Cell Biol 145:795
21. Lemasters JJ, Nieminen AL, Qian T, Trost LC, Elmore SP, Nishimura Y, Crowe RA, Cascio WE, Bradham CA, Brenner DA, Herman B (1998) Biochim Biophys Acta 1366:177
22. Scaduto RC, Jr., Grotyohann LW (1999) Biophys. J. 76:469
23. Bernardi P, Petronilli V, Di Lisa F, Forte M (2001) Trends Biochem Sci 26:112
24. Cortese JD (1999) Am J Physiol 276:C611
25. Zorov DB, Filburn CR, Klotz L-O, Zweier JL, Sollott SJ (2000) J Exp Med 192:1001
26. Elmore SP, Qian T, Grissom SF, Lemasters JJ (2001) Faseb J 17:17
27. Leach JK, Van Tuyle G, Lin PS, Schmidt-Ullrich R, Mikkelsen RB (2001) Cancer Res 61:3894
28. Swift LM, Sarvazyan N (2000) Am J Physiol 278:H982
29. Burkitt MJ, Wardman P (2001) Biochem Biophys Res Commun 282:329
30. Pap EH, Drummen GP, Winter VJ, Kooij TW, Rijken P, Wirtz KW, Op den Kamp JA, Hage WJ, Post JA (1999) FEBS Lett 453:278
31. Kuznetsov AV, Mayboroda O, Kunz D, Winkler K, Schubert W, Kunz WS (1998) J Cell Biol 140:1091
32. Gustafsson MG, Agard DA, Sedat JW (1999) J Microsc 195:10

Expression of Multicolor Fluorescent Fusion Proteins in Zebrafish Cell Cultures: A Versatile Tool in Cell Biology

C. K. D. Breek, F. van Iren, S. E. Wijting, N. Stuurman, and H. P. Spaink

A genetically stable zebrafish cell line was transfected with plasmids encoding fluorescent marker proteins of various colors. The markers for human tubulin, actin, ER, golgi and endosome are located at their correct position in zebrafish cells. We discus the benefits of transfected zebrafish cell lines over mammalian cells in cell biological studies.

23.1
Introduction

Cell cultures have long been used to study cell behavior under well-defined conditions. With the discovery of genetically encoded fluorescent markers, cell cultures also made it possible to visualize location and dynamics of specific proteins within living cells. Most use has been made of cell lines derived from mammals. To maintain a native environment, however, mammalian cells need to be incubated at 37°C, and often need higher than atmospheric carbon dioxide concentrations. These two requirements are difficult to combine with microscopy and require design or purchase of specialized equipment.

We therefore explored the possibility to study mammalian genes in cell lines derived from zebrafish. Zebrafish cells grow at room temperature and can be cultured under atmospheric carbon dioxide concentrations [1]. An additional benefit of these cells is that they can be re-implanted during embryogenesis [2]. Thus, zebrafish cells could potentially be transfected with genes coding for specific fluorescently marked proteins and these proteins could be studied in the living embryo. Since zebrafish embryos are small, free living, and completely transparent, they are well suited for microscopic analysis of cell biological and developmental processes.

While studies on mammalian, *Arabidopsis* and yeast systems have made use of the whole spectrum of autofluorescent proteins, until now studies on zebrafish made almost exclusive use of the green fluorescent protein (GFP). There is only one report that shows expression of blue fluorescent protein (BFP) and red fluorescent protein (DsRed) in live zebrafish. [3]. The GFP based studies in zebrafish make use of GFP as a reporter for gene expression and/or tissue specific localization, while the only report on sub-cellular localization with a zebrafish protein, using GFP, was performed in *Xenopus* oocytes and cell lines. [4]. Despite the technical advantages of cell cultures over complete organisms there are currently no reports on the expression of fluorescent proteins in zebrafish cell lines.

Here we demonstrate that zebrafish tissue culture cells can be successfully transfected with constructs coding for different fluorescent proteins. Furthermore, we show that various commercially available fusion proteins used for studies in human systems are targeted to the correct compartments in a zebrafish cell line.

23.2
Zebrafish Cell Lines

Various zebrafish cell lines derived from embryonic tissues have been described. These were either isolated from early embryonic stages, such as blastula stage, or from later stages such as 28 somite embryos. Recently, the use of zebrafish cell cultures for the production of embryonic germ-line chimeras has also been reported [5]. In this paper, we made use of the zebrafish cell lines ZF13 and ZF29 described by [1]. These cell lines were derived from 20 h old late somitogenesis embryos, are presumed to be of mesenchymal origin [6] and have the morphology of fibroblasts.

In order to test genetic stability of the cell lines we used flow cytometric analysis. Somatic cells from zebrafish muscle had a nuclear DNA content (2C) of 3.79 pg (Fig. 23.1A), close to the value that can be calculated from preliminary sequence data ([7] reported 1.9 Mbp, equivalent to 3.9 pg / 2C). However, the cell lines had a significantly lower amount of DNA, with both ZF13 and ZF29 yielding

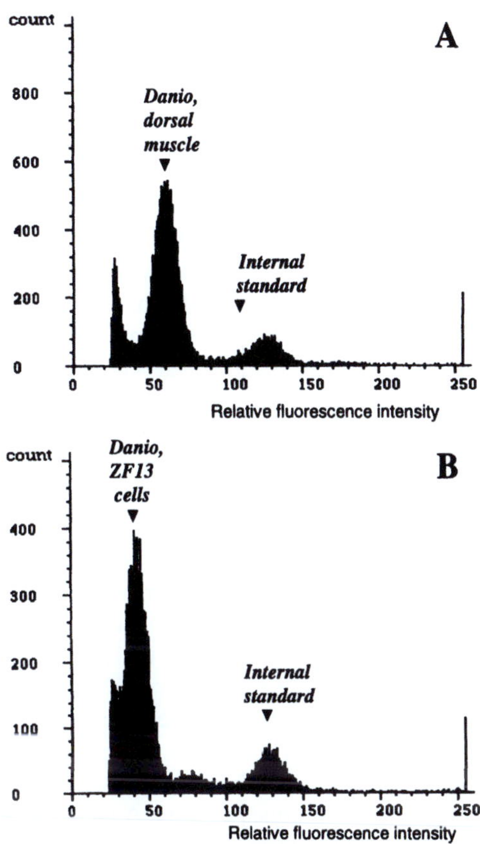

Fig. 23.1. Absolute DNA content estimations of nuclei as measured by propidium iodide staining and flow cytometry. Internal standard (*Agave stricta*, 2C = 7.84 pg, see [9]) was chopped with a razor blade, together with zebrafish cell material in a buffer (modified after [10]), and stained with propidium iodide (100 mL L^{-1} final concentration). Time curves revealed small but significant differences in staining rate of the different nuclei, so staining times had to be optimised per sample type. Because peaks were broad (coefficient of variation (CV) ca 10%) several independent measurements had to be made. We used a CAII flow cytometer of Partec GmbH, Münster, Germany. **A** Dorsal muscle nuclei from fish. Fluorescence yield per nucleus = 0,483 relative to standard. DNA content = 3.79 pg per nucleus. **B** ZF13 cell culture (P18). Fluorescence yield per nucleus = 0,346 relative to standard. DNA content = 2.71 pg per nucleus

28% less at 2.71 pg DNA per cell (Fig. 23.1B). Aneuploid DNA contents are frequently found in plant cell cultures (FVI, unpublished) and mammalian cells. In order to determine whether the loss of DNA continued during prolonged subculture, DNA content was estimated by DAPI staining of ZF13 cultures 18, 22, and 32 passages after cryostorage of the original cell line. DAPI is a very sensitive and reproducible DNA-stain [8], resulting in better peak resolution (typical coefficients of variation were 4 to 5%) and lower background. Because it is AT-specific, DAPI is not suited for absolute measurements. No differences were found (measurements relative to internal standard 3.536, 3.548, and 3.532, respectively, SE always below 0.6%).

23.3
Microscopical Analysis

Confocal laser scanning microscopy was carried out with a Leica-DMIRBE inverted fluorescence microscope (Leica, Bensheim, Germany) equipped with a Leica SP confocal scan head. Usually multicolor images were acquired by sequential scanning with settings for ECFP (excitation at 457 nm, emission detection between 470 and 490 nm) followed by settings for EYFP (excitation at 515 nm, emission detection between 530 and 550 nm) and/or DsRed2 (excitation at 568 nm, emission detection between 560 and 610 nm). For excitation of Hoechst-33258-stained DNA use was made of 2-photon excitation at 750 nm. 2-photon excitation was established by direct coupling of a TiS Mira 900F laser (Coherent Inc., Sunnyvale, CA) to the Leica SP scan head, resulting in pulse lengths of around 150 femtoseconds and power of up to 400 mW. In some cases, out of focus information in the original data was removed by deconvolution using the maximum likelihood estimate method as offered by the Huygens software package (SVI, Hilversum, The Netherlands).

Images were archived in the web-driven Scientific Image DataBase (SIDB). This system lets users upload images through the Internet, or directly through a mounted file system in the local intranet. SIDB generates thumbnail views of 2-D and 3-D (confocal) images in user-defined colours. A view on individual layers of 3-D images is provided by a MPEG format movie file. Keywords can be assigned and images can be grouped in projects. Access by other users to each image can be specified. The database can be accessed through the Internet at http://impi. leidenuniv.nl (login: guest, password: guest). More information about this web-driven image database can be found at http://sidb.sourceforge.net.

23.4
Analysis of Transfected Cell Lines

Zebrafish cell lines ZF13 and ZF29 were cultured at 25°C in medium consisting of 67% Leibovitz L-15 (Life Technologies) and 9% fetal bovine serum (Life Technologies) as described [1]. For transfections, 1 mL cell cultures were grown in medium on incubation chambers with a 0.17 mm thick glass bottom (Nunc). Medium was removed and replaced with 0.1 mL fresh medium and a mixture of 1 µg DNA and 4 % DOTAB transfection reagent in 0.1 mL medium was added. After 2 hours the transfection mixture was replaced with fresh medium. Cells were analyzed one or two days after transfection.

All plasmids used in this study were obtained from Clontech and contain fusion proteins with the enhanced form of the green fluorescent protein from *Aequorea victoria* [11] or the DsRed2 protein from *Dictosoma* [12]. These vectors have been optimized for expression and fluorescence in human cells. The EYFP-actin construct is the result of a fusion between the enhanced yellow fluorescent protein and human cytoplasmic β-actin [13]. As can be seen in Fig. 23.2 fluorescence after transfection of this construct can be detected in filamentous structures in pseudopodial and phillopodial extensions of the cells. In addition, stress-fiber like structures are often detected. These patterns suggest efficient incorporation of the EYFP-actin fusion protein in microfilaments. The EGFP-tubulin construct resulted from a fusion between the gene for enhanced green fluorescent protein and the human α-tubulin gene. As can be seen in Fig. 23.2, transfection of this construct resulted in fluorescence that is present in radiating filamentous structures characteristic for microtubules. Note the typical centrosome structure close to the nucleus. Time-lapse analysis of EGFP-tubulin transfected zebrafish cells demonstrated growing and shrinking microtubules, i.e., microtubule dynamics (not shown, but see http://rulbim.leidenuniv.nl/publications/zebrafish_cells.htm). To visualize DNA, cells were treated with 0.1 µg/mL Hoechst 33258 and visualized by 2-photon excitation at 750 nm (Fig. 23.2). Our results show that this concentration of Hoechst 33258 dye did not influence the cell division processes.

The plasmids ECFP-Endo, ECFP-Golgi and ECFP-ER encode fusion proteins with enhanced cyan fluorescent protein (ECFP) that accumulate in the different organelles. A fusion with the human rhoB GTPase is responsible for endosome targeting of the ECFP-Endo construct in human cells. Transfection of this construct in ZF13 cells (Fig. 23.3) results in typical endosomal vesicular structures of which the insides do not fluoresce. This is expected since the rhoB protein is membrane localized [14]. In the ECFP-Golgi construct the human beta 1,4-galactosyl-transferase directs the expressed protein to the Golgi apparatus [15]. The results of the transfection experiments (Fig. 23.3) show accumulation in a limited number of stacked structures that are close to the nucleus. Such a pattern is consistent with Golgi localization of the fusion protein. The construct encoded by ECFP-ER localizes to the endoplasmic reticulum (ER) lumen by the calreticulin targeting signal at amino-terminus and a ER retrieval sequence, KDEL at the C-terminus [16, 17]. Transfection leads to fluorescence in a largely spread out continuous net-

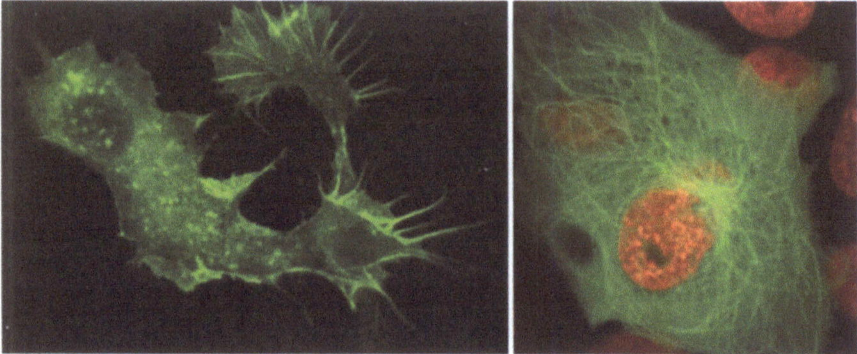

Fig. 23.2. CLSM and 2-photon analysis of cytoskeletal protein fusions. Zebrafish cell line ZF13 transfected with pYCFP-Actin (*left panel*) or pEGFP-Tubulin and a DNA staining with 0.1 μg/mL Hoechst generating a two color image (*right panel*)

work surrounding and connected to the nucleus (Fig. 23.3). As fluorescence was not found in Golgi-like structures we can conclude that the ER retention signal is efficiently recognized in zebrafish cells. Time-lapse analysis of cells transfected with the above-mentioned constructs shows the dynamic behavior of all visualized structures.

A construct similar to the ECF-Golgi was produced in which the autofluorescent protein is exchanged for the DsRed2 protein, recently commercialized by Clontech. The DsRed2 protein is an improved version of the original DsRed protein derived from *Dictosoma* species [12], differing by several point mutations. These mutations seem to improve the solubility of DsRed2 by reducing its tendency to form aggregates. The results obtained are very similar to those with ECFP-Golgi, showing that DsRed2 offers good possibilities to study protein targeting. Co-transfection experiments with CFP-ER, YFP-actin and DsRed2-Golgi show that it is now possible to distinguish three autofluorescent proteins in multicolor imaging. Transfection efficiencies were typically in the order of 2–5%. However, co-transfection efficiencies of three plasmid constructs were rather low,

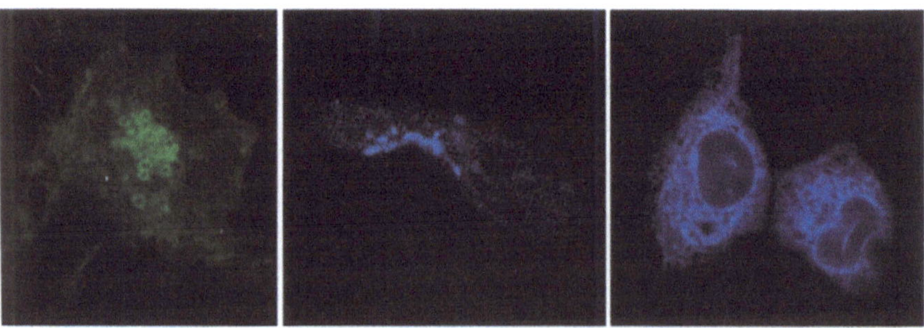

Fig. 23.3. CLSM analysis of organelle-targeted protein fusions. Zebrafish cell line ZF13 40 h. after transfection with, from left to right, pECFP-Endo, pECFP-Golgi and pECFP-ER

indicating the need for construction of plasmids coding for multiple fusion proteins to study co-localization of three proteins efficiently. Color pictures for the figures shown and complementary 2-D, 3-D and time-lapse data can be found at http://rulbim.leidenuniv.nl/publications/zebrafish_cells.htm.

23.5
Conclusions

Our work shows that zebrafish cell lines offer a versatile tool for analysis of cellular processes using fluorescence microscopy. Flow cytometric analysis shows that the used cells offer a stable genetic background at least after 18 passages of culture. Since various human fusion proteins are apparently correctly targeted to their cellular position these constructs can be used as markers for determining the cellular location of proteins of unknown function. For instance, in our current work we are using zebrafish cells for localisation of glycosyltransferases involved in hyaluronate and chitin oligosaccharide synthesis. Considering the ease of culturing and storing the used cell lines, zebrafish cells undoubtedly can be a useful tool for many researchers in cell biology. Furthermore, autofluorescence of the cells seems sufficiently low as to also be useful for specialized biophysical studies at the single molecule level.

Acknowledgements. We thank Gerda Lamers for technical advice for the confocal microscopic studies. We thank Dr. Claude Backendorf for the kind gift of DOTAB transfection agent.

References

1. Peppelenbosch MP, Tertoolen LGJ, de Laat SW, Zivkovic D (1995) Ionic responses to epidermal growth factor in zebrafish cells. Exp Cell Res 218:183–188
2. Hong Y, Chen S, Schartl M (2000) Embryonic stem cells in fish: current status and perspectives. Fish Physiol. and Biochem 22:165–170
3. Finley KR, Davidson AE, Ekker SC (2001) Three-color imaging using fluorescent proteins in living zebrafish embryos. Biotechniques 31:66–68, 70, 72
4. Watine DB, Shorte SL, Fucile S, de-Saint-Jan D, Korn H, Bregestovski P (1999) Functional integrity of green fluorescent protein conjugated glycine receptor channels. Neuropharmacol 38:785–792
5. Ma CG, Fan LC, Ganassin R, Bols N, Collodi P (2001) Production of zebrafish germline chimeras from embryo cell cultures. Proc Natl Acad Sci USA 98:2461–2466
6. Speksnijder JE, Hage WJ, Lanser PH, Collodi P, Zivkovic D (1997) In vivo assay for the developmental competence of embryo-derived zebrafish cell lines. Mol Marine Biol and Biotechn 6:21–32
7. Gabor Miklos GL, Rubin GM (1996) The role of the genome project in determining gene function: insights from model organisms. Cell 86:521–529

8. De Laat. AAM, W Göhde, MJDC Vogelzang (1987) Determination of ploidy of single plants and plant populations by flow cytometry. Plant Breeding 99:303–307

9. Zonneveld BJM, Van Iren F (2001) Genome size and pollen viability as taxonomic criteria: application to the genus *Hosta*. Plant Biol 50:176–185

10. Otto F (1990) DAPI staining of fixed cells for high-resolution flow cytometry of nuclear DNA. In: Darzynkiewicz Z, Crisman HA (eds) Methods in Cell Biology 33, Acad. Press, pp 105–110

11. Heim R, Tsien RY (1996) Engineering green fluorescent protein for improved brightness, longer wavelengths and fluorescence resonance energy transfer. Curr Biol 6:178–182

12. Matz MV, Fradkov AF, Labas YA, Savitsky AP, Zaraisky AG, Markelov ML, Lukyanov SA (1999) Fluorescent proteins from nonbioluminescent Anthozoa species. Nature Biotech 17:969–973.

13. Ponte P, Ng SY, Engel J, Gunning P, Kedes L (1984) Evolutionary conservation in the untranslated regions of actin mRNAs: DNA sequence of a human beta-actin cDNA. Nucleic Acids Res 12:1687–1696

14. Adamson P, Paterson HF, Hall A (1992) Intracellular localization of the p21rho-proteins. J Cell Biol 119:617–627

15. Roth J, Berger EG (1982) Immunocytochemical localization of galactosyltransferase in HeLa cells: codistribution with thiamine pyrophosphatase in trans-Golgi cisternae. J Cell Biol 93:223–9

16. Munro S, Pelham HR (1987) A C-terminal signal prevents secretion of luminal ER proteins. Cell 48:899–907

17. Fliegel L, Burns K, MacLennan DH, Reithmeier RA, Michalak M (1989) Molecular cloning of the high affinity calcium-binding protein (calreticulin) of skeletal muscle sarcoplasmic reticulum. J Biol Chem 264:21522–21528

Subject Index